Mechanical Performance of Advanced Composite Materials and Structures

Mechanical Performance of Advanced Composite Materials and Structures

Guest Editor

Yin Fan

MDPI

Basel • Beijing • Wuhan • Barcelona • Belgrade • Novi Sad • Cluj • Manchester

Guest Editor
Yin Fan
School of Aeronautics and
Astronautics
Shanghai Jiao Tong University
Shanghai
China

Editorial Office
MDPI AG
Grosspeteranlage 5
4052 Basel, Switzerland

This is a reprint of the Special Issue, published open access by the journal *Materials* (ISSN 1996-1944), freely accessible at: https://www.mdpi.com/journal/materials/special_issues/Mech_Perform_Adv_Compos_Mater_Struct.

For citation purposes, cite each article independently as indicated on the article page online and as indicated below:

Lastname, A.A.; Lastname, B.B. Article Title. *Journal Name* **Year**, *Volume Number*, Page Range.

ISBN 978-3-7258-7514-6 (Hbk)
ISBN 978-3-7258-7515-3 (PDF)
https://doi.org/10.3390/books978-3-7258-7515-3

Contents

About the Editor

Yin Fan

Yin Fan is a researcher specializing in the design and mechanics of advanced composite materials. His work focuses on multiscale modeling, the mechanical response of composite structures, and machine learning-assisted design of composite laminates with superior mechanical properties. His research goal is to understand the fundamental material behaviors and physical mechanisms of composite materials and structures under various loading conditions through mathematical modeling and experimental investigations, and then to use the acquired insights to guide the design and development of novel multifunctional composite materials and structures (e.g., nanostructured, bioinspired) for further improvement in mechanical performance, as well as to guide the development and optimization of advanced design methods for composite structures. Dr. Yin Fan has authored more than 40 peer-reviewed publications in journals such as *Communication Engineering* (Nature portfolio journal), *Research* (Science partner journal), *Advanced Functional Materials*, and *Composites Part B*. His recent work explores the intelligent inverse design of bioinspired layered composites with enhanced auxetic and multifunctional properties. He has also contributed to several interdisciplinary projects combining materials science, solid mechanics, and artificial intelligence for performance optimization of next-generation composites.

Preface

The rapid advancements in experimental techniques and computational methods have led to significant progress in the study and application of advanced composite materials and structures. These materials, known for their superior strength-to-weight ratios and tailored mechanical properties, are now at the forefront of engineering innovation. From aerospace to civil engineering, mechanical engineering to naval architecture, the scope of composite materials continues to expand, offering new possibilities for designing more efficient and durable systems.

This work is dedicated to exploring the mechanical performance of advanced composite materials, from fundamental principles to real-world applications. The scope of this collection encompasses experimental investigations, mechanical and numerical analyses, multiscale modeling, and the evaluation of damage and failure mechanisms. Furthermore, the study extends to the emerging field of next-generation composites, including nanocomposites and metamaterials, which hold the promise of revolutionizing industries with their enhanced properties.

The primary aim of this volume is to provide a comprehensive understanding of how advanced composites perform under various loading conditions, how they can be optimally designed for specific applications, and how their potential can be realized in practice. The motivation behind compiling this work stems from the increasing demand for high-performance materials in modern engineering, driven by industries seeking lighter, stronger, and more resilient solutions. The development of advanced composites is essential to meet these challenges, and this work seeks to shed light on the current state of research and the future directions for these materials.

This collection is primarily addressed to researchers, engineers, and professionals working in the fields of materials science, mechanical engineering, aerospace engineering, civil engineering, and related disciplines. It is also intended for students and academic researchers who wish to gain a deeper understanding of the fundamental and applied aspects of composite materials. By providing an up-to-date resource on the mechanical performance and design considerations of advanced composites, this work aims to contribute to the continued development of materials that can meet the demands of next-generation engineering systems.

In conclusion, the study of advanced composite materials is a dynamic and evolving field, and the knowledge contained within this volume is intended to inspire further research, collaboration, and innovation. As we continue to push the boundaries of material performance, it is our hope that this work serves as both a comprehensive guide and a catalyst for the next wave of discoveries in the field.

Yin Fan
Guest Editor

materials MDPI

Editorial

Mechanical Performance of Advanced Composite Materials and Structures

Yin Fan

School of Aeronautics and Astronautics, Shanghai Jiao Tong University, Shanghai 200240, China;
yfan1987@sjtu.edu.cn

1. Introduction and Summary

In the realm of material science and engineering, the pursuit of lighter, stronger, and more durable materials has been an enduring quest. In recent decades, advanced composite materials have emerged as frontrunners in revolutionizing various industries, from aerospace and automotive to construction and sports equipment. Undoubtedly, mechanical performance is one of the most important attributes of composite materials and structures. On the one hand, the mechanical properties are expected to be fully explored and utilized to meet engineering requirements. In addition, improved process levels for existing composite materials and new fabrication methods for novel composite materials are other effective approaches to enhance their mechanical behaviors, leading to an industrial revolution.

This Special Issue on "Mechanical Performance of Advanced Composite Materials and Structures" is devoted to the most relevant studies related to advanced reinforced composites and their mechanical performances. The Guest Editor of the Special Issue, Dr. Yin Fan, has proposed to collate submissions on the following topics according to different types of materials: carbon fiber-reinforced polymer, metal matrix composite, nano reinforcement, auxetic lattice, and other composites. The suggested application areas include aerospace, civil engineering, sensors, and medical equipment. Fourteen research articles from China, America, Germany, Italy, Portugal, Poland, Austria, and Saudi Arabia were selected, peer-reviewed, and accepted for publication. These articles cover all aspects of the mechanical performance study of composite materials, such as numerical modeling, lay-up design, detection, mechanical tests, and fabrication. The range of research scales varies from nanometers to meters.

2. Contributions

2.1. Carbon Fiber-Reinforced Polymer

As is known, carbon fiber-reinforced polymer (CFRP), which is lightweight but high in strength, has been widely applied in aerospace engineering. To take full advantage of the potential of CFRPs, accurate and precise mechanical models are required in the design stage. In the paper entitled "Crash Performance of Inward-Inverting Composite Tubes Filled with Foam: Experimentation and Simulation", Yu et al. [1] proposed a novel finite element model (FEM) adapted for shock absorber with an inward-inverting composite foam-filled tube to assess crashworthiness. The results revealed that the mean crush force from the simulation aligned well with the experimental data, with errors of the specific energy absorption being less than 7%. This indicates that the method is appropriate for the crush simulation of the proposed shock absorber.

CFRP joints have attracted the attention of engineers as they generally play an important role in aircraft structures. Two papers entitled "Study of Hybrid Composite Joints with Thin-Ply-Reinforced Adherends" and "Tests and Numerical Study of Single-Lap Thermoplastic Composite Joints Bolted by Countersunk" by Ramezani et al. [2] and Zhang et al. [3], respectively, explored the performance of CFRP joints using different methods of testing and FEM. However, the joint type was different: the first focused on bonded joints, whereas the second examined bolted joints. Both studies successfully established effective FEMs to simulate the damage propagation process of CFRP joints and obtained valuable results for practical engineering applications.

The designing of the stacking sequence of CFRP composite laminates inspired by nature may improve their desired mechanical properties. Yu et al. [4] investigated helicoidal laminates inspired by naturally occurring impact-resistant micro-structures in a research article entitled "Increasing the Compressive Strength of Helicoidal Laminates after Low-Velocity Impact upon Mixing with 0° Orientation Plies and Its Analysis". In this study, they employed both numerical and experimental methods and found that laminates with pitch angles of 10° and 20° exhibited higher compression-after-impact (CAI) strengths, whereas using 0° layer could lead to large indentation areas. Hence, the authors concluded that the key parameters for CAI strength are pitch angles and the arrangement of 0° plies. Another paper on helicoidal CFRP laminate, titled "Optimization Design and Nonlinear Bending of Bio-Inspired Helicoidal Composite Laminated Plates" by Lu et al. [5], focused on the bending characteristics of helicoidal CFRP laminates and utilized FEM as an analytical tool. A parametric study revealed that single-form and combination-form helicoidal layups with lower rotation angles could decrease bending deflection under the boundary condition of four fixed edges.

Designing CFRP laminate layups can not only improve mechanical properties but also achieve abnormal auxeticity. Wang [6] contributed a paper entitled "Auxetic Composite Laminates with Through-Thickness Negative Poisson's Ratio for Mitigating Low Velocity Impact Damage: A Numerical Study", which proposed layups of CFRP composite laminates that produce negative Poisson's ratios in the thickness direction. To directly exhibit the auxetic effect on the low-velocity impact of CFRP laminates, the author also identified two layups of non-auxetic CFRP counterpart laminates whose elastic moduli in three directions were very close to those of the auxetic laminate. Numerical comparison revealed that a negative Poisson's ratio significantly influences the global impact response and damage behavior. Auxetic laminates with a through-thickness negative Poisson's ratio consistently demonstrated higher impact forces, shorter impact times, reduced maximum displacements, and lower dissipated energies at elevated impact energy levels.

The use of CFRP is also garnering attention in medical apparatus and instruments due to its safety and comfort. Golewski et al. [7] conducted a study entitled "Composite Medical Tabletops Made of CFRP with Different Cross-Sections: Numerical Analysis and Laboratory Testing", which was the first attempt to design CFRP medical tabletops. Numerical simulations and laboratory testing methods were used in the study for design and verification purposes, respectively. Four variants of CFRP medical tabletops were presented in the study, as well as three types of ergonomic tops. This work provides a reference for both researchers and designers involved in the medical application of CFRP.

Although most aerospace-grade CFRPs are made using thermoset polymers, thermoplastic polymers, such as PEEK (polyether ether ketone), are being considered as better substitutes due to their high mechanical, thermal, and chemical performance. Hümbert et al. [8] investigated the effect of processing parameters on in situ bonding between continuous fiber-reinforced PEEK laminates and short fiber-reinforced PEEK 3D printing during overprinting in a study entitled "Mechanical Characterisation of Bond Formation during Overprinting of PEEK Laminates". The study found that the bonding strength primarily depended on temperature, with a maximum value reaching up to 15 MPa. The study results can guide future processes and the upscaling of the overprinting process using PEEK.

2.2. *Metal Matrix Composite*

Compared to CFRPs, metal matrix composites (MMCs) can serve at higher temperatures. Petrus et al. [9] proposed a novel alumina matrix composite reinforced with MAX phases (titanium aluminum carbide Ti_3AlC_2) in the paper entitled "Novel Alumina Matrix Composites Reinforced with MAX Phases—Microstructure Analysis and Mechanical

Properties". The authors emphasized that spark plasma sintering technology effectively reduces the decomposition of the used MAX phase. The application of MAX phases as a reinforcing phase in the alumina allows one to obtain sinters with high hardness, similar to the hardness of pure sinter while improving the fracture toughness. The composites demonstrated almost 20% higher fracture toughness while maintaining high hardness.

Currently, high-entropy alloys (HEAs) have attracted the attention of the research community owing to their excellent mechanical, wear, thermal, fatigue, and surface properties. Sivasankaran et al. [10] studied the influence of Al_2O_3 in CrFeCuMnNi high-entropy alloy matrix composites (HEMCs) on their microstructure, phase changes, and mechanical and wear performances in the paper "Effect of Al_2O_3 (x = 0, 1, 2, and 3 vol.%) in CrFeCuMnNi-x High-Entropy Alloy Matrix Composites on Their Microstructure and Mechanical and Wear Performance". HEMCs were synthesized via mechanical alloying (MA) followed by hot compaction (550 °C at 550 MPa), medium frequency sintering (1200 °C), and hot forging (1000 °C at 50 MPa). The study found that the hardness and mechanical strength gradually increased, and the VHS and UCS of the CrFeCuMnNi-3 vol.% Al_2O_3 HEAMC sample were around 1.21 times higher than those of the unreinforced CrFeCuMnNi HEA matrix, correlating with the increase in the Al_2O_3 content.

2.3. Nano Reinforcement

Carbon nanotubes (CNTs), as one of the ideal potential fillers for nanocomposites [11], exhibit superior excellent mechanical properties compared to carbon fiber. In the paper "Chirality-Dependent and Intrinsic Auxeticity for Single-Walled Carbon Nanotubes", Zhang et al. [12] discovered the auxetic nature of CNTs. They employed the molecular dynamics (MD) method to analyze the effect of chiral indices on the mechanical behavior, especially Poisson's ratio, of CNTs. It was found that various chiral CNTs exhibited different critical strains, at which the incremental Poisson's ratio could shift from negative from positive. Their study showed that the auxeticity of CNTs was mainly determined by the stretch of C-C bonds, and the critical strain was achieved when the negative contribution of bond stretch to Poisson's ratio counteracted the positive contribution of angle variation. In their subsequent work, they proposed the atomic structures of auxetic nanocomposites [11] and verified their negative Poisson's ratios using MD methods. Moreover, the mechanical performance of auxetic nanocomposites greatly surpassed that of existing man-made reentrant auxetic materials [13,14]. Thus, auxetic nanomaterials hold promise for future industrial applications.

2.4. Auxetic Lattice

Man-made auxetic materials exhibit densification behavior under compression or impact, making them suitable as cores in sandwiches with composite face sheets. In their article entitled "Crashworthiness of Additively Manufactured Auxetic Lattices: Repeated Impacts and Penetration Resistance", Franzosi et al. [15] first selected and tested representative specimens, and then proceeded with an experimental and numerical study of repeated impact behavior and penetration resistance. Finally, they proposed a new design of a metallic auxetic absorber optimized for additive manufacturing and aimed at high-performance crash applications. Their work successfully demonstrated the potentiality of additively manufactured auxetic cellular structures for impact absorption, marking an advancement over previous literature in terms of empirical demonstrations of repeated impacts and penetration behavior.

2.5. Other Composites

In a paper entitled "Filament-Reinforced 3D Printing of Clay", Jauk et al. [16] developed a method for filament-reinforced 3D printing of clay, along with relevant supporting hardware and software components. Mechanical tests showed that filament-reinforced 3D printing of clay is a viable solution that increases the tensile strength of unfired clay in both wet and dry states. Furthermore, when comparing unfired clay test specimens without and with filament reinforcement during printing, an increase of approximately 15% in tensile strength values along the axis of extrusion was observed.

Peng et al. [17] contributed a research article entitled "Assessing Effects of van der Waals Corrections on Elasticity of $Mg_3Bi_{2-x}Sb_x$ in DFT Calculations". They conducted a systematic investigation into the mechanical properties of the room-temperature thermoelectric material $Mg_3Bi_{2-x}Sb_x$ ($0 \leq x \leq 2$) through first-principles calculations within the framework of density functional theory. The Perdew–Burke–Ernzerhof (PBE) exchange–correlation functional was used in calculations, and the mechanical properties, such as elastic constant and bulk modulus, were generally found to be within a range of 10%, except for a few outliers.

3. Outlook

This Special Issue of *Materials* was well supported by numerous submissions, and the final publication consists of 14 high-quality peer-reviewed articles. As the Guest Editor, I appreciate every author's contribution to this Special Issue. It is anticipated that due to this success, a new Special Issue ("Mechanical Performance of Advanced Composite Materials and Structures (Second Edition)"; website: https://www.mdpi.com/journal/materials/special_issues/Q464B4HKW8, accessed on 22 November 2023) will be commissioned as a follow-up to accept global contributions.

Funding: The work is sponsored by the National Natural Science Foundation of China (NSFC) under Grant No. 12102255 and the Natural Science Foundation of Shanghai under Grant No. 22ZR1429500.

Conflicts of Interest: The author declares no conflict of interest.

References

1. Yu, P.; Yu, Z.; Zhou, X.; Xu, W. Crash Performance of Inward-Inverting Composite Tubes Filled with Foam: Experimentation and Simulation. *Materials* **2023**, *16*, 6378. [CrossRef]
2. Ramezani, F.; Carbas, R.J.C.; Marques, E.A.S.; da Silva, L.F.M. Study of Hybrid Composite Joints with Thin-Ply-Reinforced Adherends. *Materials* **2023**, *16*, 4002. [CrossRef]
3. Zhang, J.; Chen, X.; Tian, A.; Fan, Y. Tests and Numerical Study of Single-Lap Thermoplastic Composite Joints Bolted by Countersunk. *Materials* **2022**, *15*, 2386. [CrossRef] [PubMed]
4. Yu, Z.; Du, X.; Liu, R.; Xie, Q.; Zhang, X.; Zhu, Q. Increasing the Compressive Strength of Helicoidal Laminates after Low-Velocity Impact upon Mixing with 0° Orientation Plies and Its Analysis. *Materials* **2023**, *16*, 4599. [CrossRef] [PubMed]
5. Lu, T.; Shen, H.-S.; Wang, H.; Chen, X.; Feng, M. Optimization Design and Nonlinear Bending of Bio-Inspired Helicoidal Composite Laminated Plates. *Materials* **2023**, *16*, 4550. [CrossRef] [PubMed]
6. Wang, Y. Auxetic Composite Laminates with Through-Thickness Negative Poisson's Ratio for Mitigating Low Velocity Impact Damage: A Numerical Study. *Materials* **2022**, *15*, 6963. [CrossRef] [PubMed]
7. Golewski, P.; Pietras, D.; Sadowski, T.; Wit-Rusiecki, A.M. Composite Medical Tabletops Made of CFRP with Different Cross-Sections: Numerical Analysis and Laboratory Testing. *Materials* **2023**, *16*, 7574. [CrossRef] [PubMed]
8. Hümbert, S.; Atzler, F.; Voggenreiter, H. Mechanical Characterisation of Bond Formation during Overprinting of PEEK Laminates. *Materials* **2024**, *17*, 161. [CrossRef]
9. Petrus, M.; Wozniak, J.; Cygan, T.; Pawlak, W.; Olszyna, A. Novel Alumina Matrix Composites Reinforced with MAX Phases—Microstructure Analysis and Mechanical Properties. *Materials* **2022**, *15*, 6909. [CrossRef] [PubMed]
10. Sivasankaran, S.; Ammar, H.R.; Sherif, E.-S.M.; Alaboodi, A.S.; Mekky, A.-b.H. Effect of Al_2O_3 (x = 0, 1, 2, and 3 vol.%) in CrFeCuMnNi-x High-Entropy Alloy Matrix Composites on Their Microstructure and Mechanical and Wear Performance. *Materials* **2023**, *16*, 3672. [CrossRef] [PubMed]
11. Zhang, H.-N.; Fan, Y.; Shen, H.-S. Prediction of Temperature-Dependent Mechanical Properties for SWCNT/Cu Nanocomposite Metamaterials: A Molecular Dynamics Study. *Nanomaterials* **2023**, *13*, 1885. [CrossRef] [PubMed]
12. Zhang, H.-N.; Fan, Y.; Shen, H.-S. Chirality-Dependent and Intrinsic Auxeticity for Single-Walled Carbon Nanotubes. *Materials* **2022**, *15*, 8720. [CrossRef] [PubMed]
13. Fan, Y.; Xiang, Y.; Shen, H.-S. Temperature-Dependent Mechanical Properties of Graphene/Cu Nanocomposites with In-Plane Negative Poisson's Ratios. *Research* **2020**, *2020*, 5618021. [CrossRef] [PubMed]
14. Fan, Y.; Shen, H.-S. Non-symmetric stiffness of origami-graphene metamaterial plates. *Compos. Struct.* **2022**, *297*, 115974. [CrossRef]
15. Franzosi, P.; Colamartino, I.; Giustina, A.; Anghileri, M.; Boniardi, M. Crashworthiness of Additively Manufactured Auxetic Lattices: Repeated Impacts and Penetration Resistance. *Materials* **2024**, *17*, 186. [CrossRef]

16. Jauk, J.; Gosch, L.; Vašatko, H.; Königsberger, M.; Schlusche, J.; Stavric, M. Filament-Reinforced 3D Printing of Clay. *Materials* **2023**, *16*, 6253. [CrossRef]
17. Peng, Q.; Ma, X.; Yang, X.; Zhao, S.; Yuan, X.; Chen, X. Assessing Effects of van der Waals Corrections on Elasticity of $Mg_3Bi_{2-x}Sb_x$ in DFT Calculations. *Materials* **2023**, *16*, 6482. [CrossRef]

materials

MDPI

Article

Crashworthiness of Additively Manufactured Auxetic Lattices: Repeated Impacts and Penetration Resistance

Paolo Franzosi [1], Ivan Colamartino [2,*], Alessandro Giustina [1], Marco Anghileri [1] and Marco Boniardi [2]

[1] Department of Aerospace Science and Technology, Politecnico di Milano, Via La Masa 34, 20156 Milan, Italy
[2] Department of Mechanical Engineering, Politecnico di Milano, Via La Masa 1, 20156 Milan, Italy
* Correspondence: ivan.colamartino@polimi.it

Abstract: Auxetic materials have recently attracted interest in the field of crashworthiness thanks to their peculiar negative Poisson ratio, leading to densification under compression and potentially being the basis of superior behavior upon impact with respect to conventional cellular cores or standard solutions. However, the empirical demonstration of the applicability of auxeticity under impact is limited for most known geometries. As such, the present work strives to advance the investigation of the impact behavior of auxetic meta-materials: first by selecting and testing representative specimens, then by proceeding with an experimental and numerical study of repeated impact behavior and penetration resistance, and finally by proposing a new design of a metallic auxetic absorber optimized for additive manufacturing and targeted at high-performance crash applications.

Keywords: auxetic; crashworthiness; meta-materials; impact; compression; penetration

1. Introduction

Auxetic materials are characterized by a negative Poisson's ratio, responding to a compressive load with a transverse contraction: such peculiar property potentially leads to significant enhancement in terms of impact, shear, penetration, and fracture resistance. The definition was introduced in the scientific literature by Evans [1].

While auxeticity is present in nature under different forms [2], it is common today to artificially produce cellular structures [3–6], foams [7], composite materials [8–10], and cellular geometries with globally auxetic behavior; the latter are often called auxetic meta-materials. Among the several cellular meta-materials [11], auxetic materials have recently attracted a lot of interest in multiple fields of engineering [12]. Focusing only on studies related to impact engineering and crashworthiness in particular, as such is the field on which the present study is centred, 2D auxeticity was extensively analyzed in the literature in analytical, numerical, and experimental terms; multiple novel geometries have been designed, modeled, and produced, usually in the form of sandwich panels [13] or foam-filled tubes [14–16]: auxetic cores mostly consisted of planar extrusion of 2D geometries, providing useful practical demonstration of auxeticity applicability in real case studies. Such a procedure is advantageous as related core geometries are relatively easy to produce, especially with classical additive processes, while numerically, they can be treated with shell elements. On the other hand, in-plane performances of 2D cores are inevitably low with respect to the normal direction, especially in largely unstable loading conditions such as crushing impacts.

While 2D auxeticity has been comprehensively investigated by multiple authors, 3D auxetic structures have been featured less in the literature, and little experimental data can be found for what concerns impact properties. The first notable study was published in 2015 by Yang et al. [17], who experimentally and numerically investigated the mechanical properties of 3D re-entrant Honeycombs, as theorized previously by Almgren et al. [18]. Ebrahimi et al., in 2018, proposed a novel anti-chiral topology with a tunable Poisson ratio [19]. Between 2019 and 2020, Novak et al. thoroughly studied the dynamic and impact properties of metallic chiral (often addressed as "sinusoidal") structures, additively manufactured via SEBM (Selective Electron Beam Melting) and SLM (Selective Laser Melting) [20–22]. In 2020, Jenett et al. proposed a novel assembly mechanism targeting the mass production of large geometries [23]. Similarly, Balaji et al. presented a new production method based on the modification of the re-entrant cubic topology [24]. In 2022, Novak et al. [25] proposed a new gradable, axisymmetric chiral absorber. Recently, Varas et al. proposed a new re-entrant topology produced via SLA (stereolithography) techniques, evaluating its properties under static and dynamic compression [26]. Finally, it is worth citing the recent work of Galea et al., who proposed a novel method to produce 3D auxetic metamaterials [27].

In this context, the aim of the study was to advance the investigation of three-dimensional auxetic reticula and analyze their impact behavior with primary focus on the experimental feasibility and applicability of such structures for crashworthiness purposes. In particular, properties commonly shown in the literature as potentially present and advantageous in auxetic lattices are a peculiar triangular load curve, resistance to repeated impacts, and superior penetration resistance. However, these features, for most lattices, have not been fully investigated; the literature is lacking, especially in the experimental/numerical evaluation of low-speed repeated impacts and penetration performances, and the extension of auxetic behavior under such loading conditions. On the numerical side, it is worth noting that solid-meshed explicit models were developed, which is a strategy rarely implemented in the literature for strut-based lattices due to the difficulty in obtaining high-quality meshing and to the higher computational costs; here, such a strategy was considered necessary for the reliable construction of predictive models to be used in the penetration analyses.

First, preliminary investigations were performed on SLA-produced specimens, via static compression tests on four significant topologies, of which the two best performing were chosen to proceed with the study. Then, these two topologies were produced via SLS, and the impact properties of samples manufactured with both the technologies were evaluated. Subsequently, numerical strategies were exploited to investigate the penetration resistance. Last, the most promising topology in terms of energy absorption potential, the planar anti-chiral, was modified to allow for high-quality printing through SLM stainless steel to target applications in high-performance energy absorption.

2. Materials and Methods

2.1. Cellular Cores

Heretofore, there are several potential designs for auxetic reticula, and multiple analytical and numerical procedures have been presented for the identification and generation of novel geometries. Among such countless designs, the present study preliminary focused on four representative topologies and then proceeded with the two best-performing ones. Geometric parameters were chosen to allow for manufacturability via SLA and because they conformed to literature-proven choices; in particular, the common parameter for each topology is a strut radius of 0.5 mm.

Tables 1 and 2 provide a brief overview of the four selected geometries together with the parameters chosen for preliminary testing and finalized choices.

In the following paragraphs, the four geometries are outlined with a comprehensive definition of their geometric parameters. Clarifying figures, inserted to provide an immediate understanding of the topological features of the structures, are mostly taken from the relevant literature, and a few annotations are added to describe the geometric parameters used. Analytical equations for relative density, a non-dimensional parameter representing the volume fraction of the material, have been computed and are reported for each topology.

Table 1. Geometric parameters. h, l, θ, A, and D, are defined below as a function of the topologies; r represents the strut radius, kept at 0.5 mm for every topology; ρ/ρ_0 is the relative density, expressed as ratio between the density of the cellular solid and the one of the related base material.

Geometry	h (mm)	l (mm)	θ (deg)	r (mm)	A (mm)	D (mm)	ρ/ρ_0
IT	3.5	2.64	11	0.5	-	-	0.18
RH	3.75	2.58	76	0.5	-	-	0.28
SIN	-	5	-	0.5	1.75	-	0.11
PAC	5	10	90	0.5	-	6	0.10

Table 2. Geometric parameters for preliminary testing and finalized geometries. Except for the inverted tetrapod geometry, for which the numbers reported in the present table are treated in Section 2.1.1, all lattices are defined as cuboid-bounded repeatable shapes.

Geometry	Unit Cell Dimensions (mm)	NCells (-)	Dimensions (mm)
IT	3.5×2.64	$7 \times 9 \times 6$	$35 \times 40 \times 18$
REH	$10 \times 10 \times 10$	$4 \times 4 \times 4$	$40 \times 40 \times 40$
SIN (preliminary)	$20 \times 20 \times 10$	$4 \times 4 \times 2$	$40 \times 40 \times 30$
PAC (preliminary)	$10 \times 10 \times 10$	$2 \times 2 \times 2$	$40 \times 40 \times 20$
SIN (finalized)	$20 \times 20 \times 10$	$4 \times 4 \times 2$	$40 \times 40 \times 30$
PAC (finalized)	$10 \times 10 \times 10$	$2 \times 2 \times 3$	$40 \times 40 \times 30$

2.1.1. Inverted Tetrapods (ITs)

Inverted tetrapods (ITs) were initially designed by Wei et al. [28] as a micro-scale polymeric network and brought to a macro-scale lattice by Schwerdtfeger et al. [29]. It is defined by four parameters: length of the vertical strut (h), length of the oblique beams (l), angle of the overhangings from the horizontal plane (θ) and strut radius (r).

Figure 1 shows details of the present geometry. Inverted tetrapods are anisotropic with a negative Poisson's ratio along the axes X1 and X2; Schwerdtfeger experimentally identified the ratio for a Ti-6Al-4V alloy produced with EBM (Electron Beam Melting), resulting in values between -0.2 and -0.4, respectively. The relative density can be calculated analytically via Equation (1), and for parameters set at h = 3.5 mm, l = 2.64 mm, θ = 11°, and r = 0.5 mm, the result is 0.18.

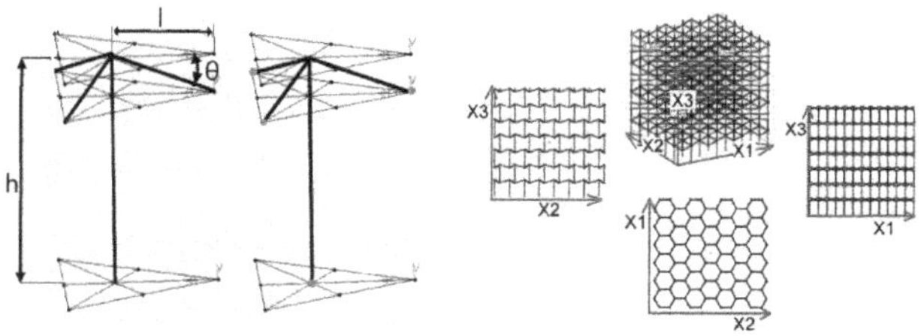

(a) (b)

Figure 1. Inverted tetrapods (ITs): unit cell (**a**) and cellular structure (**b**). Pictures were taken from [29] and [30], respectively.

$$\frac{\rho}{\rho_0} = \frac{8\pi r^2(h/l+3)}{l(h-l\sin\theta)11\sqrt{3}} \tag{1}$$

The unit cell and cellular parameters reported in Table 2 are defined as follows: for what concerns the unit cell dimensions, parameters refer to $h \times l$, as defined above; the cell count in directions X1 and X2 is measured as the maximum number of complete hexagons; in direction X3, the cell count is measured as the maximum number of vertical layers; overall dimensions are intended as finalized nominal values of the specimen.

2.1.2. Re-Entrant Hexagons (REHs)

Re-entrant hexagons (REHs) were firstly proposed by Evans et al. [7,31] in the 1990s, while comprehensive analysis of their mechanical properties was provided by Yang et al. [17]. The parameters defining this lattice are shown in Figure 2: length of the vertical struts (h), length of the re-entrant struts (l), angle of the re-entrant struts with respect to the vertical direction (θ) and radius of the struts (r).

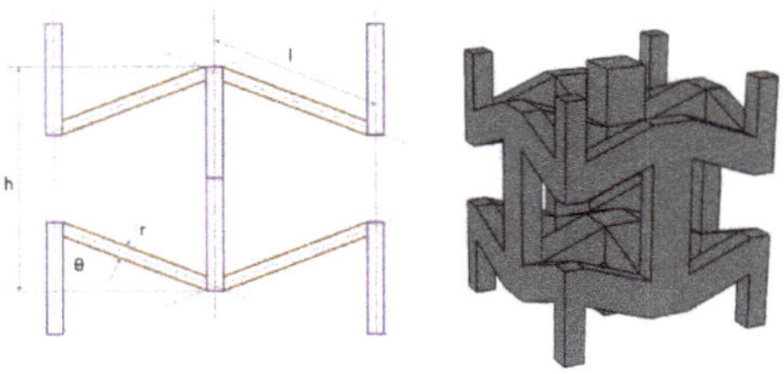

(a) (b)

Figure 2. Re-entrant hexagons (REHs) lattice: 2D view (**a**) and 3D unit cell (**b**). Pictures were taken from [32].

With h = 3.75 mm, l = 2.58 mm, θ = 76°, and h = 0.5, the relative density is ρ/ρ_0 = 0.28, which is calculated through Equation (2).

$$\frac{\rho}{\rho_0} = \frac{\pi r^2 h/l}{2l^2 sin\theta(h/l - cos\theta)} \tag{2}$$

The expected maximum Poisson's ratio is −0.31 (Equation (3)).

$$\nu_{zy} = \frac{\varepsilon_y}{\varepsilon_z} = \frac{cos\theta(h/l - cos\theta)}{sin\theta^2} \tag{3}$$

2.1.3. Sinusoidal (SIN)

The sinusoidal (SIN), or chiral lattice, shown in Figure 3, is based on the 10th eigenmode of the cube, three-dimensional evolution of the well-known 2D square chiral lattice structure. Among the topologies here evaluated, it is the only symmetric—and thus auxetic—in all the three directions. Hence, the Poisson's ratio is equal in all directions and was computed at −0.4 for the geometric parameters evaluated by [33].

The parameters of the structure are the following: distance between the nodes of the cube d, amplitude of the sinusoidal waves A and radius of the strut r.

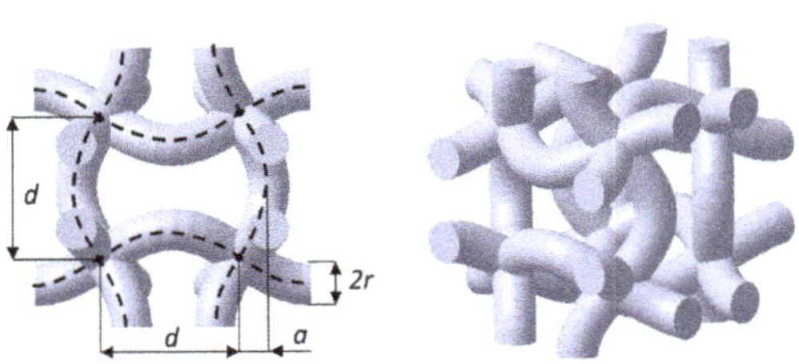

(a) (b)

Figure 3. Sinusoidal (SIN) lattice: unit cell geometric parameters (**a**) and perspective view (**b**). Pictures were taken from [20].

With parameters set as d = 5 mm, A = 1.27 mm, and r = 0.5 mm, the relative density of the lattice is 0.11, which is calculated as per Equation (4).

$$\frac{\rho}{\rho_0} = \frac{12\int_0^{2d}\sqrt{1+\left(A\frac{\pi}{5}sin\left(x\frac{\pi}{d}\right)\right)^2}dx}{(2d)^2} \tag{4}$$

2.1.4. Planar Anti-Chiral (PAC)

As anticipated in Section 1, the planar anti-chiral (PAC) lattice was first proposed by Ebrahimi et al. (2018) [19]. The structure can be decomposed in two distinct parts: a planar anti-chiral geometry and a set of oblique ligaments (Figure 4): when subjected to vertical compression (that is, out-of-plane with respect to the anti-chiral plane) forces the oblique ligaments to bend, causing the rotation in two opposite direction of the planar nodes.

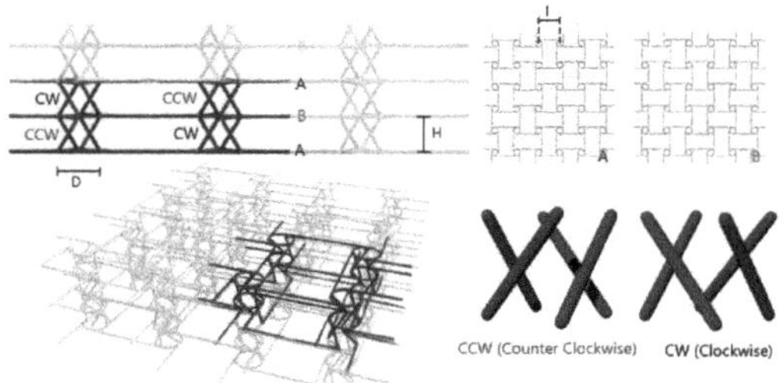

Figure 4. Planar anti-chiral (PAC) geometry and cellular structure.

The parameters of the structure are ligaments length l, diameter of the central node D, distance between the planes H and radius of the struts r. With l = 10 mm, D = 6 mm, h = 5 mm, and r = 0.5 mm, chosen to obtain the lowest Poisson's ratio, and as such, the maximum auxeticity, the relative density is computed at 0.1 from Equation (5).

$$\frac{\rho}{\rho_0} = \frac{4\pi r^2\left(\sqrt{\frac{D^2}{2}+h^2}+\frac{l}{2}+\frac{\pi D}{4}\right)}{l^2 h} \tag{5}$$

The Poisson's ratio of this structure has been investigated by Ebrahimi et al. [19] and is expected to be -0.9.

2.2. Manufacturing

After the quick discarding of classical filament technologies, such as FDM (Fused Deposition Modeling), due to their inability to manage low resolutions, manufacturing was initially approached via SLA, in particular through a Formlabs 3L and with its proprietary material, T2K (or Tough 2000). The first manufacturing campaign targeted the initial screening of manufacturability and mechanical properties, leading to the choice of the two best-performing topologies for comprehensive impact analysis.

SLA is an additive process based on the polymerization of photo-sensitive polymers; briefly, the technology consists of a tank filled with liquid resin, that is solidified layer by layer through a laser. The process is known to guarantee high resolution and surface finish and to offer high-performance materials in terms of toughness and impact strength. These peculiarities were seen as indisputable advantages for an initial investigation, in particular related to powder-based techniques such as Selective Laser Sintering (SLS) and Multi-Jet Fusion (MJF), which are able to manage high resolutions as well but are known for high brittleness due to intrinsic porosity issues. On the other hand, one fundamental advantage of powder-based technologies is the capability of managing overhangs (shapes that are not supported or are partially supported by previous layers), which is an issue that is instead suffered by SLA. Nonetheless, it was decided to proceed with SLA and evaluate its capability to tolerate the problem.

Preliminary printability investigations resulted in the identification of the lower limit of a three-dimensional feature at 0.5–0.6 mm, which is a dimension that drove the design parameters and in particular the choice of $r = 0.5$ mm for all topologies. Specimens were designed through the software Grasshopper Intralattice with parameters as defined in the previous section. Overhangs had a major role in determining failure of 20–25% of the prints. A set of printed samples is shown in Figure 5.

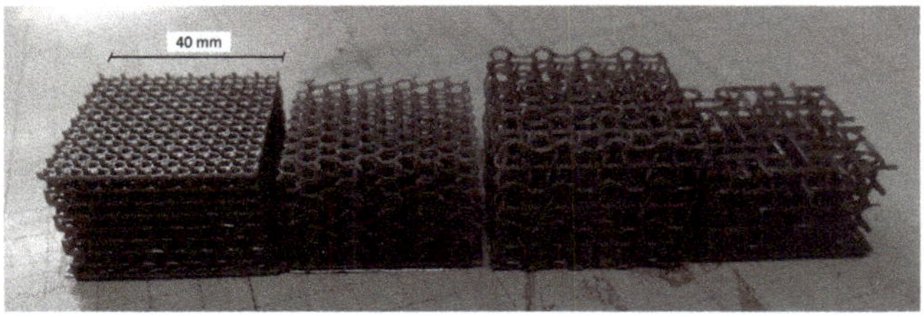

Figure 5. SLA samples for the preliminary quasi-static tests, left to right: re-entrant hexagons (REHs), inverted tetrapods (ITs), sinusoidal (SIN), planar anti-chiral (PAC).

SLA specimens production and testing resulted in two finalized geometries (the selection process and its outcomes are presented in Section 3), which were later also printed in Nylon PA12 (Figure 6) via SLS technology. After preliminary manufacturability investigations, the same geometric parameters were confirmed to be over the identified lower limit for optimal printability, reported at 0.7–0.8 mm, which is slightly higher with respect to SLA's but nonetheless sufficiently low to allow reliable production and testing.

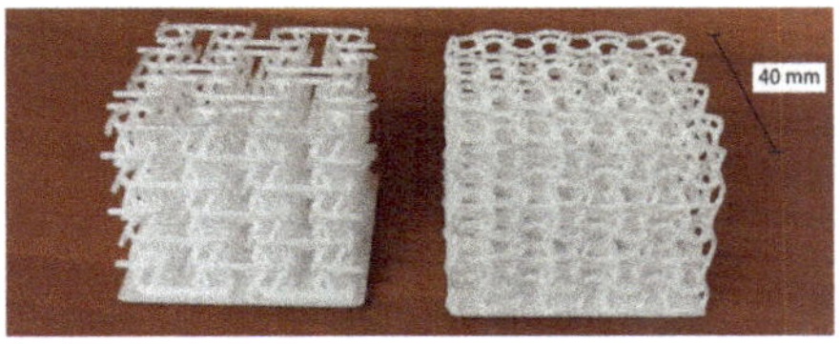

Figure 6. PA12 samples, left to right: planar anti-chiral (PAC) and sinusoidal (SIN).

The two basic materials have similar mechanical properties except for elongation at break, as visible in Table 3, thus allowing for a comparison on the effect of brittleness on the lattices' performances.

Table 3. Materials' mechanical properties. E is Young's modulus, σ_Y is the yield stress, UTS is the ultimate tensile stress, ε_F% is the failure strain, ρ is the density.

Material	E (MPa)	σ_Y (MPa)	UTS (MPa)	ε_F (%)	ρ (kg/m^3)
Tough 2000	2000	38	40	50	1300
PA22	1500–1650	28	48	18	930
AlSi316L	180,000	530	500–600	30–40	7900

2.3. Experimental Methods

Experimental activities may be divided in two phases: controlled compression and impact tests.

Controlled compression was carried out through an MTS-370.10 at displacement rates of 5 mm/min, which was traceable to a quasi-static compression. The acquired signals were imposed displacement, compressive force and videos of the frontal view of the specimen.

Impact tests were performed internally on laboratory drop towers, loaded with a steel, cylindrical impactor of 1.205 kg weight (Figure 7) able to reach impacting energies of 0.66,

2.3 and 7 J. To distribute the impact load evenly across the impacting surface, a rigid steel plate was positioned over the specimens. Acceleration data were collected with a uniaxial piezoresistive accelerometer mounted on top of the cylinder recording at 10 kHz. The impact was further recorded by means of a high-speed camera at 10,000 fps.

A second set of repeated tests was conducted with a semi-automatic drop tower, a StepLab DW100, which was equipped with a steel cylindrical impactor of total mass of 5.6 kg (Figure 7). Data were collected through a high-performance load cell acquired at 3 MHz; an identical high-speed camera set up with respect to previous impact tests was used. The specimens were tested three consecutive times.

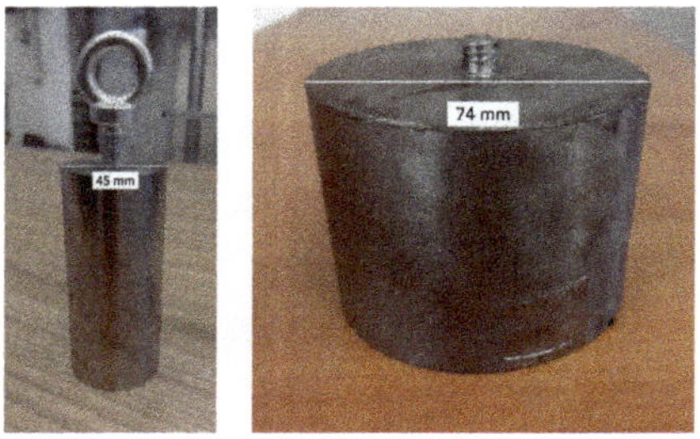

(**a**) (**b**)

Figure 7. Impactors: 1.205 kg cylinder (**a**), cylindrical anvil of the 5.6 kg impactor (**b**).

In the experiments, the measurement of the Poisson's ratio was carried out assuming transverse isotropicity via the analysis of the videos. In particular, the maximum transverse contraction, always recorded at mid-height due to the friction acting on the top and bottom of the samples, was manually measured on one of the two auxetic planes. Subsequently, the ratio was calculated as per Equation (6).

$$\nu_{zt} = \frac{\Delta l_z / L_z}{\Delta l_x / L_x} \tag{6}$$

2.4. Numerical Methods

Numerical models were constructed after the initial selection for both the chosen geometries, starting with simple beam-based models and scaling up with advanced material implementation including strain-rate sensitivity and a triaxiality-dependent failure model. Numerical models were developed for the Formlabs Tough 2000 material only, as proper consideration of the impact behavior of powdered materials is still today an open question in scientific research, and related implementation was beyond the scope of the work. The software used for analysis was Ansys LS-DYNA, which is the standard choice for most crashworthiness applications.

The beam models were implemented directly in LS-DYNA through Belytschko–Schwer elements with full cross-section integration. Sensitivity analyses lead to choose meshes with 20 elements per wave for the sinusoidal lattice, one element per strut and 20 per torus for the planar anti-chiral. The purpose of these models was preliminary analysis of the topologies to better design the test campaign.

Solid meshes were then implemented with two different strategies. The planar anti-chiral geometry was meshed by Python scripting through the API of Gmsh; the sinusoidal was treated with Altair Hypermesh. The resulting meshes, after sensitivity analyses, resulted in 6244 and 10,232 tetrahedral elements per cell.

The material model for the Tough 2000 was constructed and validated through dedicated tests, which is not of interest here and as such is presented in Appendix A. Finalized models exploited a piecewise linear elastic–plastic law implemented through *MAT24 with triaxiality-dependent failure to consider physically coherent struts breakage. The triaxiality curve, not available in the literature for the material under study, was taken from an internal study on a S275JR steel and linearly scaled referencing to the tensile failure strain of the material.

All analyses were run with the explicit scheme. For compression and impact analyses, displacement and initial velocity were, respectively, imposed to a rigid flat plate with properly adjusted mass and in contact with the lattice; the contact implemented was penalty-based and the featured friction coefficient of 0.3. As shown in Figure 8, beam-based models presented contact issues toward densification, which were successfully overcome by the adoption of solid-meshed models.

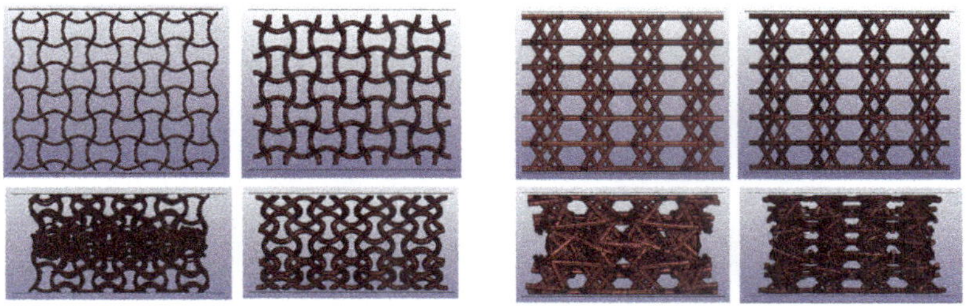

Figure 8. Numerical impact analyses: on the top, the meshed geometries prior to compression, on the bottom, the same geometries during compression and shortly before densification, with related contact issues visible in the beam-based models. (**a**) Sinusoidal lattice: beam (**left**) and solid meshes (**right**). (**b**) Planar anti-chiral lattice: beam (**left**) and solid meshes (**right**).

The measurement of Poisson's ratio was carried out evaluating the maximum transverse contraction in both the auxetic directions. Subsequently, the finalized value was calculated by taking the average of the two, as per Equation (7).

$$\nu_{zt} = \frac{\Delta l_z / L_z}{\frac{1}{2}\left(\frac{\Delta l_x}{L_x} + \frac{\Delta l_y}{L_y}\right)} \quad (7)$$

For penetration, multiple solid models were constructed; then, they were reduced to a quarter exploiting symmetry, and initial velocities were imposed to the rigid impactor (Figure 9). The impactor was always centred with respect to the cellular geometry with nodal displacements perpendicular to the two symmetry planes fixed at zero. The contact implemented was, as for the impact analyses, a penalty-based contact with a friction coefficient set at 0.3.

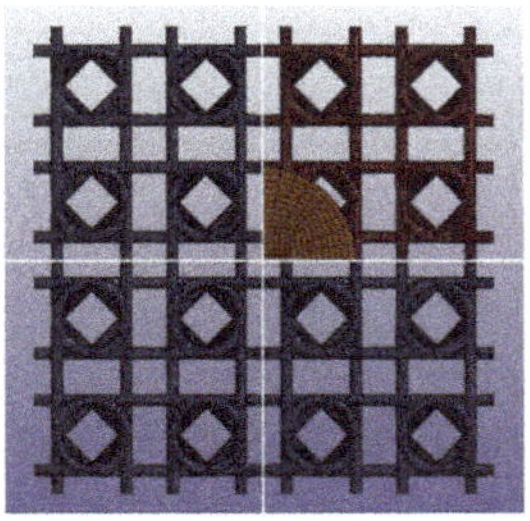

Figure 9. Low-speed penetration models, top view: the whole geometry (blue) and its symmetric portion (red).

3. Results

3.1. Quasi-Static Tests

Static compression tests, performed with T2K resin specimens as described in Section 2.3, involved all of the four presented geometries to compare the performances of the designs (Figure 10). Since this was a preliminary test campaign, intended to lead to a first initial screening and selection of the two best-performing structures, only one sample per geometry was tested.

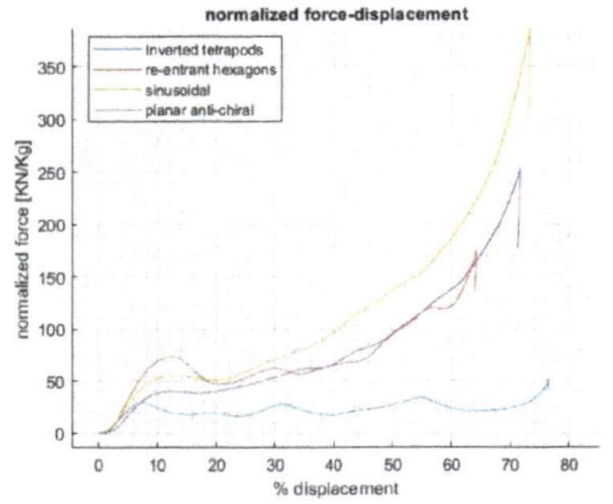

Figure 10. Quasi-static test results.

Stress–strain curves of the selected lattices highlighted two main results: the first is that the reaction force is increasing along the whole deformation due to the shrinking-induced interactions between the lattice beams in contrast to conventional non-auxetic lattices, which offer a plateau or a succession of peaks and valleys in the central section of the curve; the second is that toward densification, a bottoming point is not easily discernible as in conventional lattices with a gradual steepness increase. Such behavior, peculiar of auxetic structures, was however not observed for the inverted tetrapods geometry due to multiple local failures of the struts within contraction.

IT and REH showed a sequential failure layer by layer, originating the oscillations observable in the curve, which is behavior found also in most of the studies in the literature, while SIN and PAC reported an almost simultaneous bending of the vertical beams leading to optimal auxetic deformation. These results led to choosing PAC and SIN to proceed with the investigation. Further tests were carried out in parallel to numerical analyses, of which the final correlation is shown in Figure 11.

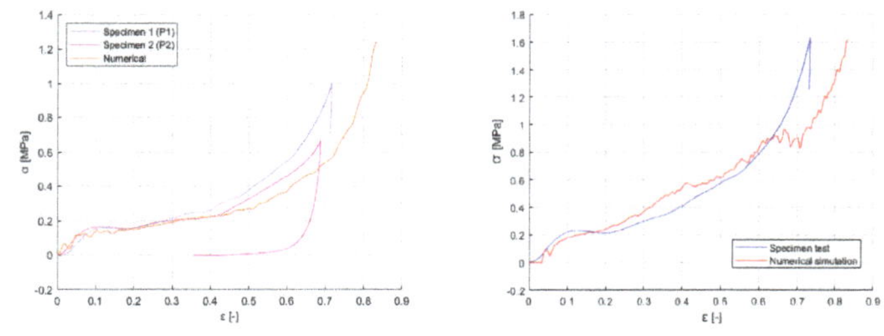

Figure 11. PAC (**left**) and SIN (**right**) quasi-static re-tests and numerical correlation; being the re-tests phase preliminary to the impact testing campaign, more demanding and relevant to the work, few samples were used: two for PAC, and one for the SIN lattice.

A set of PA12 specimen, one per each selected geometry, was then used to carry out further static tests, and properties were compared to the results of the resin ones (Table 4). The stress–strain curves for PAC were extremely coherent with the T2K behavior, while the SIN geometry showed pronounced brittleness and premature failure caused by the high deformations concentrated in the vertical beams. Full results are shown in the following section, where visual comparison of the stress–strain curves is provided between resin-based and powder-based materials, static and impact performances (Figure 12).

Table 4. Lattice mechanical properties. SEA is the specific energy absorption, E is the elastic modulus, σ_Y is the collapse stress, ν is Poisson's ratio.

-	SEA (KJ/Kg)	E (MPa)	σ_Y (MPa)	ν (-)
Sinusoidal T2K	2.24	3.43	0.232	−0.37
Sinusoidal PA12	2.13	2.1	0.169	−0.3
Anti-chiral T2K	1.61	2.4	0.16	−0.81
Anti-chiral PA12	1.74	2.47	0.15	−0.83

3.2. Impact Tests

Impact test results are shown in Figure 12.

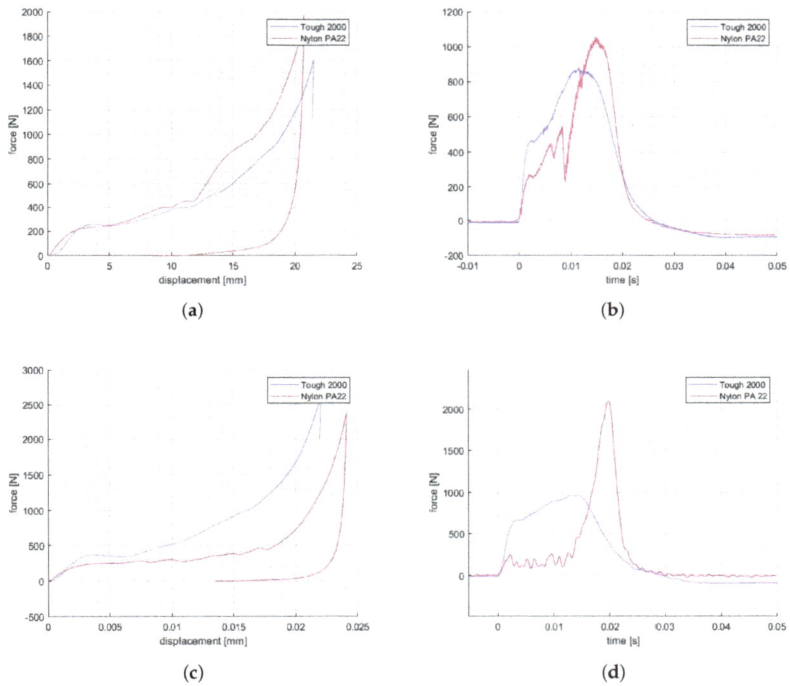

Figure 12. Static (**a**) and impact (**b**) compression of PAC lattices, static (**c**) and impact (**d**) compression of SIN lattices.

Concerning the resin-based specimens, the auxetic behavior observed under quasi-static conditions was effectively translated to low impact velocities as both the selected geometries, PAC and SIN, showed a pronounced transverse contraction under all tested speed up to 5 m/s. Considerable strength increase, close to 50%, was recorded with respect to the static counterpart in accordance with the high strain-rate dependency observed during material characterization (Appendix A). The shape of the curve, and thus the lattices behavior, was coherent with what was described in the previous section, and the peculiar force response was observed. Ultimately, the resin specimens improved performances under impact with no discernible weakening of their auxetic features. A few shots of the high-speed videos are reported in Figures 13 and 14.

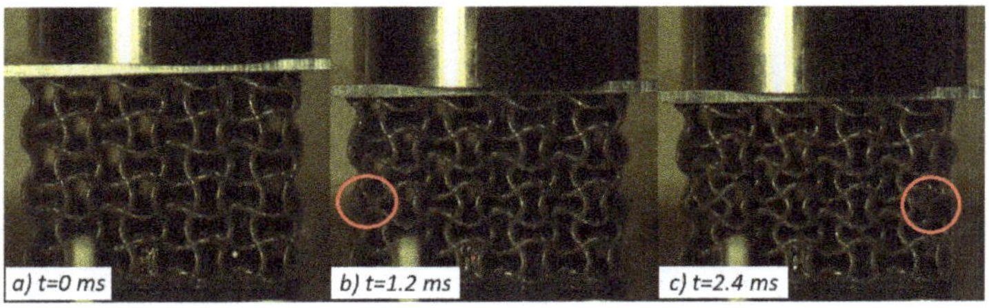

Figure 13. Impact test of SIN-T2K at 2.3 J with visible auxetic contraction (**c**); local failures are highlighted in frames (**b**,**c**).

Figure 14. Impact test of PAC-T2K at 2.3 J, with visible auxetic contraction (**c**); no local failures were reported.

On the contrary, nylon cores gave radically different results with respect to the static tests: the SIN specimen underlined notable brittleness, which was further worsened by the high deformation rate, causing total failure of the structure layer by layer and complete loss of the auxetic properties; PAC geometries, instead, were able to sustain the load without complete collapse of every layer and maintaining the auxeticity with the sole exception of the layer in immediate contact with the printing base.

All the geometries, except for SIN manufactured with nylon powder, recovered a notable portion of the deformation up to 70%. Specimens were then subjected to multiple impact tests in a row to investigate the structure capability of sustaining repeated impacts. The results, at each repetition, showed progressively lower force responses in the initial elastic phase, higher slopes in the subsequent crushing phase, and higher force peaks at densification. For the nylon specimens, such a trend is the most pronounced, as highlighted by the data collected in Table 5.

Table 5. Multiple impact load peak.

-	Impact 1 (F (N))	Impact 2 (F (N))	Impact 3 (F (N))
Sinusoidal T2K	983.6	1372.2	1698.5
Sinusoidal PA12	2102.7	-	-
Planar Anti-Chiral T2K	845.5	1031.7	1238.2
Planar Anti-Chiral PA12	1057.4	1796.3	2737.6

Of the three tested cases, the best-performing result was the PAC-T2K, able to sustain impacts with increasing peaks at an average rate of 23% per impact (Figure 15), which is in opposition to SIN-T2K and PAC-PA12, with rates of 37 and 80%, respectively.

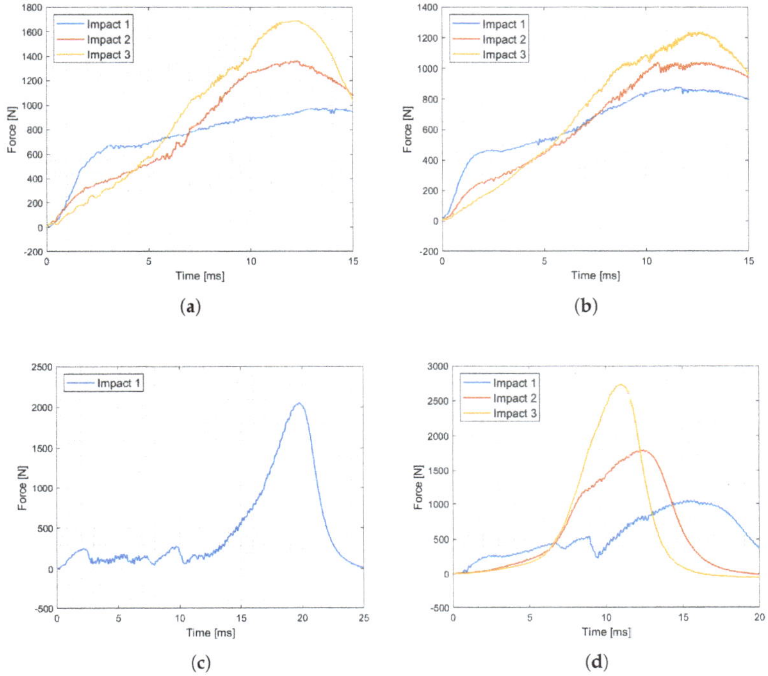

Figure 15. Repeated impacts results: PAC Tough 2K (**a**), SIN Tough 2K (**b**), PAC PA12 (**c**), SIN PA12 (**d**).

3.3. *Penetration Analyses*

Penetration is a complex phenomenon that is intrinsically local and dependent on many factors: the shape dimension of the impactor, material properties, impact energy and speed. For a cellular material, further complication comes with the relation between the cell and impactor dimension as well as the geometric and topological configuration.

As a consequence, it was decided to study the penetration behavior as a function of significant geometric parameters for both the planar anti-chiral and the sinusoidal lattices; investigations were performed numerically as reported in Section 2.4, and for both cases, the T2K-validated resin material was used with triaxiality dependency in order to capture the material behavior up to failure. The impactor was a rigid cylinder with a hemispherical tip of radius 8 mm.

3.3.1. Sinusoidal

The base sinusoidal geometry was the one reported in Section 2.1.3: from that configuration, it was decided to variate the sine amplitude, which is expressed as the percentage ratio between sine height and cell dimension and ranged between 0 and 20% with the former value equivalent to an FCC (face-centered cubic) lattice. Parallelly, the variation of impact speed was carried out with extraction of the maximum penetration depth of the impactor.

The results illustrated in Figure 16 reported that between amplitudes of 0% and 5%, straight (or slightly curved) beams break in bending, showing little to no auxeticity and ultimately low penetration resistance. Similar outcomes were obtained with amplitudes toward 20%: in such a case, the low clearances between the struts prevent auxetic contraction, and Poisson's ratio, theoretically the highest, dramatically increases while penetration depths reach maximum values. The optimum was observed for amplitudes between 12 and 14%, where a compromise between large clearances and a high Poisson's ratio allows reaching peaks of penetration resistance.

For what concerns the dependency on impact speed, an increase of penetration depth with impacting energy was reported, as reasonably expected; however, in qualitative terms, the results reported above were not affected by such energy.

Final outcomes are summarized in Figure 16.

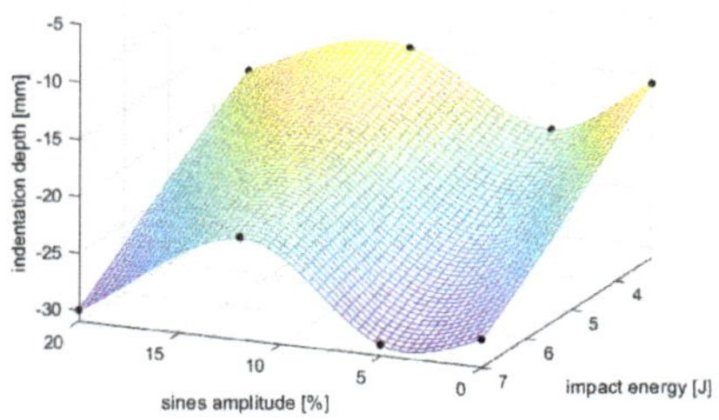

Figure 16. Penetration analysis of sinusoidal lattice; final results.

3.3.2. Planar Anti-Chiral

For the planar anti-chiral, the chosen variating parameters were the diameter of the toruses and the number of vertical layers of the lattice. The first was chosen due to its importance for the lattice mechanical properties and deformation mechanism; the second was chosen to evaluate the influence of multiple layers in the penetration response. Given that no significant dependence on the impact speed was shown for the analysis of the SIN lattices, the impact energy was kept constant at 3.2 J.

The radius of the node ranged from 2 to 3.5 mm, and each geometry iteration was tested for 3 and 6 layers of cells. As for the analyses of the sinusoidal presented in Section 3.3.1, the parameters are expressed as a percentage of the cell's characteristic length.

The results shown in Figure 17 indicated that the lattice performances are inversely proportional to the node radius increase. Such results were reasonably anticipated: higher radii translate in higher angles between the oblique beams and the horizontal plane, lower resisting torques of the plastic hinge at the base of the oblique ligaments and ultimately lower auxetic contraction capabilities. Since the reduction of penetration resistance is directly related to the benefits of the auxetic behavior, there was an evident decrease in the penetration resistance for low values of node radius.

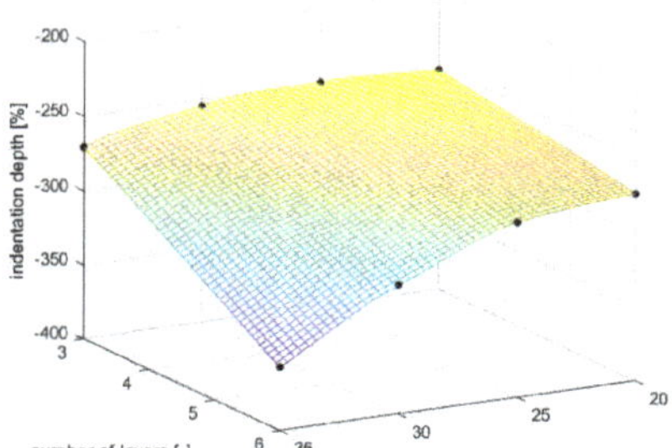

Figure 17. Penetration analysis of planar anti-chiral lattice, final results.

For what concerns the number of layers, the six-layers geometry presented a steeper increase in penetration resistance from 35% to 20% of the radius, while the three-layer models reported a maximum at 25%; this result suggests that the optimal value of the toroidal radius is not constant as a function of the number of vertical layers but shifts backwards as the number grows.

3.3.3. Non-Auxetic Geometries and Final Comparison

The unoptimized PAC and SIN were further used in a numerical comparison with non-auxetic lattices. In particular, two geometries were used: a regular BCC (Body-Centred Cubic), which was chosen as one of the most known and easily additively manufacturable geometries studied in the literature, and a stochastic Voronoi, which is surely amongst the most studied geometries in the field of mechanics of meta-materials. Geometric parameters were chosen targeting for both relative densities of 10% to allow a direct comparison in terms of penetration resistance, which is measured as the maximum vertical displacement of the impactor.

The BCC presented a cell dimension of L = 10 mm, and a strut radius r = 0.7 mm, which was set up to reach a relative density of 10%. The model was built by using Altair Hypermesh exploiting cell repetition schemes with an average element dimension of 0.5 mm.

The Voronoi, on the other hand, is a stochastic geometry and requires dedicated procedures to be properly designed and meshed for explicit analysis. For what concerns the design, parameters were chosen by exploiting recent work of the authors on stochastic Voronoi lattices [34], in particular by using the analytical equations relating relative density to cell dimension and strut diameter. Finalized meshes presented a strut radius of 0.5 mm and cell dimension of 4.7 mm. CAD geometries were automatically obtained through the related free software (https://github.com/ivncl/lattice300, 6 December 2023). Meshes were then constructed via the Altair Hypermesh Shrink Wrap feature, which is a method capable of computing high-quality meshes from complex components by approximation of the geometry; this method has recently been used by multiple researchers in the meshing of lattices [35,36] and is particularly suited for the treatment of stochastic geometries for which repetition schemes cannot be exploited [37].

Both BCC and Voronoi models were studied alongside PAC and SIN with identical setups as reported in Section 2.4. Impact energies were 3 and 7 J with lattice thicknesses set at 30 and 60 mm.

The results, reported in Table 6 and Figure 18, showed the best performances for the SIN lattice, followed by the Voronoi, at once confirming the benefits of the auxetic behavior for said lattice and questioning the effectiveness of the auxetic contribution for the PAC in this specific loading condition.

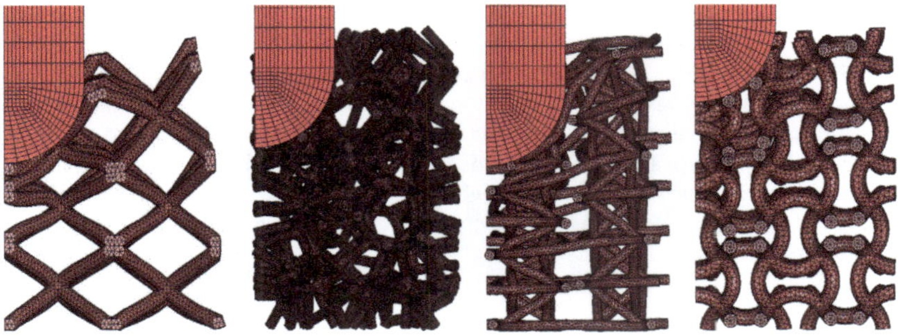

(**a**) Non-auxetic: BCC (**b**) Non-auxetic: VORO (**c**) Auxetic: PAC (**d**) Auxetic: SIN

Figure 18. Penetration analyses at 3 J impact energy. The images capture the maximum vertical displacement of the impactor, showing the penetration resistance of the non-auxetic structures (**a**,**b**) with respect to the auxetic ones (**c**,**d**).

Table 6. Final penetration results: maximum vertical displacement of the impactor (mm).

Lattice	30 mm–3 J	30 mm–7 J	60 mm–7 J
BCC	13.6	20.3	23.2
VORO	12.3	17.1	13.2
PAC	14.1	20.5	24.1
SIN	7.9	11.5	12.8

3.4. Modified Topology Investigations

After thorough analysis of polymeric lattices, it was decided to scale up the investigation in terms of absorbing performances of metallic auxetic structures and in particular additively manufactured stainless steel samples. Between the sinusoidal and the planar anti-chiral, the first has been already investigated experimentally and numerically [20–22], while to the authors' knowledge, the literature still lacks proper treatment of the impact properties of the planar anti-chiral topology; furthermore, the higher crushing force carried by the sinusoidal lattice is associated with higher local stresses and lower elasticity, possibly leading to fragile rupture in impulsive conditions such as impact. Consequently, the focus of this last stage of the present research was apply the second geometry.

When manufacturing was approached, it was immediately clear that the production of metallic anti-chiral structures via SLM is problematic: the main issue was identified in the horizontal ligaments, being at the highest possible overhang with short bridging, which is a well-known complication for most additive technologies, SLM included. Even if instances of considerably overhanged cellular structures can be found in the literature, studies suggest that suboptimal results are often obtained [38], and being fundamental for the objectives of the present work analyzing practical manufacturability and design robustness, it was decided to modify the planar topology to comply to the best practices of industrial manufacturing.

As such, horizontal ligaments were tilted 45°, as shown in Figure 19. The new design was assumed to cause lower values of Poisson ratio: in the planar topology, horizontal ligaments are in fact mainly responsible for auxetic contraction, and the design modification considerably changes the state of stress from purely compression/tension to bending, which is known to be less efficient. On the other hand, the overall force performances, mostly driven by the oblique ligaments' plastic rotation, were not likely to change. The new design was applied to specimens with identical dimensions and geometric parameters as per Section 2.2.

Once manufactured, the specimens highlighted issues on the first of the three planar layers, resulting in immediate breakage of the V-shaped, former horizontal, ligaments (Figure 19). The defect emerged due to the immediate contact between the first printed layer and the printing plate, resulting in unsupported ligaments. Prior to testing, numerical analysis of the samples taking into account the defect were carried out, resulting in almost superimposable force signals and a notable Poisson ratio decrease, as expected due to the absence of auxeticity for 1/3rd of the samples' height (Figure 20). These results were considered acceptable as no loss of crushing performances was evident, and proper treatment of the undefective geometry for a comprehensive overview of the absorbing properties was later addressed numerically.

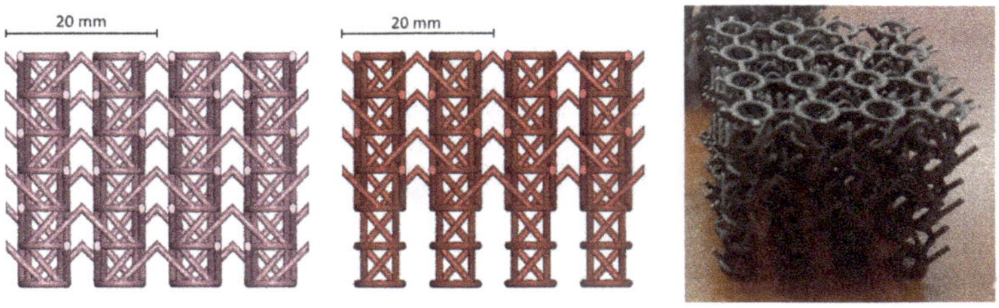

(a) (b) (c)

Figure 19. Modified planar topology, original design (**a**), defected geometry (**b**), printed samples (**c**); as easily seen in (**b**), the printed samples and numerical models are characterized by missing 45° ligaments on the whole first layer.

Impact tests were performed with a drop tower equipped with 90 kg mass dropped from a height of 400 mm, resulting in 353 J of energy and an impact speed of 2.8 m/s. Acquisition consisted of the measurement of vertical acceleration history on the impactor, to extract the force–displacement compressive behavior, and high-speed videos to evaluate crushing dynamics and Poisson's ratio evolution. Results are reported in Figures 20 and 21, showing a crushing strength of 5 t/m^2, a densification strain reported around 40–50% and specific energy absorption prior to said densification of 3.3 kJ/kg. The Poisson's ratio was measured at −0.4 by analysis of the high-speed videos, resulting in a value that was slightly lower, in modulus, than the numerical prediction (−0.54), and considerably lower with respect to the original structure (−0.9).

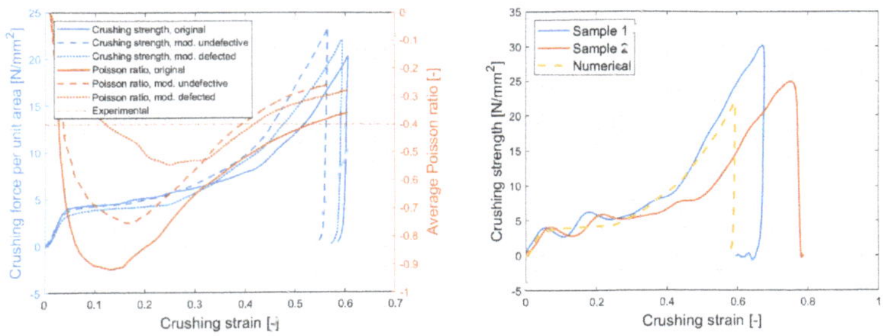

Figure 20. Numerical analyses of the original, modified defected and undefective geometries (**left**) and experimental results (**right**). As mentioned in the text, while the reduction of auxetic effect from original to final geometry is significant (respectively, −0.9, −0.75, −0.54, and −0.4 for original, modified undefected, modified defected and experimental result), notable differences in crushing strength performances were not identified.

As far as the crushing strength is concerned, it is worth noting the good correlation between the experimental and numerical results, which was obtained with little base material information for modeling build-up; this result suggests that the samples' behavior had remained in the elastic and mildly plastic regions for most of the compression, proving a major auxetic contribution up to densification. On the other hand, the reason for the suboptimal Poisson's ratio correlation can be identified in both the different measurement methods and related assumptions and the local geometric defects not implemented in the models.

Figure 21. High-speed footage of the modified PAC steel specimen subjected to impact. Prior to densification, occurring at t = 6.6 ms, pronounced transverse contraction can be appreciated at the center of the sample.

Further investigations were lastly carried out to systematically consider the influence of the new design on the crushing properties. To do so, an LS-OPT routine was exploited to parametrize the topology and analyze the effects of tilting angles lower than 45 degrees. The results, shown in Figure 22, are in good agreement with the experimental signals and confirm the negligible loss of performances in terms of crushing strength with respect to the original configuration with a slight increase of the slope of the linear pre-densification phase. On the other hand, a notable loss of Poisson's ratio was observed; in particular, the auxetic features drop to −0.75 at 45 degrees ligaments and −0.85 at 26 degrees ligaments: the trend may be approximated as parabolic, as per Equation (8).

$$\nu_{zy} = \nu_{zx} = 6.393 * 10^{-5}\,\theta^2 + 7.321 * 10^{-4}\,\theta - 0.9147 \tag{8}$$

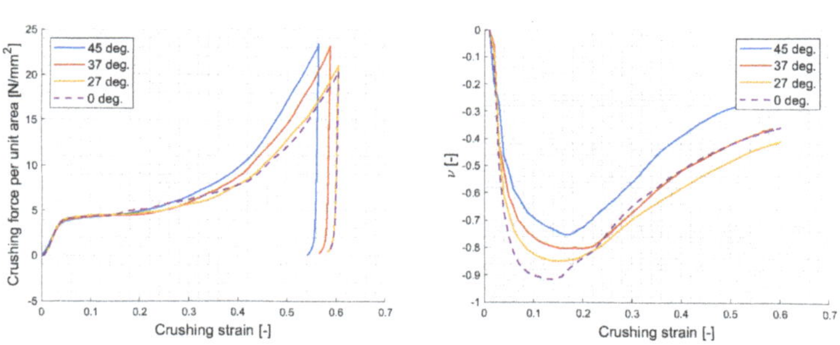

(**a**) (**b**)

Figure 22. Final numerical results of the modified planar topology: crushing strength (**a**) and Poisson ratio (**b**).

4. Summary and Discussion

The present work aimed to study the impact properties of auxetic lattices with particular attention on an empirical demonstration of manufacturability via common additive technologies, efficacy in resisting to impact loading, and validity of auxetic behavior under dynamic crushing. The work aimed to demonstrate the benefits of auxeticity in terms of peculiar crushing performances with enhanced energy absorption, resistance to repeated impacts, superior penetration behavior, feasibility of design and the use of auxetic absorbers in high-energy applications.

The work started with the selection of four of the most commonly evaluated geometries, produced via SLA with a high-performance resin, and subsequently tested in quasi-static compression conditions. After such preliminary static testing, two geometries showing the best performances in terms of resisting force and Poisson's ratio were probed under impact loading via single and repeated crushing. Following, the same geometries were produced via SLS and tested under similar conditions. Numerically, a preliminary study of behavior under penetration impact was carried out. Finally, an optimized geometry was constructed and produced via SLM for maximum performances; such geometry was then tested under impact with parallel comprehensive numerical analysis to robustly grasp its behavior.

4.1. Peculiar Crushing Behavior

In the tested structures that did not exhibit elastic instability, the load–displacement curve displayed a consistent upward "triangular" trend for the load. The steepness of this curve gradually increased until total densification was achieved: the behavior is considerably different from conventional cores, which typically have a plateau either at or below their initial response level with densification leading to a sudden load peak. This characteristic is not detrimental to the crushing response; in fact, it is a significant feature in load transmission. Comparison with conventional lattices can be made in terms of the maximum load, absorbed energy and crushing stroke. Figure 23 shows the response of the two investigated lattices and a typical honeycomb core loaded in the in-plane direction, which is implemented via shell-based numerical models using the same material relative density and core dimensions, with a numerical setup identical to the one used for the static tests described in Section 2.4.

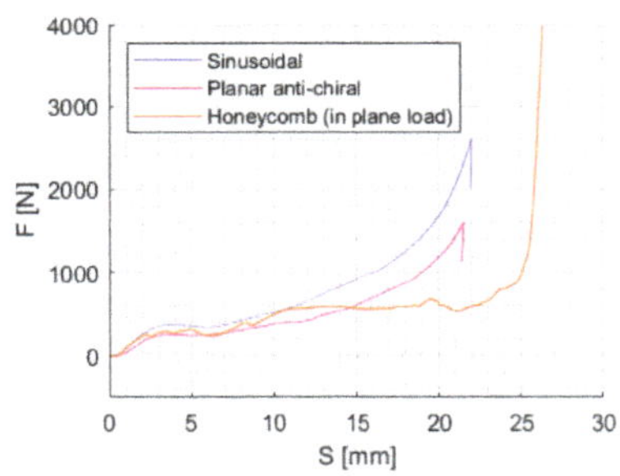

Figure 23. Comparison between auxetic lattices examined (sinusoidal and planar anti-chiral) and a non-auxetic honeycomb.

In the initial section, load levels are similar, with the honeycomb displaying lower values due to its plateau. However, at higher energy levels, the auxetic geometries show an early increase in the crushing force inevitably causing early densification though reducing the risk of bottoming phenomena.

If this unique pulse shape reduces the efficiency when compared to ideal absorbers, as expected, this peculiarity may be exploited when gradual crushing force application is needed. For example, in personal protective devices, a soft response to low-energy impacts may be desired, while a stronger response is needed for high-energy impacts. In this sense, the auxetic behavior could be an effective solution to protect from mild injuries.

4.2. Resistance to Repeated Impacts

When subjected to repeated tower drops, it was observed that for deformations up to 50% of the specimen's total height, the lattice could withstand multiple impacts with only a slight increase in peak force, particularly in the case of resin-based cores. A comparison with specimens produced from nylon powders confirmed that a low brittleness of the material is key to extend auxeticity to repeated impacts. It is apparent that the elastic–plastic behavior of the core is effective only if supported by materials with a pronounced elastic/plastic range.

In this context, the planar anti-chiral geometry showed greater resistance to brittle failure when compared to the sinusoidal one with promising versatility in various applications; the sinusoidal geometries, on the other hand, work at higher global forces and local stresses, likely leading to premature local failures even for high-elongating materials. Ultimately, the core's capability to absorb multiple impacts may be attributed to the material itself as well as to the lattice topology, and its evaluation, let alone numerical prediction, seems to require comprehensive base material characterization, including fracture and post-fracture behavior. Since such a process would be rather complex, especially for newly developed 3D-printed polymers, the authors believe that the construction of suited predictive models is necessary for the safe use of these structures.

4.3. Superior Penetration Behavior

Of the formulated hypotheses, the present is surely the most complicated to study let alone demonstrate in general terms due to the local effects of the phenomenon. However, numerical analyses here performed indubitably gave an affirmative response.

The results of the parametric analysis indicated that the sinusoidal cell reaches the optimum for sines amplitudes between 10 and 15% of the cell's characteristic length in contrast to the low performances of a FCC (Face-Centered Cubic) lattice; the planar anti-chiral, instead, presented the best results for low values of toroidal node radius, and minimum penetration depth was found for a radius close to 25% of the characteristic length.

To better appreciate the penetration resistance of PAC and SIN, performances of unoptmized auxetic lattices were compared to classical cores by means of further numerical analyses of a BCC (Body-Centered Cubic) and a Voronoi lattice of similar relative density (10%). Benefits of the auxetic behavior were evident especially for the SIN lattice, while PAC reported a penetration resistance comparable to non-auxetic cores. Such a result is directly linked to the lower force carried by the PAC, which, given its higher densification strain and limited local stresses, seems to be particularly suited for resistance to repeated loading rather than to the maximization of crushing/penetration performances.

As for the repeated impacts, it is however reasonable to hypothesize that such benefits are strongly dependent on the material properties, in particular on the toughness of the T2K resin, which is able to sustain impacts without leading to local failure detrimental to the auxetic deformation. In this sense, the present results need surely to be extended not only by experimental testing on tough materials such as the T2K, but also, and most important, by developing numerical models and experimental setups able to treat and analyze the fracture behavior of 3D-printed samples, especially for powder-based materials. In parallell, the variation of geometric parameters must be carried out for both auxetic and non-auxetic geometries, targeting a comparison of optimized configurations.

4.4. Use in High-Performance Applications

To demonstrate the possibility of constructing high-performance, metallic, auxetic absorbers, the planar anti-chiral lattice was redesigned for metal additive manufacturability and then tested under impact.

Within crushing, the results clearly showed both pronounced auxetic behavior and high performances, with the latter not invalidated by the presence of notable manufacturing defects and reported at 5 N/mm^2, which is comparable to high-resistance honeycombs. Furthermore, the absorber showed consistency for the two samples both in terms of qualitative deformation behavior and force performances, suggesting that the auxetic contraction is effective in reducing the probability of unsymmetrical or unstable compression. In this sense, this novel geometry, optimized for fast and reliable metal additive production, could be potentially a useful tool in conjunction with conventional absorbing structures—such as tubes—to be used as a multi-stage device able to respond to loads at different force levels. Of sure value would be further investigation with consideration of the post-impact response and analysis of its recovery capabilities.

5. Conclusions

The present work investigated the impact behavior of three-dimensional strut-based auxetic lattices in terms of practical manufacturability, feasibility and effective applicability for crashworthiness purposes. The activity focused at first on four representative topologies whose geometric parameters were taken from the literature; they were manufactured via resin-based additive technologies and experimentally tested under quasi-static compression. Subsequently, the two most promising geometries were thoroughly tested and modeled under impact crushing, which was followed by the evaluation of their main mechanical properties with a resin and a powder-based technology. While the powder-based samples struggled to maintain the auxetic contraction shown in quasi-static tests, the resin-based specimens reported outstanding performances and the capability of resisting multiple successive impacts. Afterwards, the response to impact penetration was studied numerically via high-quality solid-meshed explicit models, reporting high potentiality for both geometries evaluated and outperforming conventional cores. Last, the topology with higher auxetic behavior was re-designed, proposing a novel configuration optimized for metal additive manufacturing; it was then produced and tested under impact, reporting robustness and high specific crushing strength.

In conclusion, it can be surely said that, to date, the construction of effective auxetic meta-materials for impact absorption via additive techniques is a complex task; however, it is achievable. Topologies available in the literature showed particular reliability when manufactured in resin-based polymers with persistent auxeticity up to densification and the capability of sustaining repeated impact loading. Powder-based polymers presented instead multiple difficulties, demonstrating that auxetic properties are not easily transferable from static to impact compression and from one material to another: in this sense, the investigation of powder-based materials is surely needed, on both the numerical and experimental side, to design auxetic topologies suited for such technologies. However, the present work successfully demonstrated the potentiality of additively manufactured auxetic cellular structures for impact absorption, which is a step ahead of the previous literature regarding empirical demonstrations of repeated impacts and penetration behavior. Furthermore, it was shown that a robust additive production of metallic auxetic structures is achievable within standard manufacturing best practices, and high-performance devices can be obtained.

Since the present work is mostly of an empirical nature, experimental and numerical, the authors believe that future research needs to focus on the development of effective theoretical methods able to explain the repeated impact and penetration performances of auxetic lattices. Systematic numerical studies and effective experimental campaigns need to be carried out to better understand the penetration problem, targeting the reliable, generalized prediction of penetration depth as a function of geometric and material parameters. In parallel, focus must be put on the investigation of powder-based polymers with the development of robust strategies for the minimization of threatening consequences of their manufacturing technology and suited numerical models to predict their impact behavior. Further numerical and experimental studies must also be carried out to optimize the geometry of the hereby proposed metallic absorber and develop high-performant multi-stage devices.

Author Contributions: Conceptualization, I.C. and M.A.; methodology, P.F. and I.C.; validation, P.F. and I.C.; formal analysis, P.F. and I.C.; investigation, P.F. and I.C.; writing—original draft preparation, P.F.; writing—review and editing, I.C. and A.G.; supervision, M.A. and M.B.; project administration, M.A. and M.B. All authors have read and agreed to the published version of the manuscript.

Funding: This research received no external funding

Institutional Review Board Statement: Not applicable.

Informed Consent Statement: Not applicable.

Data Availability Statement: Data will be made available on request.

Acknowledgments: This research was supported by the company cirp GmbH. In particular, fundamental support was provided in samples' manufacturing by Ubiratan S. Freitas, Fatma Karayagiz and Hans Nopper.

Conflicts of Interest: The authors declare no conflicts of interest.

Abbreviations

The following abbreviations are used in this manuscript:

API	Application Program Interface
BCC	Body-Centered Cubic
EBM	Electron Beam Melting
FDM	Fused Deposition Modeling
IT	Inverted Tetrapod
PAC	Planar Anti-Chiral
RH	Re-Entrant Hexagon
SEBM	Selective Electron Beam Melting
SIN	Sinusoidal
SLA	Stereolithography
SLM	Selective Laser Melting
SLS	Selective Laser Sintering

Appendix A. Characterization of Formlabs Tough 2000

Since data on Formlabs Tough 2000 were not available in the literature, in particular as far as the dynamic properties are concerned, dedicated test campaigns were carried out to extract the material properties in order to properly set up the numerical models. The campaign was composed of static and dynamic tests. Specimens were classical dogbones as per ASTM D638-14 [39] for both campaigns (Figure A1).

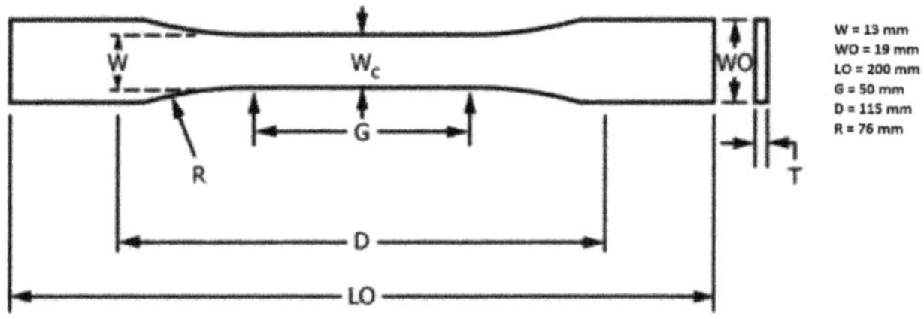

Figure A1. Dogbone specimens dimensions.

Appendix A.1. Quasi-Static Tests

Quasi-static tests were conducted on an MTS-370.10 machine, at speeds of 5 mm/min (0.0016 s^{-1}); in total, six tests were performed at this strain rate, three specimens for each printing orientation to evaluate the degree of isotropy of the printed material. A strain gauge was also applied to each specimen to record the strain history at the expected failure point. Prior to testing, the samples were treated with dedicated painted patterns to perform

DIC analyses, which were fundamental to correlate global and local strain measurement methods and calculate transverse strain and related Poisson's ratio.

The results, as shown in Figure A2, clearly showed quasi-isotropic behavior up to failure with almost superimposable curves and considerably repeatable outcomes. Even the failure strain, calculated through the strain gauge, presented notable variability between 30% and 50% and did not exhibit dependence on the printing direction.

Failure of the material was brittle, not showing prominent necking effects, while of notable importance seemed to be the layering direction with respect to the rupture dynamics and shape, as shown in Figure A3: The XZ specimen have a neat inter-layer failure, while the XY specimen have an 45° failure line with detachment of a triangular chip; such a mechanism was coherent between the two sets.

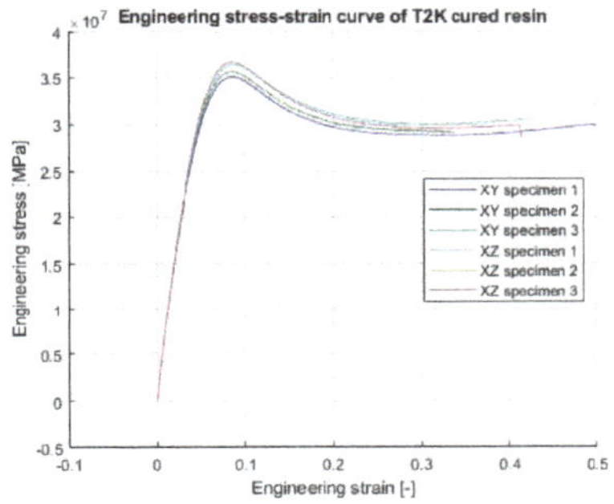

Figure A2. Tough 2000 tensile tests.

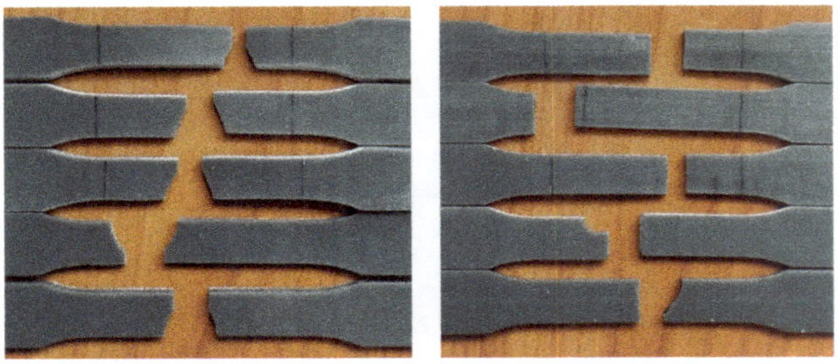

Figure A3. Tensile failure in XY (**left**) and XZ (**right**) direction.

Appendix A.2. Dynamic Tests

Dynamics tests were aimed to investigate the strain rate dependency, obtaining the variation of the tensile properties at increasing deformation rates. Low deformation rates (40, 80, 100 s^{-1}) were analyzed here, as crashworthiness properties were of interest. In order to execute the test, a Step Lab DW1000 was exploited: the lower part of the dogbone was clamped to a T-shaped grip, free to move, while the upper part was clamped to a rigid support equipped with a force sensor: in the present case, a Dytran 1210V4. The tensile dynamic load was then applied through a falling mass mounted on a vertical sled, released from the designed height, impacting the lower grip and quickly tearing the specimen.

As commonly expected from similar tests, dynamic tests reported an increasing in both yield and ultimate stress. Results are shown in Figure A4.

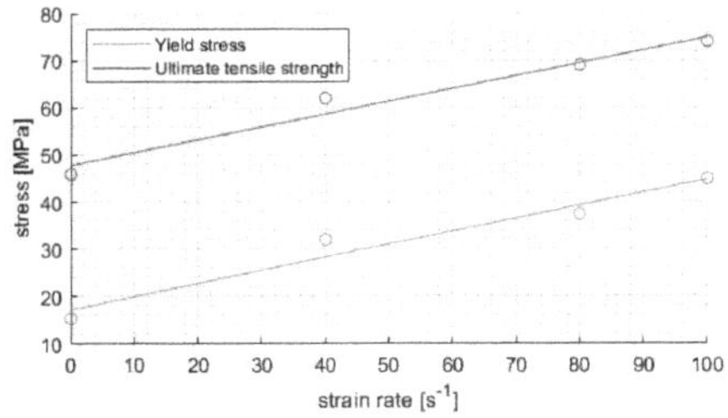

Figure A4. Dynamic tensile tests results.

References

1. Evans, K.E. Auxetic polymers: A new range of materials. *Endeavour* **1991**, *15*, 170–174. [CrossRef]
2. Carneiro, V.H.; Meireles, J.; Puga, H. Auxetic materials—A review. *Mater. Sci.* **2013**, *31*, 561–571. [CrossRef]
3. Baughman, R.; Shacklette, J.; Zakhidov, A.; Stafström, S. Negative Poisson's ratios as a common feature of cubic metals. *Nature* **1998**, *392*, 362–365. [CrossRef]
4. Kimizuka, H.; Kaburaki, H.; Kogure, Y. Mechanism for Negative Poisson Ratios over the α- β Transition of Cristobalite, SiO_2: A Molecular-Dynamics Study. *Phys. Rev. Lett.* **2000**, *84*, 5548–5551. [CrossRef] [PubMed]
5. Grima, J.N.; Winczewski, S.; Mizzi, L.; Grech, M.C.; Cauchi, R.; Gatt, R.; Attard, D.; Wojciechowski, K.W.; Rybicki, J. Tailoring Graphene to Achieve Negative Poisson's Ratio Properties. *Adv. Mater.* **2015**, *27*, 1455–1459. [CrossRef] [PubMed]
6. Poźniak, A.A.; Wojciechowski, K.W.; Grima, J.N.; Mizzi, L. Planar auxeticity from elliptic inclusions. *Compos. Part Eng.* **2016**, *94*, 379–388. [CrossRef]
7. Evans, K.; Nkansah, M.; Hutchinson, I. Auxetic foams: Modelling negative Poisson's ratios. *Acta Metall. Mater.* **1994**, *42*, 1289–1294. [CrossRef]
8. Milton, G.W. Composite materials with poisson's ratios close to -1. *J. Mech. Phys. Solids* **1992**, *40*, 1105–1137. [CrossRef]
9. Wang, Y. Auxetic Composite Laminates with Through-Thickness Negative Poisson Ratio for Mitigating Low Velocity Impact Damage: A Numerical Study. *Materials* **2022**, *15*, 6963. [CrossRef]
10. Lin, W.; Wang, Y. Low velocity impact behavior of auxetic CFRP composite laminates with in-plane negative poisson's ratio. *J. Compos. Mater.* **2023**, *57*, 2029–2042. [CrossRef]
11. Gibson, L.J. Cellular solids. *Mrs Bull.* **2003**, *28*, 270–274. [CrossRef]
12. Novak, N.; Vesenjak, M.; Ren, Z. Auxetic cellular materials—A review. *Stroj. Vestn. J. Mech. Eng.* **2016**, *62*, 485–493. [CrossRef]
13. Usta, F.; Türkmen, H.S.; Scarpa, F. Low-velocity impact resistance of composite sandwich panels with various types of auxetic and non-auxetic core structures. *Thin Walled Struct.* **2021**, *163*, 107738. [CrossRef]
14. Gao, Q.; Liao, W.H.; Huang, C. Theoretical predictions of dynamic responses of cylindrical sandwich filled with auxetic structures under impact loading. *Aerosp. Sci. Technol.* **2020**, *107*, 106270. [CrossRef]
15. Usta, F.; Ertaş, O.F.; Ataalp, A.; Türkmen, H.S.; Kazancı, Z.; Scarpa, F. Impact behavior of triggered and non-triggered crash tubes with auxetic lattices. *Multiscale Multidiscip. Model. Exp. Des.* **2019**, *2*, 119–127. [CrossRef]
16. Günaydın, K.; Gülcan, O.; Türkmen, H.S. Experimental and numerical crushing performance of crash boxes filled with re-entrant and anti-tetrachiral auxetic structures. *Int. J. Crashworthiness* **2023**, *28*, 649–663. [CrossRef]
17. Yang, L.; Harrysson, O.; West, H.; Cormier, D. Mechanical properties of 3D re-entrant honeycomb auxetic structures realized via additive manufacturing. *Int. J. Solids Struct.* **2015**, *69*, 475–490. [CrossRef]
18. Almgren, R. An isotropic three-dimensional structure with Poisson's ratio = -1. *J. Elast.* **1985**, *15*, 427–430.
19. Ebrahimi, H.; Mousanezhad, D.; Nayeb-Hashemi, H.; Norato, J.; Vaziri, A. 3D cellular metamaterials with planar anti-chiral topology. *Mater. Des.* **2018**, *145*, 226–231. [CrossRef]
20. Novak, N.; Starčevič, L.; Vesenjak, M.; Ren, Z. Blast response study of the sandwich composite panels with 3D chiral auxetic core. *Compos. Struct.* **2019**, *210*, 167–178. [CrossRef]
21. Novak, N.; Vesenjak, M.; Tanaka, S.; Hokamoto, K.; Ren, Z. Compressive behavior of chiral auxetic cellular structures at different strain rates. *Int. J. Impact Eng.* **2020**, *141*, 103566. [CrossRef]
22. Novak, N.; Krstulović-Opara, L.; Ren, Z.; Vesenjak, M. Compression and shear behavior of graded chiral auxetic structures. *Mech. Mater.* **2020**, *148*, 103524. [CrossRef]
23. Jenett, B.; Cameron, C.; Tourlomousis, F.; Rubio, A.P.; Ochalek, M.; Gershenfeld, N. Discretely assembled mechanical metamaterials. *Sci. Adv.* **2020**, *6*, eabc9943. [CrossRef]
24. Balaji, B.; Burela, R.G.; Ponniah, G. Mass production of re-entrant cubic auxetic structure. In *IOP Conference Series: Materials Science and Engineering*; IOP Publishing: Bristol, UK, 2021; Volume 1070, p. 012128.
25. Novak, N.; Nowak, M.; Vesenjak, M.; Ren, Z. Structural optimization of the novel 3D graded axisymmetric chiral auxetic structure. *Phys. Status Solidi B* **2022**, *259*, 2200409. [CrossRef]
26. Varas, D.; Pernas-Sánchez, J.; Fjeldberg, N.; Martín-Montal, J. Experimental analysis at different loading rates of 3D printed polymeric auxetic structure based on cylindrical elements. *Polym. Test.* **2023**, *119*, 107930. [CrossRef]

27. Galea, R.; Andre Farrugia, P.S.; Dudek, K.K.; Attard, D.; Grima, J.N.; Gatt, R. A novel design method to produce 3D auxetic metamaterials with continuous pores exemplified through 3D rotating auxetic systems. *Mater. Des.* **2023**, *226*, 111596. [CrossRef]
28. Wei, G. Negative and conventional Poisson's ratios of polymeric networks with special microstructures. *J. Chem. Phys.* **1992**, *96*, 3226–3233. [CrossRef]
29. Schwerdtfeger, J.; Heinl, P.; Singer, R.F.; Körner, C. Auxetic cellular structures through selective electron-beam melting. *Phys. Status Solidi B* **2010**, *247*, 269–272. [CrossRef]
30. Novak, N.; Vesenjak, M.; Ren, Z. Computational simulation and optimization of functionally graded auxetic structures made from inverted tetrapods. *Phys. Status Solidi B* **2017**, *254*, 1600753. [CrossRef]
31. Evans, K. The design of doubly curved sandwich panels with honeycomb cores. *Compos. Struct.* **1991**, *17*, 95–111. [CrossRef]
32. Chen, J.; Tao, W.; Pang, S. Impact Testing of 3D Re-Entrant Honeycomb Polyamide Structure Using Split Hopkinson Pressure Bar. *Appl. Sci.* **2021**, *11*, 9882. [CrossRef]
33. Körner, C.; Liebold-Ribeiro, Y. A systematic approach to identify cellular auxetic materials. *Smart Mater. Struct* **2014**, *24*, 025013. [CrossRef]
34. Colamartino, I.; Anghileri, M.; Boniardi, M. Investigation of the compressive properties of three-dimensional Voronoi reticula. *Int. J. Solids Struct.* **2023**, *284*, 112501. [CrossRef]
35. Peto, M.; Aguilar-Rosas, O.; Ramirez-Cedillo, E.; Jimenez, M.; Hernandez, A.; Siller, H.R. A proof of concept study of the mechanical behavior of lattice structures used to design a shoulder hemi-prosthesis. *J. Eng. Sci. Med. Diagn. Ther.* **2021**, *4*, 031004. [CrossRef]
36. Daynes, S.; Feih, S. Bio-inspired lattice structure optimisation with strain trajectory aligned trusses. *Mater. Des.* **2022**, *213*, 110320. [CrossRef]
37. Giustina, A.; Colamartino, I.; Franzosi, P.; Anghileri, M.; Boniardi, M. Parametric optimization of cellular materials through LS-OPT. In Proceedings of the 14th European LS-DYNA Conference, Baden, Germany, 18–19 October 2023.
38. Thomas, D. The Development of Design Rules for Selective Laser Melting. Ph.D. Thesis, Cardiff Metropolitan University, Cardiff, UK, 2009.
39. *ASTM D638*; Standard Test Method for Tensile Properties of Plastics. ASTM International: West Conshohocken, PA, USA, 2022.

materials

MDPI

Article

Mechanical Characterisation of Bond Formation during Overprinting of PEEK Laminates

Simon Hümbert *, Fynn Atzler and Heinz Voggenreiter

German Aerospace Center (DLR), Institute for Structures and Design (BT), 70569 Stuttgart, Germany
* Correspondence: simon.huembert@dlr.de; Tel.: +49-711-6862-8044

Abstract: The latest generation of high-temperature 3D printers enables the production of complex structural components from aerospace-grade thermoplastics such as PEEK (polyether ether ketone). However, adding long or continuous fibres is currently limited, and thermal stresses introduced during the process restrict the maximum part dimensions. Combining 3D-printed components with continuous fibre-reinforced components into one hybrid structure has the potential to overcome such limitations. This work aims to determine whether in situ bonding between PEEK laminates and PEEK 3D printing during overprinting is feasible and which process parameters are significantly responsible for the bonding quality. To this end, the bonding is analysed experimentally in two steps. Firstly, the influence of the process parameters on the thermal history and the strength of the bond is investigated. In the second step, a detailed investigation of the most critical parameters is carried out. The investigation showed the feasibility of overprinting with bonding strengths of up to 15 MPa. It was shown that the bonding strength depends primarily on the temperature in the interface. Additionally, the critical parameters to control the process were identified. The process influences that were displayed form the basis for future hybrid component and process designs.

Keywords: additive manufacturing; FDM; overprinting; PEEK; thermoplastic composites; in situ bonding; automated fibre placement

1. Introduction

In recent years, there has been notable progress in the additive manufacturing of high-temperature thermoplastics, such as PEEK (polyether ether ketone) or PEI (polyethyleneimine), through material extrusion processes like Fused Filament Fabrication (FFF) or Fused Granular Fabrication (FGF). Various solutions for thermal management and an exemplary implementation were shown by [1]. This printing process has become a standard procedure for producing complex parts. Studies have shown that high process temperatures are necessary for adequate interlayer bonding and, subsequently, proper mechanical properties of high-temperature printed parts [2,3]. To cater to this need, the newest generations of commercial, high-temperature FFF machines have heated print beds and chambers that can reach up to 300 °C, making it possible to manufacture high-temperature, aerospace-grade materials through additive manufacturing.

Although 3D-printed, high-temperature thermoplastic material has promising properties, its mechanical performance still falls short of that achieved through injection moulding [4]. Reviews by [4] and [5] have shown that the reduction in mechanical properties of 3D-printed material compared to other manufacturing processes is mainly due to low inter-layer bonding strength and porosity. Additionally, generated thermal stresses during material extrusion can result in warpage and fractures within the part, thereby limiting the part's maximum dimensions [5,6]. These factors, together with the limited ability to add long or even continuous fibres [5], still severely restrict the use of additive manufacturing of thermoplastics for aerospace applications.

To overcome these limitations in manufacturing, one possible solution is to merge thermoplastic 3D printing with automated fibre placement or compression moulding. This combination aims to combine the intricate design possibilities of 3D printing with the exceptional material properties of continuous fibre-reinforced thermoplastics and their high production output.

One approach to achieving this goal is to produce continuous fibre-reinforced laminates and 3D-printed components separately and then combine them. Li et al. [7] studied sandwich structures with complex 3D-printed core shapes joined with continuous fibre-reinforced laminates using adhesive bonding. Another option is tailored composite designs like those proposed by Janssen et al. [8], where consolidation between printed core and laminate occurs through a thermoforming process. Rakhshbahar et al. [9] proposed achieving the bonding during automated fibre placement (AFP). When using AFP to join components, the heat source of the AFP system (e.g., laser) is used to melt the surface of the substrate and thermoplastic tapes. The components are then consolidated using a compaction roller [10].

Other studies have explored the possibility of bonding 3D-printed structures directly onto laminates without additional joining steps, similar to overmoulding. In one example, Morales et al. [11] developed a process called "over 3D-printing" using PA 6 (polyamide 6) organo-sheets. A more detailed analysis of the bond strength between PA 6 laminates and 3D-printed PA 6, especially the influence of process parameters, has been shown by Penter et al. [12] and Maier et al. [13]. Boros et al. [14] conducted a study on PLA (polylactic acid), comparing the properties of overprinting with overmoulding. Raspall et al. [15] describe a facility where two cooperating robots perform the tape placement and printing simultaneously. An automated process chain for the flexible production of hybrid components using non-planar printing using industrial robots has been described by Matkovic et al. [16]. This process chain offers several features to optimise the bonding conditions, such as pre-heating, substrate heating and additional consolidation pressure.

All these studies were conducted with commodity or engineering plastics. Expanding the overprinting process to high-temperature thermoplastics will increase the demands on the process. The main requirements can be derived from the theory of layer adhesion in 3D printing. Sun et al. [17–20] described the bond formation as a diffusion bonding based on the crack healing theory by [21]. Coogan et al. [22–24] and Li et al. [25] have shown a similar approach. An adaption of such a model using PEEK was shown by Basgul et al. [26]. The main principle of these bonding mechanisms is that bond formation occurs only as long as the interface temperature is above the thermoplastic's glass transition temperature, Tg. Regarding semi-crystalline thermoplastics, the bond formations widely stopped once the interface temperature had fallen short of the re-crystallisation temperature.

The importance of interface temperatures for the 3D printing of PEEK was also shown experimentally. Zanjanijam et al. [27] give an overview of the main challenges of printing PEEK and state that, additionally to the maximum temperatures, minimising the difference between nozzle and substrate temperature is the most critical aspect for sufficient mechanical properties. Similarly, Yang et al. [2] as well as Wang et al. [28] report increasing mechanical properties and an increasing degree of crystallinity for increasing process temperatures. Yi et al. [29] investigated the correlation between crystallinity and mechanical properties and showed the importance of controlling the crystallinity. Pu et al. [30] have shown similar effects for dynamic mechanical analysis. Deng et al. [31] and Wu et al. [32] described the significant effects of layer thickness and infill. These results not only represent the main requirements for FFF printing of PEEK but can also be largely applied to the overprinting of PEEK laminates.

First experimental studies on bonding effects between printed PEEK and PEEK laminates during overprinting have been presented by Caprais et al. [33,34] and Hümbert et al. [35]. The results support the claim that high bonding temperatures are crucial for bonding strength.

The German Aerospace Center's Institute for Structures and Design (BT) is working on integrating 3D printing into the process chain for producing thermoplastic composites. The aim is to use 3D printing to increase the complexity and functionality of the structures. Overprinting of laminates made of high-performance thermoplastics plays a central role here. Ref. [36] shows an example of combining 3D printing and automated fibre placement in two robotic cells using PEEK. Other applications are the integration of electronics into sandwich structures [37,38] or the application of radiation protection for space applications [39,40].

While the basics of making hybrid components by overprinting thermoplastic laminates have been described, a deeper understanding of the process is necessary to use the method for high-performance applications. Thus, the major goal of this work is the characterisation of bonding mechanisms between the 3D-printed structure and the laminate during overprinting. Short-fibre-reinforced PEEK is considered for overprinting CF-PEEK laminates to enable the application of this process to primary aerospace structures. To achieve this goal, the investigation is divided into two steps. First, a parameter screening is carried out to identify the crucial elements that influence the bonding between the 3D print and the laminate. The bond's strength and the interface's thermal history are analysed. In the second step, a detailed characterisation of these crucial influencing factors is performed.

2. Materials and Methods

Additive manufacturing is characterised by a large number of process parameters that have a complex effect on the properties of the finished part. During overprinting thermoplastic laminates, both the requirements for the joint between the laminate and the 3D-printed component and the requirements for the 3D-printed component itself must be met. Therefore, overprinting should be carried out within the process window for PEEK 3D printing. This process window was determined experimentally in a previous study [35] and serves as a boundary condition for the experimental design space in this work. The workflow used within this work and discussed above is visualised in Figure 1.

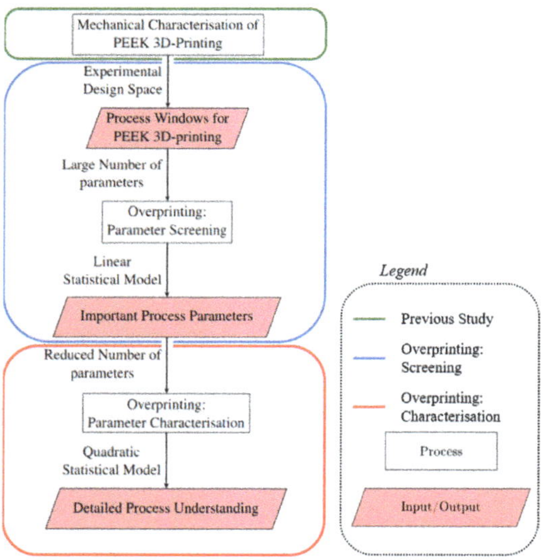

Figure 1. Flow chart of the structure of the investigation [33].

The parameter screening of the overprinting process is supposed to identify the most significant effects of process parameters on the in situ bonding of printed PEEK onto laminates using minimal testing effort. The results from the preliminary study serve as the limits for the test design.

Based on the results of the parameter screening, the test design was narrowed down for a final characterisation of the overprinting process. The characterisation described not only the main effect of the parameters but also the interactions of these parameters, providing a detailed understanding of the overprinting process.

2.1. Materials and Printer

All specimens for the screening and characterisation of overprinting were printed from Ensinger TECAFIL PEEK VX CF30 (Ensinger Plastics, Nufringen, Germany) [41], a PEEK filament filled with 30 wt% carbon fibres. The carbon fibre filling primarily improves processability. On the one hand, a stiffer filament and a higher melt viscosity ensure a stable material extrusion. On the other hand, the fibre content reduces heat shrinkage. At the same time, the fibre content also increases the mechanical properties of the printed component. The material properties, as well as the recommended printing parameters from the data sheet, are summarised in Table 1. A microsection of the used filament is shown in Figure 2. The laminate used for overprinting was a Toray Cetex 1200 UD laminate (Toray Advanced Composites, Morgan Hill, CA, USA) in the configuration $[45/0/-45/90]_s$.

Table 1. Material properties of pure PEEK and carbon fibre-reinforced PEEK filament [41].

Property	PEEK-CF30
Density [g/cm^3]	1.38
Young's modulus [MPa]	17,500
Ultimate strength [MPa]	190
Glass transition temperature [°C]	143
Melting temperature [°C]	343
Service temperature [°C]	260

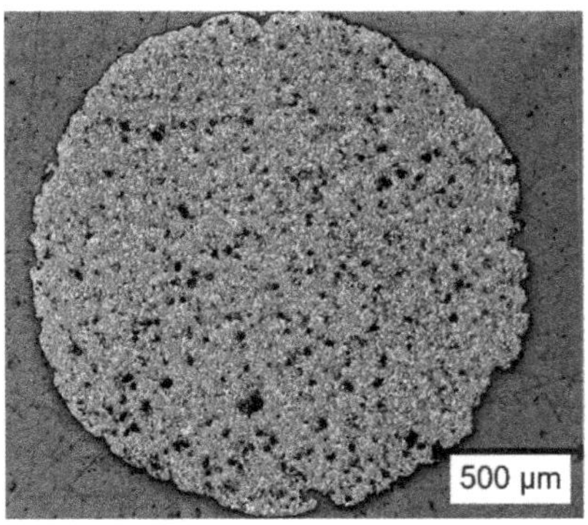

Figure 2. Microsection of PEEK-CF30 filament.

For this study, all specimens were printed using a GEWO HTP-260 (GEWO3D, Wörth, Germany) [42] high-temperature FFF printer. The printer is equipped with a heated bed that can reach up to 270 °C, a heated chamber that can reach up to 260 °C, and a nozzle that can be heated up to 450 °C.

2.2. Experimental Design

In a previous study [35], the influence of process temperatures on the mechanical properties of printed PEEK was investigated. It was shown that a build chamber temperature of at least 200 °C and a nozzle temperature of at least 430 °C are necessary for acceptable print quality with the given printer. These values serve as lower limits for the experimental design in this work.

The overprinting screening and characterisation test plans were designed using the DoE software Design Expert 13. The software optimises the tested factor levels and fits a surface response model to the experimental results. ANOVA (analysis of variance) was performed to evaluate the fitted model and find the significant effects. Model parameters and effects were considered significant with a p-value below 0.05. Results are represented by the model prediction at the centre point of the design and the 95% confidence level. The intensity of effects is described by the Pearson correlation r. Pearson correlation r describes the nature and strength of a linear effect with a single value between −1 and 1.

2.2.1. Overprinting: Parameter Screening

The screening consists of 12 runs fitting a linear surface response model. Response Surface Methodology (RSM) allows a large number of process parameters to be evaluated with relatively few tests. This linear model is designed to find the main effects of the process parameters on the bond strength. Process temperatures were investigated along with print speed, extrusion factor and the height of the first printing layer, as these parameters will likely impact the thermal history in the bonding interface. Additionally, an over-extrusion of the first printing layer was expected to increase the pressure and surface wetting during bond formation. The used parameter ranges and the parameters for each run for printing are shown in Table 2. The entire test design is shown in Table A1 in Appendix A. The bond strength was measured using guided compression shear tests. Additionally, microsections were taken from one specimen per parameter set. The specimen preparation is described in Section 2.3.

Table 2. Process parameters of the parameter screening and characterisation.

Process Parameter	Screening	Characterisation
Nozzle Temperature [°C]	430–450	430–450
Substrate Temperature [°C]	200–270	200–260
Chamber Temperature [°C]	200–260	200–260
Speed [mm/s]	5–20	12.5
First Layer Height [mm]	0.1–0.3	0.1–0.3
Extrusion Factor [%]	100–200	100
Nozzle Diameter [mm]	0.4	0.4
Layer Height Other Layers [mm]	0.2	0.2
Wall line count	2	2
Infill Raster [°]	±45	±45

2.2.2. Overprinting: Parameter Characterisation

The overprinting characterisation consists of 15 test runs fitting a quadratic surface response model. This model allows the analysis of main effects and interaction effects. A quadratic surface response model is also suitable for optimising the process. However, the optimisation is not the subject of this work. Since the separation of substrate temperature and chamber temperature has turned out unsuitable with the given printer setup, both temperatures are summarised into ambient temperature, with both temperatures at the same value. The investigated parameters were the nozzle temperature, ambient temperature and the height of the first printing layer. The print parameters are shown in Table 2. The entire test plan is shown in Table A2 in Appendix A. Specimen preparation microsections were carried out as described in Section 2.3. Additionally, the first layer height was measured using a 3D optical profilometer.

2.3. *Test Procedure and Equipment*

2.3.1. Compression Shear Testing

To qualify the bond strength between 3D printing and laminates, the inter-layer shear strength was measured using single lap shear specimens based on ASTM D3846-08 [43]. In contrast to other frequently used ISO standards, this ASTM standard provides a fixture to guide the specimens. The guided compression test was used to avoid buckling and peel loads in the interface. Due to the asymmetric stiffness of the specimens created by the short fibre-reinforced 3D print and the continuous fibre-reinforced laminate, non-guided specimens like single-lap tensile tests have proven unsuitable to characterise the bond strength. To further reduce the effect of residual stresses and warping in the specimens, the dimensions were scaled down to the values shown in Figure 3.

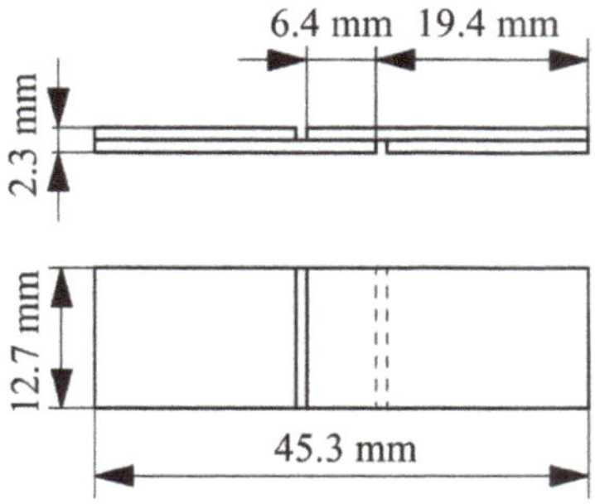

Figure 3. Geometry of compression shear specimens based on ASTM D3846-08 [43].

The testing machine RetroLine 1494 from ZwickRoell GmbH & Co. KG (Ulm, Germany) was used to determine the mechanical characteristics. Tensile tests were performed using mechanical clamps, and the compression shear tests used a fixture, according to [43].

Using this geometry, six specimens per configuration were printed onto one laminate, from which one was used for temperature measurement and the remaining five were used for mechanical testing. The interface temperature was measured using a single thermocouple embedded in the laminate. For this purpose, the laminate was notched halfway and the thermocouple was soldered in so that it was flush with the surface of the laminate. The print setup is shown in Figure 4.

Figure 4. Print setup with embedded thermocouple.

Using this setup, the laminate, including the embedder thermocouple, was overprinted with the six specimens. The specimens were then separated and notched with a disk saw. The overprinted laminate with six specimens is shown in Figure 5.

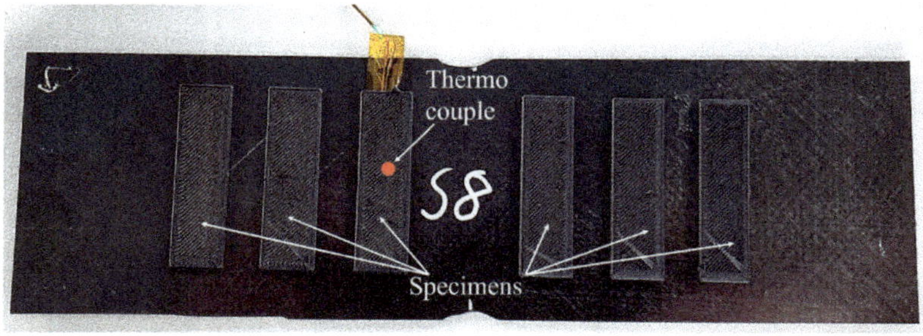

Figure 5. Printed specimens on laminate.

2.3.2. Microanalysis

The light microscopic examination was performed using the digital VHX-5000 from Keyence Deutschland GmbH (Leinfelden-Echterdingen, Germany) [44]. The objective used was the dual zoom objective VH-ZST with 20 to 200× or 200 to 2000× magnification.

The height profiles were measured using the Keyence vr-5200 optical 3D profilometer. The device achieves a measurement accuracy of ±2.5 µm [45].

3. Results

Test results are presented in two steps: parameter screening and characterisation.

3.1. Overprinting: Parameter Screening

The main effects of the tested process parameters and the 95% confidence band of the used linear effect model are shown in Figure 6. Similar to the material properties of printed PEEK, higher process temperatures increase the bond strength. A significant effect can be observed for the nozzle temperature (p = 0.036, r = 0.127) and the substrate temperature (p = 0.0001, r = 0.537) generated by the heated print bed. The effect of the chamber temperature is statistically not significant, with p = 0.071. Besides the process temperatures, print speed (p = 0.002, r = −0.135) and first layer height (p < 0.0001, r = −0.460) are significant and thereby, the distance between the hot nozzle and the bonding zone shows significant effects. The effect of the extrusion factor is not significant at p = 0.193.

To better understand the impact of process parameters on the thermal conditions in the bonding zone and the correlation between bonding temperature and bonding strength, the temperature on the top surface of the laminate was measured during each print job. The maximum temperatures reached in the bonding zone are shown in Figure 7. Like the effects on shear strength, process temperatures strongly affect the bonding temperature (nozzle temperature: p = 0.036, r = 0.200; substrate temperature: p < 0.0001, r = 0.064; chamber temperature: p = 0.015, r = 0.236). Additionally, reducing the print speed increases the bonding temperature (p < 0.0001, r = −0.362), as well as smaller first layer heights (p < 0.0001, r = −0.312). A significant effect can also be shown for the extrusion factor (p = 0.014, r = 0.361).

Overall, the measurements are subject to large fluctuations within the individual factor combinations. This is shown, among other things, by the wide confidence intervals. The microsections produced provide an indication of a possible cause of the fluctuations. The actual layer height of the first layer does not always correspond to the set target value. This is shown as an example in Figure 8.

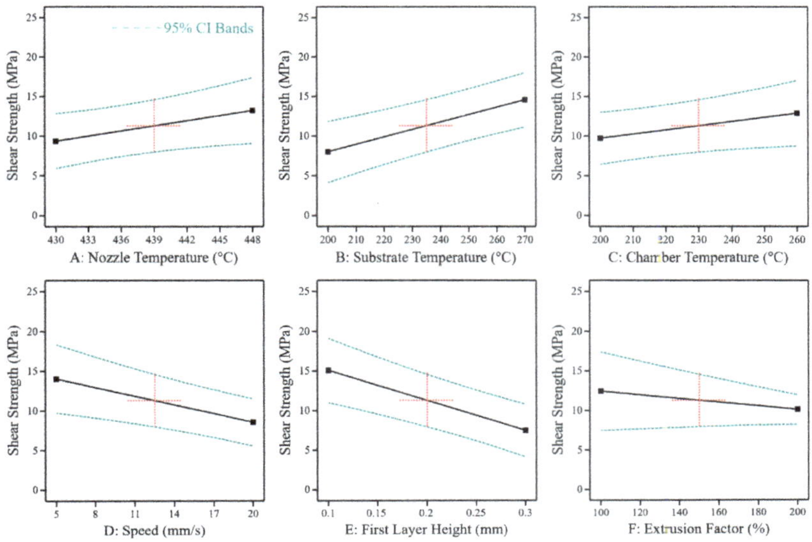

Figure 6. Main effects of process parameters on the shear strength of bonding between PEEK 3D print and PEEK laminate (extended from [35]). The black line shows the predicted effect. The blue lines show the 95 % confidence interval. The model was evaluated at the centre points of the test room (red crosses).

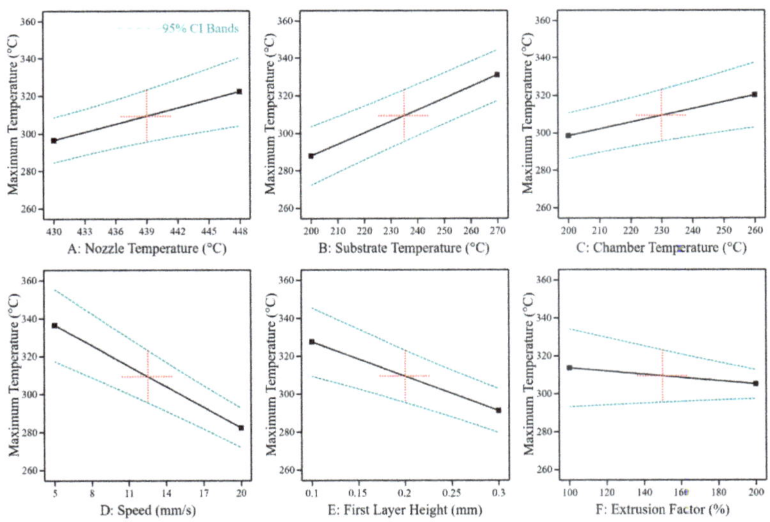

Figure 7. The main effects of process parameters on the maximum bonding temperature between PEEK 3D print and PEEK laminate. The black line shows the predicted effect. The blue lines show the 95 % confidence interval. The model was evaluated at the centre points of the test room (red crosses).

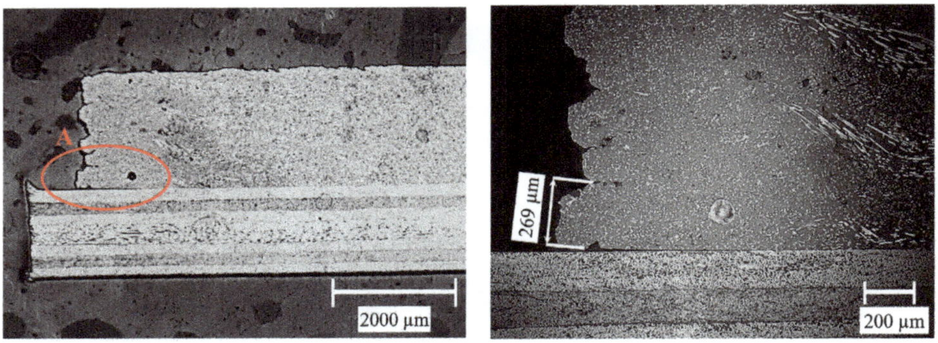

Figure 8. Microsection of a shear specimen with measured first layer height in section A.

The deviation in the first layer height is due to the fact that the bed levelling, i.e., the setting of the distance between the print bed and the nozzle, is subject to a certain tolerance. On the one hand, this concerns the setting of the zero point of the Z-coordinate. With the printer used, this tolerance is in the range of 0.05 mm and can significantly influence the layer height. Secondly, the parallelism between the print bed and the movement plane of the nozzle plays an important role. Due to a tolerance of this parallelism, the layer height of the first layer can fluctuate across a laminate. An example of the effect of a non-parallel laminate is shown in Figure 9. In this example, the first layer height for each specimen of a configuration was determined using microsections and displayed together with the corresponding strengths of the specimen. A correlation between the position of the sample and the strength is clearly recognisable.

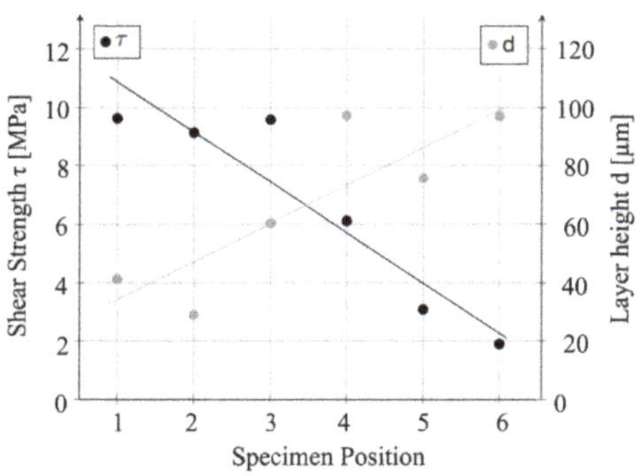

Figure 9. Correlation between bonding strength and tolerance of first layer height depending on the specimen position: 1—left to 6—right (compare Figure 5).

3.2. Overprinting: Parameter Characterisation

Based on the results from the screening, three parameters were chosen for closer investigation. The substrate temperature showed the strongest correlation overall, with a Pearson coefficient of $r = 0.537$. Together with the nozzle temperature, these two factors directly influence the temperature in the bonding zone and, thus, the strength. Therefore, nozzle and bed temperatures were included in the characterisation. The screening tests have shown that individual settings for substrate (print bed) temperature and chamber temperature are very limited with the given 3D printer, as the heating power does not allow large differences between the two parameters. Therefore, the chamber temperature is set to the same value as the substrate temperature and both parameters are combined as one ambient temperature. The second strongest effect was measured for the first layer height with $r = -0.460$. Consequently, the third factor investigated is the first layer height.

Furthermore, the screening has shown that the tolerance of the first layer height is difficult to control. It therefore makes sense to measure the first layer height of each specimen individually in order to avoid errors in the evaluation. However, the microsections did not prove to be a reliable measurement method, as the boundary between the first and second layer is not always clearly recognisable. For this reason, the first layer height was measured using a 3D optical profilometer. For this purpose, a reference line with a single layer height was printed along all specimens and analysed at the position of each sample. The measurement concept is shown in Figure 10.

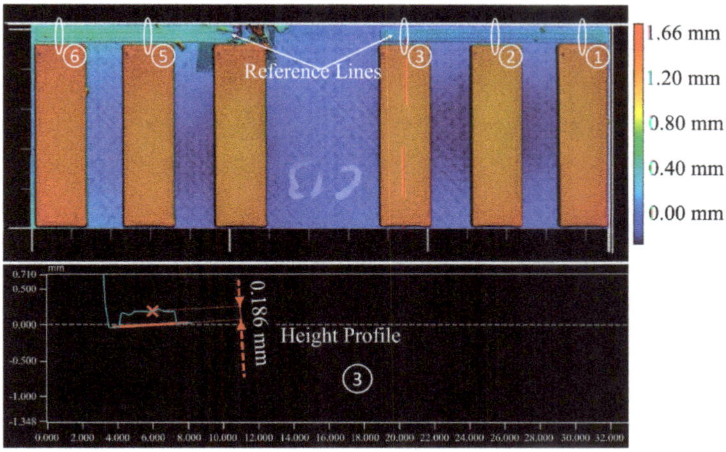

Figure 10. Measurement of first layer height using a 3D optical profilometer.

The measurements with individually adjusted layer height were then used to create a quadratic model of the effects of the three factors. The model shows significant effects for all three factors (nozzle temperature: $p = 0.0057$; ambient temperature: $p < 0.0001$; first layer height: $p = 0.0002$). The effect plots are shown in Figure 11. In particular, the ambient temperature and the layer height have a non-linear effect. An increase in the ambient temperature, therefore, increases the shear strength disproportionately, while an increase in the layer height reduces the strength disproportionately. Regarding the Pearson correlation, the effects of the ambient temperature (r = 0.567) and the layer height (r = −0.579) are similarly strong. On the other hand, the effect of nozzle temperature is notably weaker (r = 0.325).

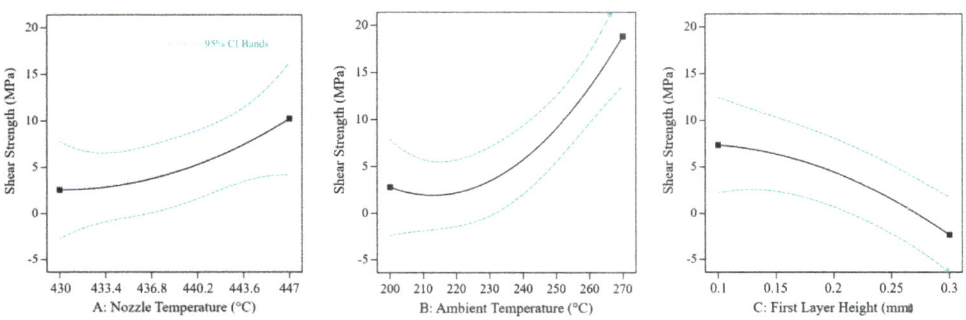

Figure 11. Main effect plots of the characterisation on shear strength. The black line shows the predicted effect. The blue lines show the 95% confidence interval.

In addition to the main effects, the quadratic model also allows the effect interactions to be analysed. The interactions between nozzle temperature and ambient temperature ($p = 0.2934$), as well as nozzle temperature and layer height ($p = 0.1165$), are not significant. Only the interaction between ambient temperature and layer height is statistically significant ($p < 0.0001$). This interaction is shown in Figure 12. At high first layer heights, the ambient temperature has no clear effect. However, the effect of the ambient temperature is clearly recognisable at lower first layer thicknesses.

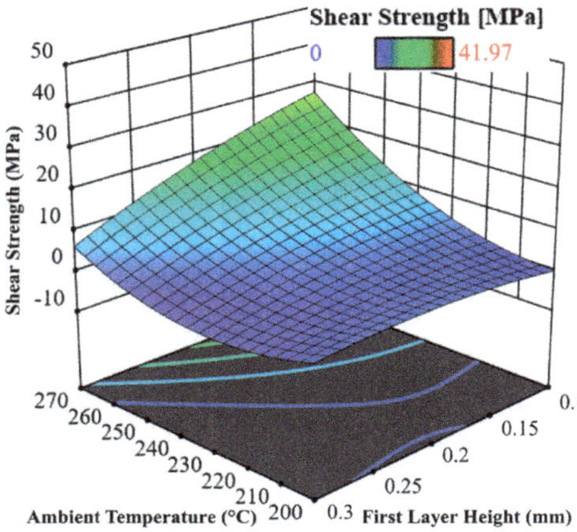

Figure 12. 3D plot of the interaction effect between first layer height and ambient temperature.

In addition to the shear strength, the temperature in the interface between the laminate and 3D print was measured, as in the screening. As the sample size for the temperature measurement was significantly smaller than for the shear tests (only one sample per configuration), only a linear statistical model was used for the maximum temperatures. It was also not possible to analyse the first layer height of the sample for temperature measurement, as the thermocouple interrupts the reference line at this point. Therefore, the first layer height of the neighbouring sample (position 5 in Figure 10) was used for the evaluation. In the resulting model, only the ambient temperature shows a significant effect ($p = 0.0005$; $r = 0.879$). The effects of the nozzle temperature ($p = 0.5528$) and the first layer height ($p = 0.4830$) are not significant. The effect plot for the maximum interface temperature is shown in Figure 13.

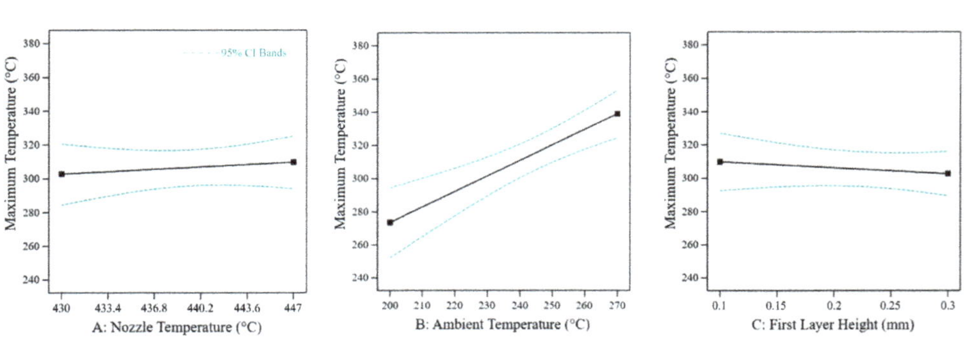

Figure 13. Main effect plots of the characterisation on maximum temperature. The black line shows the predicted effect. The blue lines show the 95 % confidence interval.

4. Discussion

The parameter screening shows significant effects of the process temperatures, print speed, and first layer height on the bond strength between printed PEEK and the laminate. Furthermore, all parameters show an effect on the maximum bonding temperature with very similar magnitudes compared to the effects on bond strength. This indicates that the bonding temperature is the dominant factor for the bond formation between laminates and 3D print. The influence of the process temperatures on the interface temperature thus corresponds to the expectations and observations in the literature on PEEK 3D printing, as these process parameters directly affect the interface temperature. The influence of printing speed and first layer height, however, is indirect. In FFF printers, the print head with the heated nozzle and a heating block forms a strong heat source that affects the process. At a slow printing speed, the heated nozzle remains in the area of influence of the freshly extruded material for longer, which increases the interface temperature. The situation is similar to the layer height. At low layer heights of the first layer, the heated nozzle is closer to the interface, which increases the interface temperature. Increasing the bonding pressure by introducing an over-extrusion showed no additional effect on the bond strength, supporting this theory. The dominant influence of the bonding temperature in the strength is also shown statistically, with a correlation coefficient of $p = 0.707$. These results are in agreement with the bonding theory for 3D printing used by the group of Sun et al. [17,18] and Coogan et al. [22], as well as experimental studies of PEEK 3D printing [27].

The detailed characterisation of the three factors—nozzle temperature, ambient temperature and first layer height—has enabled a better understanding of the effects. The non-linearity of the effects also fits in with the theoretical models mentioned, which are based on crack healing, according to Wool et al. [21]. According to this theory, the temperature of the bonding zone influences the development of strength to the fourth power.

While in the screening all effects on the bonding strength can be attributed to the effect of the process parameters on the interface temperature, this statement cannot be made so easily in the characterisation. In this case, the nozzle temperature and the layer height have a significant effect on the bonding strength but not on the interface temperature. The cause of this observation cannot be conclusively assessed within the scope of this work. While five samples were tested for each parameter combination in the mechanical tests, only one measured value is available for the temperature in the interface. At the same time, the first layer height at the temperature measurement point cannot be determined with sufficient accuracy using the profilometer. The lack of significance could, therefore, be attributed to the small sample size.

The observed interaction effect between ambient temperature and first layer height does not initially match the expected results. The data indicate that this effect is primarily due to the fact that the highest first layer height of 0.3 mm and the lowest ambient temperature of 200 °C are both at the very edge of the process window. The interaction effect is presumably caused by the fact that the main effects in the edge area of the process window can no longer be mapped cleanly.

Overall, the shear tests show high deviations, compromising the quality of the used statistical models. Especially the first layer height, or in other words, the distance of the nozzle from the substrate, has turned out to be a critical factor for the overall process. While the first layer height is a major factor impacting the bond strength, the factor is hard to control. Typical set values of 0.1–0.3 mm of standard printers are within the tolerance range of print bed calibration of the used 3D printer. Additionally, the thickness tolerance of the laminate adds to the tolerance chain of the first layer height. Controlling the tolerance chain of the first layer must be a primary requirement for overprinting laminates, especially when upscaling to large structures.

Another hard-to-control error source is the porosity of the printed material. The filament used to print the samples already shows a porosity of approx. 20% (see Figure 2). In the printed samples, a porosity of a comparable order of magnitude can be observed (see Figure 8). The exact influence of this porosity on the bonding strength cannot be determined within the scope of this work.

5. Conclusions

This work shows an analysis of bond formation during the overprinting of continuous fibre-reinforced PEEK laminates with short fibre-reinforced PEEK filament. The analysis is carried out in two steps. Firstly, the influence of the six most important process parameters (nozzle temperature, substrate temperature, chamber temperature, print speed, first layer height, and extrusion factor) on the bonding strength was analysed in an experimental screening. In the second step, the three parameters with the greatest influence were analysed in detail.

The study shows a strong correlation between the process temperatures and the bonding strength. At the same time, it was shown that the toolpath generation, e.g., speed and layer height, also influences the bonding strength. Overall, the temperature and strength measurements showed that the bonding temperature in the interface is the critical factor for successful bonding. This interface temperature is, in turn, influenced by the aforementioned process parameters.

While the concept of overprinting PEEK laminates was proven with an inter-layer shear strength of up to 15 MPa, four major findings can be highlighted as guidance for future process development:

- Bonding temperature in the interface is the primary factor influencing the bond strength.
- Nozzle temperature, substrate temperature, and first layer height are the most important parameters for optimising the bonding strength.
- Substrate pre-heating by means of print bed and chamber heating above 200 °C is required to enable in-situ bonding.
- Controlling the tolerance chain of the first layer height is a crucial requirement to ensure a stable process.

Overall, the results shown can form the basis for future process development and upscaling of the overprinting process using PEEK.

Author Contributions: Conceptualization, S.H. and H.V.; methodology, S.H.; validation, S.H. and F.A.; formal analysis, S.H. and F.A.; investigation, S.H. and F.A.; resources, S.H.; data curation, S.H. and F.A.; writing—original draft preparation, S.H.; writing—review and editing, S.H.; visualization, S.H. and F.A.; supervision, S.H. and H.V.; project administration, S.H. and H.V.; funding acquisition, H.V. All authors have read and agreed to the published version of the manuscript.

Funding: This research was by the Ministry of Economic Affairs, Labour and Tourism Baden-Württemberg reference numbers 3-4332.62-DLR/49 (IRAS), 3-4332.62-DLR/56 (IRAS II) and 3-4332.62-DLR/59 (IRAS III).

Institutional Review Board Statement: Not applicable.

Informed Consent Statement: Not applicable.

Data Availability Statement: The data presented in this study are available on request from the corresponding author.

Conflicts of Interest: The authors declare no conflicts of interest.

Appendix A

Table A1. Experimental design parameter screening.

Run	Nozzle Temp. [°C]	Substrate Temp. [°C]	Chamber Temp. [°C]	Speed [mm/s]	Layer Height [mm]	Extrusion Factor [%]
1	430.0	200	200.0	8.6	0.3	100
2	430.0	235	260.0	12.5	0.3	150
3	447.0	230	252.5	5.0	0.3	200
4	447.0	240	200.0	5.0	0.3	200
5	430.0	200	200.0	5.0	0.1	200
6	447.0	235	260.0	5.0	0.1	100
7	447.0	270	252.5	20.0	0.3	100
8	447.0	235	260.0	20.0	0.1	200
9	434.5	230	252.5	20.0	0.1	100
10	430	244	200	20	0.1	100
11	448	200	200	20	0.3	100
12	439	250	207.5	5	0.2	100

Table A2. Experimental design parameter characterisation.

Run	Nozzle Temperature [°C]	Ambient Temperature [°C]	Layer Height [mm]
1	439	150	0.2
2	432	260	0.1
3	447	260	0.2
4	432	260	0.3
5	430	150	0.3
6	447	161	0.1
7	447	162	0.3
8	430	206	0.2
9	442	226	0.1
10	430	150	0.1

References

1. McNiffe, E.; Ritter, T.; Higgins, T.; Sam-Daliri, O.; Flanagan, T.; Walls, M.; Ghabezi, P.; Finnegan, W.; Mitchell, S.; Harrison, N.M. Advancements in Functionally Graded Polyether Ether Ketone Components: Design, Manufacturing, and Characterisation Using a Modified 3D Printer. *Polymers* **2023**, *15*, 2992. [CrossRef] [PubMed]
2. Yang, C.; Tian, X.; Li, D.; Cao, Y.; Zhao, F.; Changquan, S. Influence of Thermal Processing Conditions in 3D Printing on the Crystallinity and Mechanical Properties of PEEK Material. *J. Mater. Process. Technol.* **2017**, *248*, 1–7. [CrossRef]
3. Xiaoyong, S.; Liangcheng, C.; Honglin, M.; Peng, G.; Zhanwei, B.; Cheng, L. Experimental Analysis of High Temperature PEEK Materials on 3D Printing Test. In Proceedings of the 2017 9th International Conference on Measuring Technology and Mechatronics Automation (ICMTMA), Changsha, China, 14–15 January 2017; pp. 13–16. [CrossRef]
4. Cuan-Urquizo, E.; Barocio, E.; Tejada-Ortigoza, V.; Pipes, R.B.; Rodriguez, C.A.; Roman-Flores, A. Characterization of the Mechanical Properties of FFF Structures and Materials: A Review on the Experimental, Computational and Theoretical Approaches. *Materials* **2019**, *12*, 895. [CrossRef] [PubMed]
5. Vyavahare, S.; Teraiya, S.; Panghal, D.; Kumar, S. Fused Deposition Modelling: A Review. *Rapid Prototyp. J.* **2020**, *26*, 176–201. [CrossRef]
6. Arif, M.F.; Kumar, S.; Varadarajan, K.M.; Cantwell, W.J. Performance of Biocompatible PEEK Processed by Fused Deposition Additive Manufacturing. *Mater. Des.* **2018**, *146*, 249–259. [CrossRef]
7. Li, T.; Wang, L. Bending Behavior of Sandwich Composite Structures with Tunable 3D-Printed Core Materials. *Compos. Struct.* **2017**, *175*, 46–57. [CrossRef]
8. Janssen, H.; Peters, T.; Brecher, C. Efficient Production of Tailored Structural Thermoplastic Composite Parts by Combining Tape Placement and 3d Printing. *Procedia CIRP* **2017**, *66*, 91–95. [CrossRef]
9. Rakhshbahar, M.; Sinapius, M. A Novel Approach: Combination of Automated Fiber Placement (AFP) and Additive Layer Manufacturing (ALM). *J. Compos. Sci.* **2018**, *2*, 42. [CrossRef]
10. Lukaszewicz, D.H.-J.A.; Ward, C.; Potter, K.D. The Engineering Aspects of Automated Prepreg Layup: History, Present and Future. *Compos. Part B Eng.* **2012**, *43*, 997–1009. [CrossRef]

11. Morales, U.; Esnaola, A.; Iragi, M.; Aretxabaleta, L.; Aurrekoetxea, J. Over-3D Printing of Continuous Carbon Fibre Composites on Organo-Sheet Substrates. *AIP Conf. Proc.* **2019**, *2113*, 020015. [CrossRef]
12. Penter, L.; Maier, J.; Kauschinger, B.; Lebelt, T.; Modler, N.; Ihlenfeldt, S. 3D Printing Technology for Low Cost Manufacturing of Hybrid Prototypes from Multi Material Composites. In *Lecture Notes in Production Engineering, Proceedings of the Production at the Leading Edge of Technology, Dresden, Germany, 23–24 September 2020*; Behrens, B.-A., Brosius, A., Hintze, W., Ihlenfeldt, S., Wulfsberg, J.P., Eds.; Springer: Berlin/Heidelberg, Germany, 2021; pp. 396–405.
13. Maier, J.; Vogel, C.; Lebelt, T.; Geske, V.; Behnisch, T.; Modler, N.; Gude, M. Adhesion Studies during Generative Hybridization of Textile-Reinforced Thermoplastic Composites via Additive Manufacturing. *Materials* **2021**, *14*, 3888. [CrossRef] [PubMed]
14. Boros, R.; Kannan Rajamani, P.; Kovacs, J.G. Combination of 3D Printing and Injection Molding: Overmolding and Overprinting. *Express Polym. Lett.* **2019**, *13*, 889–897. [CrossRef]
15. Raspall, F.; Velu, R.; Vaheed, N.M. Fabrication of Complex 3D Composites by Fusing Automated Fiber Placement (AFP) and Additive Manufacturing (AM) Technologies. *Adv. Manuf. Polym. Compos. Sci.* **2019**, *5*, 6–16. [CrossRef]
16. Matkovic, N.; Kupzik, D.; Steidle-Sailer, C.; Friedmann, M.; Fleischer, J. Novel Robot-Based Process Chain for the Flexible Production of Thermoplastic Components with CFRP Tape Reinforcement Structures. *Procedia CIRP* **2022**, *106*, 21–26. [CrossRef]
17. Sun, Q.; Rizvi, G.M.; Bellehumeur, C.T.; Gu, P. Effect of Processing Conditions on the Bonding Quality of FDM Polymer Filaments. *Rapid Prototyp. J.* **2008**, *14*, 72–80. [CrossRef]
18. Bellehumeur, C.; Li, L.; Sun, Q.; Gu, P. Modeling of Bond Formation Between Polymer Filaments in the Fused Deposition Modeling Process. *J. Manuf. Process.* **2004**, *6*, 170–178. [CrossRef]
19. Li, L.; Sun, Q.; Bellehumeur, C.; Gu, P. Composite Modeling and Analysis for Fabrication of FDM Prototypes with Locally Controlled Properties. *J. Manuf. Process.* **2002**, *4*, 129–141. [CrossRef]
20. Sun, Q.; Rizvi, G.M.; Bellehumeur, C.T.; Gu, P. Experimental Study of the Cooling Characteristics of Polymer Filaments in FDM and Impact on the Mesostructures and Properties of Prototypes. In Proceedings of the 2003 International Solid Freeform Fabrication Symposium, Texas, Austin, TX, USA, 4–6 August 2003. [CrossRef]
21. Wool, R.P.; O'Connor, K.M. A Theory of Crack Healing in Polymers. *J. Appl. Phys.* **1981**, *52*, 5953–5963. [CrossRef]
22. Coogan, T.J.; Kazmer, D.O. Healing Simulation for Bond Strength Prediction of FDM. *Rapid Prototyp. J.* **2017**, *23*, 551–561. [CrossRef]
23. Coogan, T.J.; Kazmer, D.O. Bond and Part Strength in Fused Deposition Modeling. *Rapid Prototyp. J.* **2017**, *23*, 414–422. [CrossRef]
24. Coogan, T.; Kazmer, D. Modeling of Interlayer Contact and Contact Pressure during Fused Filament Fabrication. *J. Rheol.* **2019**, *63*, 655–672. [CrossRef]
25. Li, H.; Wang, T.; Sun, J.; Yu, Z. The Effect of Process Parameters in Fused Deposition Modelling on Bonding Degree and Mechanical Properties. *Rapid Prototyp. J.* **2018**, *24*, 80–92. [CrossRef]
26. Basgul, C.; Thieringer, F.M.; Kurtz, S.M. Heat Transfer-Based Non-Isothermal Healing Model for the Interfacial Bonding Strength of Fused Filament Fabricated Polyetheretherketone. *Addit. Manuf.* **2021**, *46*, 102097. [CrossRef] [PubMed]
27. Zanjanijam, A.R.; Major, I.; Lyons, J.G.; Lafont, U.; Devine, D.M. Fused Filament Fabrication of PEEK: A Review of Process-Structure-Property Relationships. *Polymers* **2020**, *12*, 1665. [CrossRef] [PubMed]
28. Wang, P.; Zou, B.; Ding, S.; Li, L.; Huang, C. Effects of FDM-3D Printing Parameters on Mechanical Properties and Microstructure of CF/PEEK and GF/PEEK. *Chin. J. Aeronaut.* **2020**, *34*, 236–246. [CrossRef]
29. Yi, N.; Davies, R.; Chaplin, A.; McCutchion, P.; Ghita, O. Slow and Fast Crystallising Poly Aryl Ether Ketones (PAEKs) in 3D Printing: Crystallisation Kinetics, Morphology, and Mechanical Properties. *Addit. Manuf.* **2021**, *39*, 101843. [CrossRef]
30. Pu, J.; McIlroy, C.; Jones, A.; Ashcroft, I. Understanding Mechanical Properties in Fused Filament Fabrication of Polyether Ether Ketone. *Addit. Manuf.* **2021**, *37*, 101673. [CrossRef]
31. Deng, X.; Zeng, Z.; Peng, B.; Yan, S.; Ke, W. Mechanical Properties Optimization of Poly-Ether-Ether-Ketone via Fused Deposition Modeling. *Materials* **2018**, *11*, 216. [CrossRef]
32. Wu, W.; Geng, P.; Li, G.; Zhao, D.; Zhang, H.; Zhao, J. Influence of Layer Thickness and Raster Angle on the Mechanical Properties of 3D-Printed PEEK and a Comparative Mechanical Study between PEEK and ABS. *Materials* **2015**, *8*, 5834–5846. [CrossRef]
33. Caprais, I.; Joyot, P.; Duc, E. Bonding between Thermostable Polymers Processed by FFF and PEEK/Carbon Fiber Laminate. In Proceedings of the ADDFABCOMP—Additive Fabrication of Composite (2), Online Conference, 23–24 November 2021.
34. Caprais, I.; Joyot, P.; Duc, E.; Deseur, S. Bonding between High-Performance Polymer Processed by Fused Filament Fabrication and PEEK/Carbon Fiber Laminate. In Proceedings of the 24th International Conference on Material Forming, Liège, Belgium, 14 April 2021. [CrossRef]
35. Hümbert, S.; Schmidt, I.; Atzler, F.; Lengowski, M. Mechanical characterisation of in-situ bonding between peek filaments and laminates in the FFF process. In Proceedings of the 20th European Conference on Composite Materials—Composites Meet Sustainability (Volume 1–6), Lausanne, Switzerland, 26–30 June 2022. [CrossRef]
36. Atzler, F.; Hümbert, S.; Schiel, I.; Tröger, S.; Raps, L. Combining Fused Granular Fabrication and Automated Fibre Placement for the Rapid Production of Complex Sandwich-Structures. In Proceedings of the International Conference on Composite Materials (ICCM23), Belfast, UK, 2 August 2023.
37. Hümbert, S.; Gleixner, L.; Arce, E.; Springer, P.; Lengowski, M.; Sakraker, I. Material Characterization of Additively Manufactured PA12 and Design of Multifunctional Satellite Structures. In Proceedings of the ECSSMET 2018, Noordwijk, The Netherlands, 1 May 2018.

38. Stäbler, T.; Echsel, M.; Ehresmann, M.; Fugmann, M.; Galla, D.; Gottschalk, N.; Hümbert, S.; Lengowski, M.; Marigo, G.; Müller, I.; et al. Integrated Research Platform for Affordable Satellites. In Proceedings of the 70th International Astronautical Congress, Washington, DC, USA, 21–25 October 2019.
39. Echsel, M.; Springer, P.; Hümbert, S. Production and Planned In-Orbit Qualification of a Function-Integrated, Additive Manufactured Satellite Sandwich Structure with Embedded Automotive Electronics. *CEAS Space J.* **2020**, *13*, 111–118. [CrossRef]
40. Hümbert, S.; Meth, J.; Echsel, M.; Lengowski, M.; Stäbler, T. Additive Manufacturing of Radiation Shielding for Small Satellites. In Proceedings of the 72nd International Astronautical Congress, Dubai, United Arab Emirates, 25–29 October 2021.
41. Datasheets Plastics | Ensinger. Available online: https://www.ensingerplastics.com/en-us/datasheets (accessed on 9 January 2022).
42. Überblick. Available online: https://gewo3d.com/de/htp-260/ueberblick.html (accessed on 26 September 2021).
43. *ASTM D3846-08(2015)*; Standard Test Method for In-Plane Shear Strength of Reinforced Plastics. ASTM International: West Conshohocken, PA, USA, 2015.
44. Keyence, U.K. Digital Microscope—VHX-5000. Available online: https://www.keyence.co.uk/products/microscope/digital-microscope/vhx-5000/models/vhx-5000/?search_dl=1 (accessed on 10 January 2023).
45. Wide-Area 3D Measurement System Head—VR-5200 | KEYENCE America. Available online: https://www.keyence.com/products/microscope/macroscope/vr-3000/models/vr-5200/ (accessed on 15 November 2023).

materials

MDPI

Article

Composite Medical Tabletops Made of CFRP with Different Cross-Sections: Numerical Analysis and Laboratory Testing

Przemysław Golewski [1], Daniel Pietras [1], Tomasz Sadowski [1,*] and Albin Michał Wit-Rusiecki [2]

[1] Department of Solid Mechanics, Faculty of Civil Engineering and Architecture, Lublin University of Technology, Nadbystrzycka 40 Str., 20-618 Lublin, Poland; p.golewski@pollub.pl (P.G.); d.pietras@pollub.pl (D.P.)

[2] WIT-Composites Ltd., Bohdana Dobrzańskiego 3 Str., 20-262 Lublin, Poland; wit@wit-composites.com

* Correspondence: sadowski.t@gmail.com

Abstract: This paper presents the results of laboratory tests of CFRP (carbon fiber-reinforced polymer) laminates, which allowed the development of numerical material models. The obtained data were used in a further stage to perform numerical simulations of four variants of medical tabletops, differing, among other features, in the shape of the cross-section. Maximum deflections and effort in the composite material were analyzed. The final step was to perform a laboratory test for one of the tabletop versions, the results of which confirmed the correctness of the numerical calculations. This work is aimed at both researchers and designers involved in the practical application of fiber-reinforced polymer matrix composites.

Keywords: operating table; CFRP composite; numerical simulations

1. Introduction

Basic equipment in the operating room includes a basin stand, trolleys, Mayo table, run-about buckets, diathermy apparatus, suction apparatus, anesthesia machine, drip stand, operating lights, viewing screens, swab rack, and an operating table [1]. There are many types of operating tables, but from the point of view of designing the tabletop itself, the following features are important:

- The tables are usually divided into three sections to support the body, allowing the patient to assume a bent or lying position,
- Extensions for the head are also used, e.g., for neurosurgical or endoscopic examinations,
- Many accessories are available, such as a hand support that attaches to movable clamps on the rail [2].

Another solution is the Jackson table method [3], which allows the patient to be transferred from the bed to the table, which is equipped with appropriate inserts. The frame is made of CFRP and sits above the patient. Once the patient is strapped to the table, the frame can be rotated to any angle depending on the examination being performed. This type of table provides flexibility in patient positioning, which is important, for example, in examinations for oblique lumbar intervertebral fusion [4].

Adjustable carbon-fiber operating tables used for lumbar spine surgery have a similar design. The flexion angle of the table should allow for increasing or decreasing the lumbar lordosis to access the disk space [5].

In addition to the design of the tabletop, a mechanism that allows it to tilt to a lateral position is also important. The authors in the article [6] presented the results of a study of the biomechanical effort of personnel performing a patient transfer from one table to another. With the use of a tilted tabletop, the effort of the pushing nurse was eliminated and the activity of most of the muscles of the pulling nurse decreased. Thus, the use of lateral tilt tops should also be included in numerical simulations, with attention paid to the mounting locations where the complex stress state occurs.

Tabletop joint failure situations can occur in practice; for example, when a patient is morbidly obese. Such an incident is reported in [7]. While transferring a patient weighing 130 kg from a surgical table to a bed, two bolts of the turning mechanism failed, almost causing the patient to fall. This type of situation is likely to increase as the incidence of obese patients in the operating room increases. This situation once again underscores that one of the important design points is the joint between the operating tabletop and the tilt mechanism column. Both pull-out and shear tests are required [8].

Another important design issue is the occurrence of pressure from different parts of the body. If a patient lies in one position for dozens of hours, pressure injuries called pressure sores develop. This topic was presented in [9], where pressure relieving pads were used at the interface between the patient's body and the medical tabletop. Pressure at the junction of both heels and sacrum was measured in the supine position. Comfort was assessed using a visual analogue scale. The use of polyurethane foam helped reduce pressure and improve patient comfort. Thus, when designing the upper part of the medical top, some curvatures can be preliminarily introduced for greater patient comfort and against pressure sores.

A similar topic related to surface pressure is presented in [10]. Experimental studies were conducted on a total of 72 patients, divided into three groups differing in the material used to reduce pressure. The materials used were a gel support surface, a viscoelastic support surface, and a standard operating table. The pressures exerted on the patient's body were lowest with the viscoelastic material, and this type of material is recommended to minimize the risk of pressure injury in the operating room.

In conclusion, composite medical tabletops are an important component of operating room equipment. Their design is related to both safety and comfort, as well as the prevention of overpressure injuries in contact with the patient's body. However, there is a lack of articles in the literature on the process of designing and testing such composite products. The presented article will help fill this gap.

The presented results are related to a scientific project with the title Composite system for fast mechanical connections for the medical industry. Using CAD modeling and numerical simulations, three types of original tabletops were proposed, which are relatively thin, ergonomically shaped, and free of filling. In the final analysis, a solution was chosen that represents a compromise: the tabletop is flat and has infill only near the edges and the attachment zone. The purpose of the work was to test the technology of making such a tabletop and compare the laboratory results with a numerical model.

2. Materials and Methods

Polymeric materials have been used in medicine for more than 150 years. Their first appearance dates back to 1862 when Alexander Parker first presented a modified thermoplastic material at an exhibition in London [11]. The rapid dissemination of these type of materials in medicine was possible due to their advantages such as low cost, excellent mechanical properties, ease of sterilization, and bactericidal properties. In addition, they have weak magnetic properties and much less absorption of X-rays, which makes them irreplaceable in certain applications [12,13]. To increase their strength, the structure is reinforced, which leads to the formation of a composite. The literature [14] lists several types of composites that can be used in medical devices. These include composites reinforced with short fibers or flakes spaced irregularly or in a specific direction, glass beads, powders, or the use of continuous fibers resulting in laminates. Each of these reinforcing materials causes changes in the properties of the polymer material, e.g., resulting in reduced shrinkage, and increased strength and stiffness.

Medical equipment is divided into 3 classes [14,15] depending on the application and the safety and risks involved. The first class applies to such devices as tongue depressors, bandages, gloves, and bedpans. The second class includes, for example: wheelchairs, X-ray machines, MRI machines, surgical needles, catheters, and diagnostic equipment. The last group includes such products as heart valves, stents, implanted pacemakers, silicone implants, and hip and bone implants. Thus, operating tables and their tops will fall into the second class of medical devices, while the materials from which they are made must be tested for their strength.

2.1. Laboratory Tests

Laboratory tests were performed to determine the elastic and strength properties of composite materials to build their numerical model. The tests were made for two types of materials:

- Composite reinforced with unidirectional fibers—KORDCARBON-CPREG-200-UD-3K-EP-120-38 (UD),
- Composite reinforced with twill weave fabric—KORDCARBON-CPREG-245-T-3K-EP1.5-42-T (TW),
- Both types of materials use an epoxy matrix and carbon fibers and were produced by Fiberpreg GmbH, Carl Zeiss Str. 7, 89231 Neu-Ulm, Germany.

All test samples were made by Wit Composites using a vacuum bag autoclave technology [16] and epoxy matrix prepregs [17].

Tensile and compression tests were made for both types of materials. A summary of all tests, taking into account the angle of the fibers concerning the direction of loading (0°, 45°, 90°), is presented in Table 1.

Table 1. Summary of samples used in laboratory tests.

	UD		TW	
	Tension	**Compression**	**Tension**	**Compression**
0°	3 pcs.	3 pcs.	6 pcs.	3 pcs.
45°	3 pcs.	-	6 pcs.	-
90°	3 pcs.	3 pcs.	-	-

The tests were made by the following standards:

- D3518/D3518M—13 Standard Test Method for In-Plane Shear Response of Polymer Matrix Composite Materials by Tensile Test of a 45° Laminate [18],
- Designation: D3039/D3039M—14 Standard Test Method for Tensile Properties of Polymer Matrix Composite Materials [19],
- D6641/D 6641M—Standard Test Method for Determining the Compressive Properties of Polymer Matrix Composite Laminates Using a Combined Loading Compression (CLC) Test Fixture [20].
- The shapes of the used samples with dimensions are shown in Appendix A, Figures A1 and A2.

Tensile tests were made using an MTS810 machine equipped with a 250 kN force measuring head. Loads were applied at a constant speed of 2 mm/min, whose strain rate was 0.6 min^{-1} (less than 2.5 min^{-1}, which satisfies the requirement of quasi-static loading [21]). Strain measurements were made with a biaxial extensometer. The test stand is shown in Figure A3a.

For compression tests, MTS810 testing machine equipped with a special platform for clamping specimens was also used. Loading forces were measured using a load cell with a range of 250 kN. The test was made under conditions of constant displacement speed. The speed of 0.4 mm/min (strain rate 0.18 min^{-1}), was selected to meet the test-to-failure time condition, which should be in the range of 1 to 10 min. In this case, electro-resistance strain gauges with a base of 3 mm were used for strain measurements. Strain gauge measurements were made using a HBM MGC PLUS type amplifier with CATMAN application. The test stand is shown in Figure A3b.

Normal stresses σ [MPa] were determined from the following equation:

$$\sigma = \frac{F}{h \cdot w} \tag{1}$$

where: F—loading force [N], h—sample thickness [mm], w—sample width [mm]. The tangential stress τ_{12} [MPa] was determined from the following equation:

$$\tau_{12} = \frac{F}{2 \cdot h \cdot w} \tag{2}$$

The longitudinal modulus of elasticity E [GPa] was determined as follows:

$$E = \frac{\sigma_0 - \sigma_1}{\varepsilon_{a0} - \varepsilon_{a1}} \tag{3}$$

where ε_{a0} denotes the initial strain of the elastic modulus determination range of 1000 μm/m, while ε_{a1} denotes the final strain of the elastic modulus determination range of 3000 μm/m. The symbols σ_0 and σ_1 are the stress values corresponding to strains with the same subscript from the beginning and end of the range in which the longitudinal elastic modulus was determined.

The Poisson's ratio was determined as the ratio of the difference between transverse and longitudinal strains in the same range used to determine the longitudinal modulus of elasticity.

$$\nu = \frac{\varepsilon_{t0} - \varepsilon_{t1}}{\varepsilon_{a0} - \varepsilon_{a1}} \tag{4}$$

For ±45° samples, the shear strain of γ_{12} was determined as the difference of the measured longitudinal and transverse strains as follows:

$$\gamma_{12} = \varepsilon_a - \varepsilon_t \tag{5}$$

The modulus of elasticity G_{12} was determined as the ratio of the increase in shear stress to the corresponding increase in shear strain (6) according to document D3518/D3518M—13, this modulus was determined in the range of shear strain 2000 μm/m–6000 μm/m.

$$G_{12} = \frac{\tau_{12}^0 - \tau_{12}^1}{\gamma_{12}^0 - \gamma_{12}^1} \tag{6}$$

The shear strength of the tested materials was determined as the stress that occurred at a shear strain equal to 5%.

The standard deviation was calculated using Equation (7):

$$\sqrt{\frac{\sum (x - \overline{x})^2}{n - 1}} \tag{7}$$

where: x is the value of a parameter for a given sample, $\overline{x}$ is the average value of a given parameter for all samples, and n is the number of samples tested.

The differences between the parameters were determined as follows:

- If an increase in the parameter is observed:

$$\left(\frac{x_k}{x_p} - 1\right) \cdot 100\% \tag{8}$$

- If a decrease in the parameter is observed:

$$\left(1 - \frac{x_k}{x_p}\right) \cdot 100\% \tag{9}$$

The tests were performed at an average temperature of 27.7 °C and a humidity range of 49.9–51.4%. Measurements of environmental parameters were made once a day around 8:30 a.m.

2.2. Numerical Simulations

Numerical simulations were made using Abaqus software 6.16. S4R-type shell parts were used for the CFRP laminate. S4R element is the reduced integrated quadrilateral finite-membrane-strain element which adopts reduced integration theory advocated to treat the shear lock. There were also solid parts in model 4, e.g., foam filling or aluminum mounting plate for which C3D8R-type elements were used. The C3D8R element is a general purpose linear brick element, with reduced integration (1 integration point).

In the case of shell parts for laminates, it was possible to apply the Tsai–Hill composite strength criterion [22]. Before that, it was necessary to define the following values:

- Tensile strength in the fiber direction (X_t),
- Compressive strength in the fiber direction (X_c),
- Tensile strength in the direction perpendicular to the fibers (Y_t),
- Compressive strength in the direction perpendicular to the fibers (Y_c),
- Shear strength (S).

The Tsai–Hill hypothesis has the following form:

$$\frac{\sigma_{11}^2}{X^2} - \frac{\sigma_{11}\sigma_{22}}{X^2} + \frac{\sigma_{22}^2}{Y^2} + \frac{\tau_{12}^2}{S^2} = 1 \tag{10}$$

where:

σ_{11}—normal stress in the direction of the fibers,
σ_{22}—normal stress in the direction perpendicular to the fibers,
τ_{12}—tangential stress,
X—strength in the fibers' direction,
Y—strength in the direction perpendicular to the fibers,
S—shear strength.

If a value of 1 is reached in the material, the element will thus be completely efforted. The criterion in the form (10) also remains valid when the composite material has different tensile and compressive strength characteristics. Then, the modification of this criterion is to insert in place of X and/or Y, depending on the sign of the stresses σ_{11}, σ_{22}—values of tensile strength X_t, Y_t or compressive strength X_c, Y_c.

Figure 1 shows the values of loads that are taken for strength calculations of tabletops. Based on them, the corresponding partitions at the top of the tabletop were separated, which made it possible to set loads.

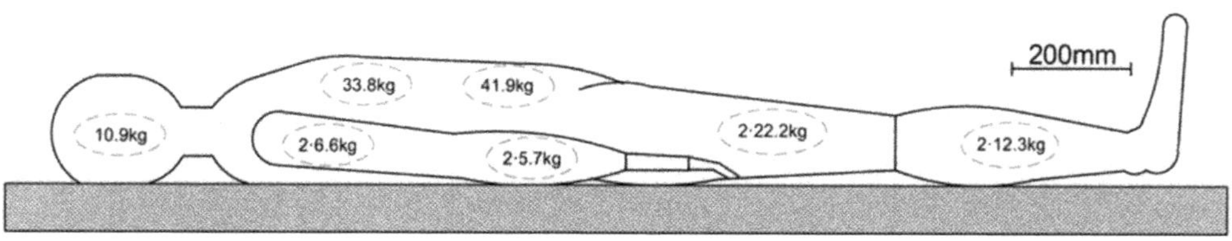

Figure 1. Distribution of weights from the patient's body.

The calculations were performed for a total of four tabletops, differing in cross-section, layer arrangement, and external dimensions. The simulations were divided into two main groups:

- Ergonomic tabletops (without making a demonstrator),
- Tabletop with a rectangular cross-section (made demonstrator).

To better illustrate the scope of performed work, it is shown in the diagram in Figure 2.

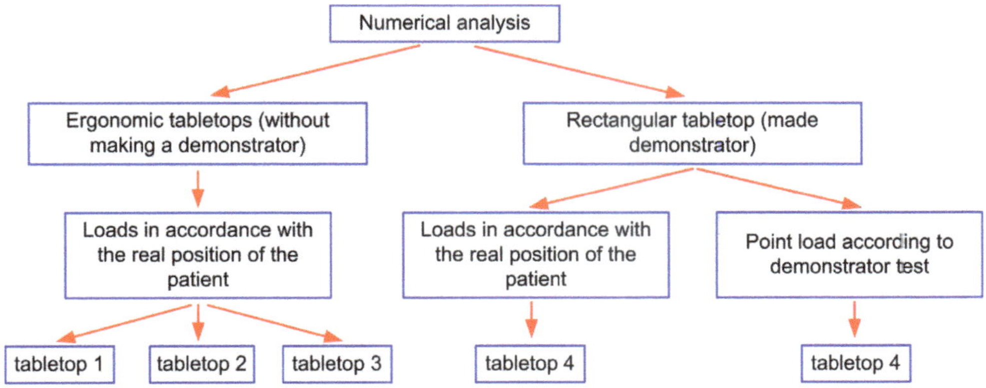

Figure 2. Classification of numerical simulations.

The external dimensions of the ergonomic tabletops (1–3) were 2100 mm × 650 mm, while the external dimensions of the rectangular tabletop 4 were 2300 mm × 600 mm. Because the tabletop should be as transparent to X-rays [23] as possible, models 1 through 3 were made as no-fill. Partial filling was used in tabletop 4. Usually, polyurethane foam is used for this purpose, which adversely affects the translucency of the tabletop during imaging with the "C" arm [24], hence the width of the filling at the edges of the tabletop is only 50 mm. Cross-sections of the analyzed tabletops are shown in Figure 3.

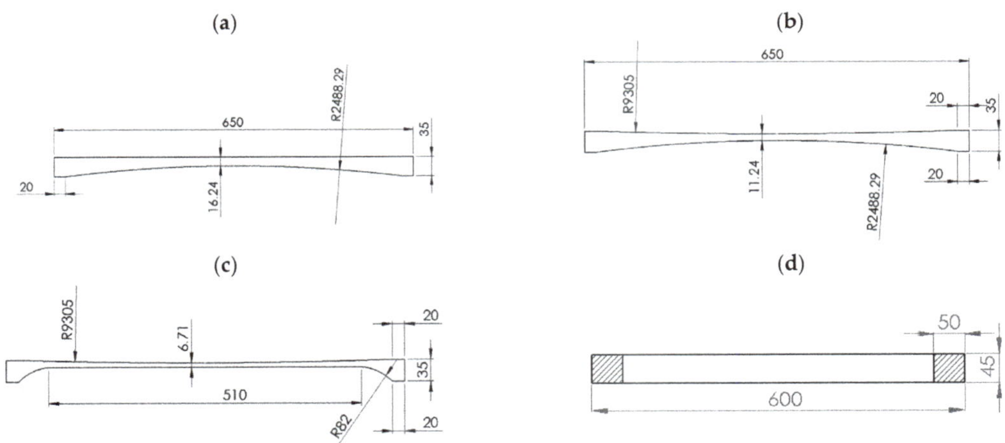

Figure 3. Cross-sections of the analyzed tabletops: (**a**) Tabletop 1, (**b**) Tabletop 2, (**c**) Tabletop 3, (**d**) Tabletop 4.

Total filling was used in a section of tabletop 4 (Figure 4) because of the mounting to the column. Depending on the imaging site, the patient can be positioned either head over the column or legs over the column, whereas higher loads will be achieved in the second position.

The same cross-section of the wall was used for models 1 through 3. The wall consisted of 15 layers. Both twill fabric (TW) and unidirectional (UD) fibers were used. In both cases, these were carbon fibers. The arrangement of the layers was as follows: 3 × TW, 2 × UD, 1 × TW, 3 × UD, 1 × TW, 2 × UD, 3 × TW, with the UD fibers arranged along the longer edge of the tabletop.

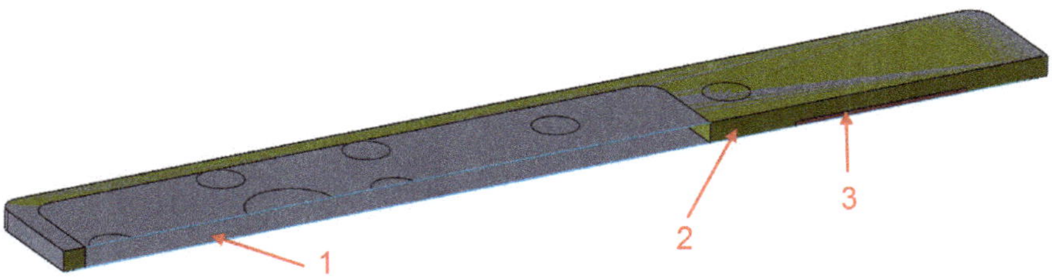

Figure 4. Longitudinal cross-section of tabletop 4. 1—CFRP laminate, 2—Polyurethane foam, 3—Aluminum plate.

For model 4, a different arrangement of layers was used: 1 × TW, 4 × UD, 2 × TW, 4 × UD, 1 × TW, making a total of 12 layers.

The purpose of simulations 1 through 3 was to determine what effect the introduction of an ergonomic shape has by using a radius to keep the patient's body on the axis of the tabletop. Demonstrators were not made for this type of tabletop, since the introduction of radiuses requires the manufacture of appropriate molding tools, which significantly increases costs.

Tabletop number 4 has a rectangular cross-section and is therefore much easier to manufacture. A demonstrator was made for this case to verify the numerical calculations.

Because the weight of the body is distributed unevenly over the entire surface, a partition was made as in Figure 5a.

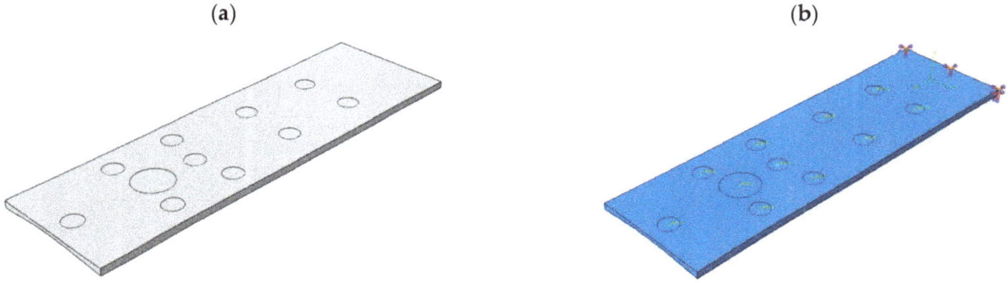

Figure 5. Models 1–3: (**a**) Performed partitions, (**b**) Boundary conditions.

The top of the operating table can be treated as a cantilever plate. For models 1 through 3, one end (on the lower limb side) was fixed—all degrees of freedom were taken away, while the other end was free. Reference points were placed in the made partitions and connected to them. This was to make it easier to apply loads in the form of concentrated forces (Figure 5b). Each layer of laminate was arranged along the axis of the tabletop.

For model 4, as mentioned above, a demonstrator was made, so the method of attachment cannot be theoretical as in the case of models 1–3. For model 4, an aluminum plate was used at the bottom (Figure 4), the purpose of which was to transfer the load from the tabletop to the point fasteners to the column. The boundary conditions used for the model can be seen in Figure 6 (orange shapes). All degrees of freedom have been taken away in the four holes of the aluminum plate; this is analogous to the use of bolts at these locations.

After the necessary partitions were made, FEM meshes were applied as shown in Figure 7. For models 1 to 3, about 42,000 S4R elements were used, while for model 4 a total of 91,703 elements were used, including 36,635—S4R, 1022—S3 (CFRP laminate), and 54,046—C3D8R (polyurethane foam and aluminum plate).

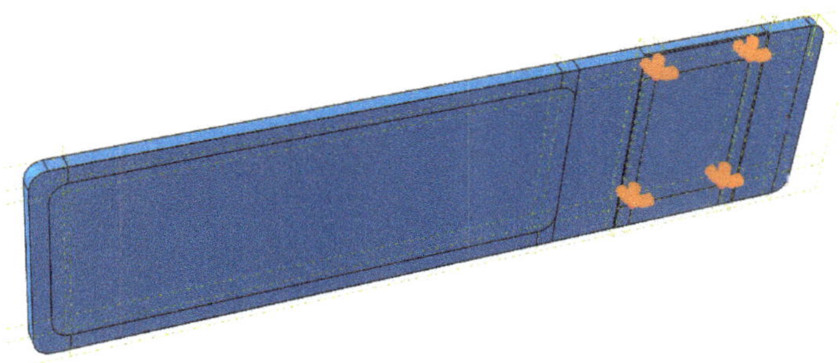

Figure 6. Method of mounting tabletop 4.

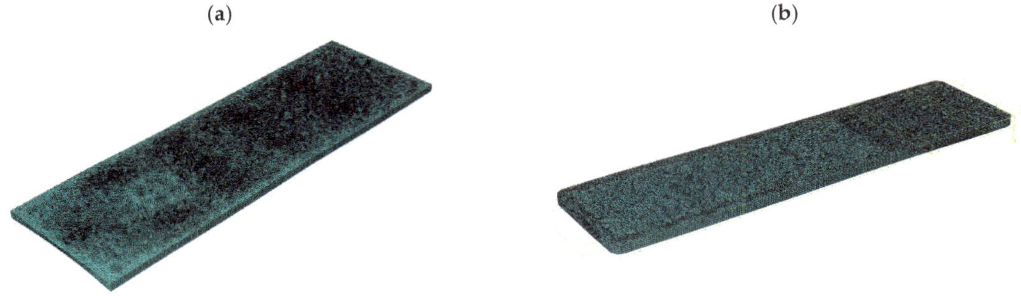

Figure 7. Finite element meshes: (**a**) Models 1–3, (**b**) Model 4.

3. Results of Material Testing

Material research for uniaxial tensile tests for samples made of unidirectional fiber-reinforced prepreg was performed on a batch of three samples. The tests were performed in three directions: 0, 90, and 45 degrees. Force-displacement diagrams are shown in Figure 8.

This type of test made it possible to determine the following elastic properties:

E_{t0} = 121.965 GPa—Young's modulus in the direction of fiber orientation,

E_{t90} = 7.817 GPa—Young's modulus transverse to the direction of fiber orientation,

G_{12} = 3.434 GPa—Kirchhoff's modulus,

ν_{12} = 0.298—Poisson's ratio,

And strength properties:

X_t = 2003.119 MPa—tensile strength in the direction of fiber orientation,

Y_t = 45.12 MPa—tensile strength transverse to fiber orientation,

S = 58.749 MPa—shear strength.

The relative standard deviation analysis is noteworthy. For samples with a fiber arrangement along the tensile direction, a value in the range of 6.37–6.5% was obtained for tensile strength and Young's modulus, and a much lower value of 2.68% was obtained for Poisson's ratio. For samples with a fiber orientation perpendicular to the tensile direction, low repeatability of results was obtained for maximum force, and thus for tensile strength. The relative standard deviation was as high as 29.2%, while for Young's modulus, it was 1.67%. Considering the tests for samples with fibers oriented at an angle of 45 degrees, the values of relative standard deviation were in the range of 1.15–3.6% for shear strength and Kirchhoff modulus, respectively.

For samples made of twill fabric, two tests were necessary for the direction along the warp orientation and at an angle of 45 degrees. The results are presented in Figure 9.

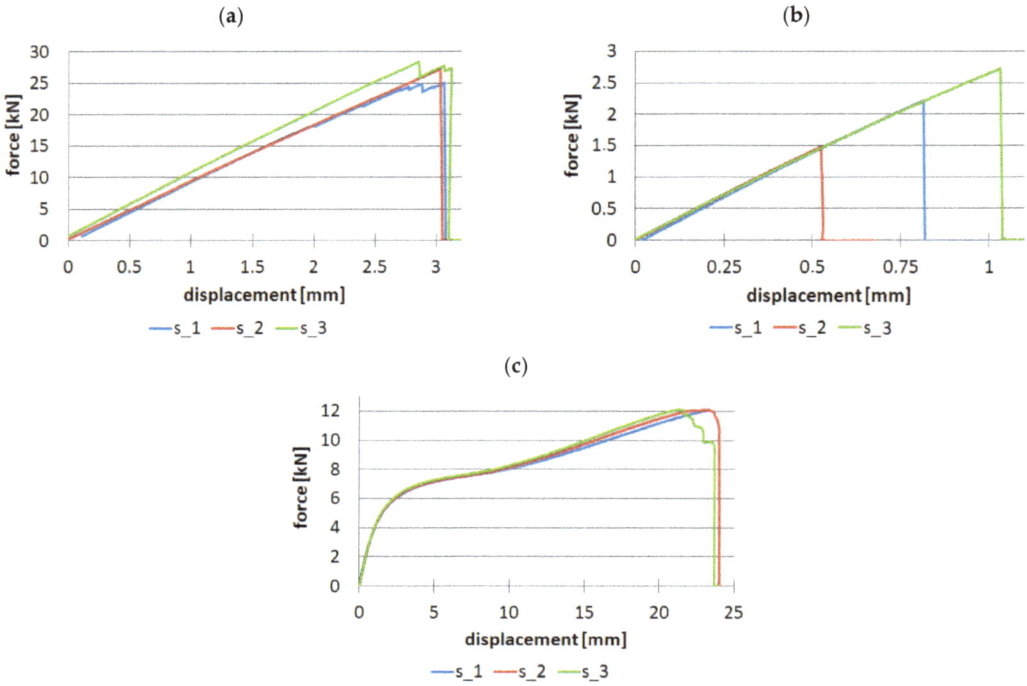

Figure 8. Tensile results for unidirectional reinforced samples: (**a**) 0° direction, (**b**) 90° direction, (**c**) 45° direction.

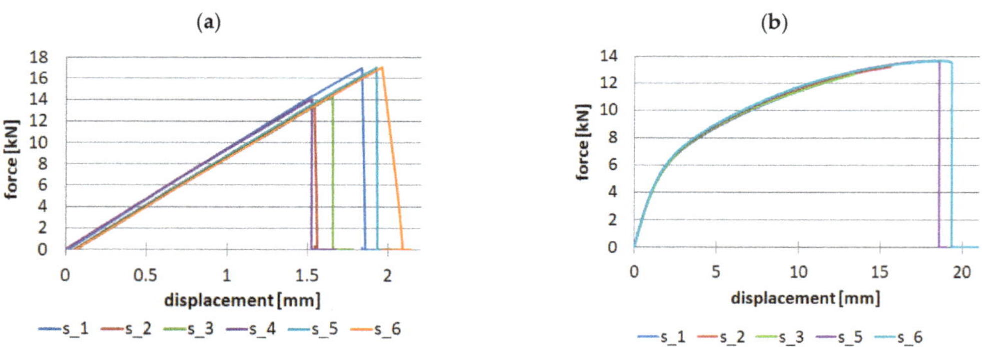

Figure 9. Tensile results for the fabric: (**a**) Warp direction, (**b**) 45° direction.

This type of test made it possible to determine the following elastic properties:
E_{t0} = 48.069 GPa—Young's modulus in the direction of warp orientation,
G_{12} = 2.652 GPa—Kirchhoff's modulus,
ν_{12} = 0.064—Poisson's ratio,
And strength properties:
X_t = 453.385 MPa—tensile strength in the direction of warp orientation,
S = 58.730 MPa—shear strength.

For fabric-reinforced samples, it was assumed that $E_0 = E_{90}$ and $X_t = Y_t$. For performed tests, the maximum value of relative standard deviation was obtained for tensile strength—10.86%. The smallest values were obtained for tests with a 45-degree fiber orientation: for shear strength 1.64% and Kirchhoff modulus 1.58%.

The composite tabletops work similarly to a cantilevered plate, so the upper part is in tension and the lower part is in compression. Hence, to make the numerical simulations more in line with reality, compression tests were also carried out, the results of which for the composite reinforced with unidirectional fibers are shown in Figure 10.

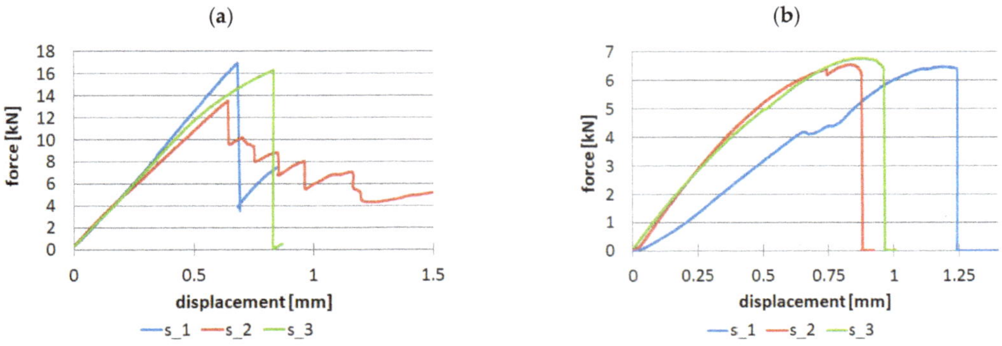

Figure 10. Results for unidirectional fibers for compression: (**a**) 0° direction, (**b**) 90° direction.

Performed tests allowed us to determine the following elastic properties:

Ec_0 = 111.564 GPa—Young's modulus in the direction of fiber orientation,

Ec_{90} = 8.094 GPa—Young's modulus transverse to the direction of fiber orientation,

And strength properties:

X_c = 793.504 MPa—compressive strength in the direction of fiber orientation,

Y_c = 131.592 MPa—compressive strength transverse to the direction of fiber orientation.

Analysis of the relative standard deviation showed that the largest value (10.9%) is for compressive strength for the direction along the fiber orientation, while the smallest value (0.69%) is for Young's modulus values also for the direction along the fiber orientation. For the perpendicular direction, the relative standard deviation is in the range of 2.5–3.78% for Young's modulus and compressive strength, respectively.

The compression test was also carried out for the composite reinforced with twill fabric, only for warp direction, since we assume the same values for weft direction. The results for the batch of samples are shown in Figure 11.

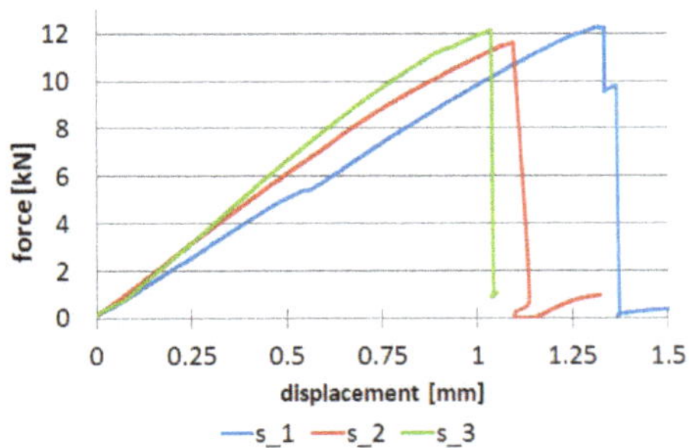

Figure 11. Results for fabric in compression.

As a result of the tests, Young's modulus in compression Ec = 43.4 GPa and compressive strength Xc = 208.99 MPa were determined. The relative standard deviations for both of the above-mentioned parameters were 0.53% and 6.27%, respectively.

A summary of the results for all tests can be found in Appendix B.

4. Results of Numerical Simulations

The results of the numerical simulations are presented in the form of resultant displacement maps "U, Magnitude" and the achievement of the Tsai–Hill strain criterion. The resultant displacement is calculated as the square root of the sum of the squares of the respective displacements along the *x*, *y*, and *z* axes. For the displacement maps, the state before deformation was also shown, and the deflection of the tabletops was not scaled. Figure 12a shows the displacement map for tabletop 1, with a maximum value of 70.31 mm.

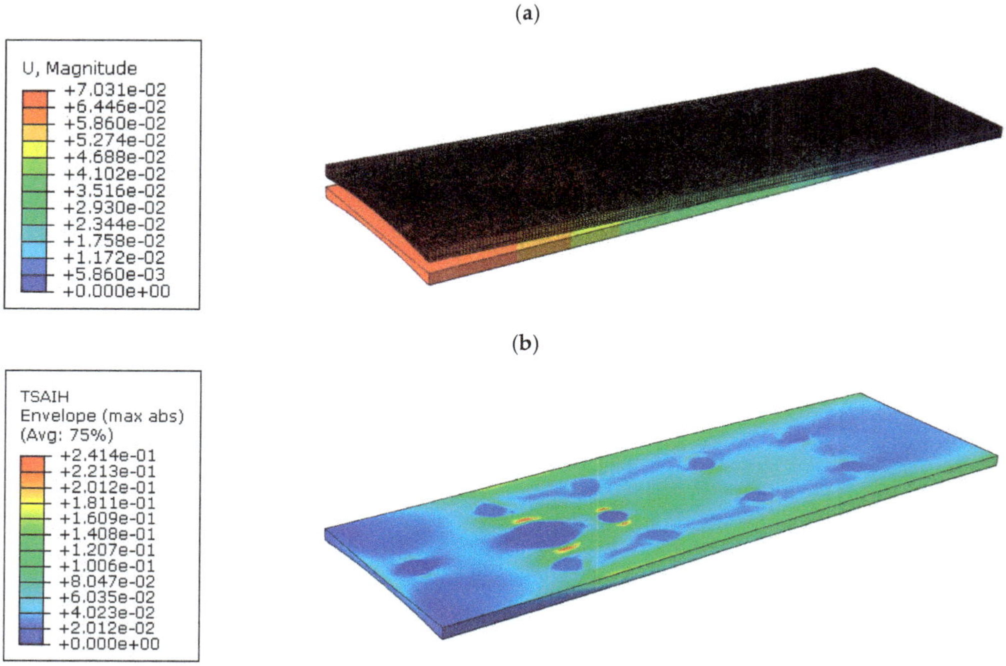

Figure 12. Results for tabletop 1: (**a**) Deflection, (**b**) Tsai–Hill criterion.

Figure 12b shows the effort of the composite material for tabletop 1. The maximum value was 24.14%.

Tabletop 2 had two radiuses of curvature compared to the previous tabletop 1, hence the thickness thinning was even greater. This resulted in an increase in maximum deflection to a value of 85.14 mm (Figure 13a).

However, the introduction of two radiuses of curvature does not result in a significant increase in material effort. For tabletop 2, a maximum value of 26.57% was obtained, as shown in Figure 13b.

Tabletop 3 had one side flat, while the other side was shaped as a curve; the minimum thickness was only 6.71 mm and was almost constant across the width of the tabletop, hence the maximum displacement was as much as 139.6 mm (Figure 14a).

However, such a large displacement did not translate into a significant increase in material effort, which in this case was 38.44% (Figure 14b).

In all three cases, the effort of the material was not exceeded, but due to technological reasons, a rectangular cross-section was finally adopted for tabletop 4. The method of mounting is also different, as can be seen in the displacement map in Figure 15a.

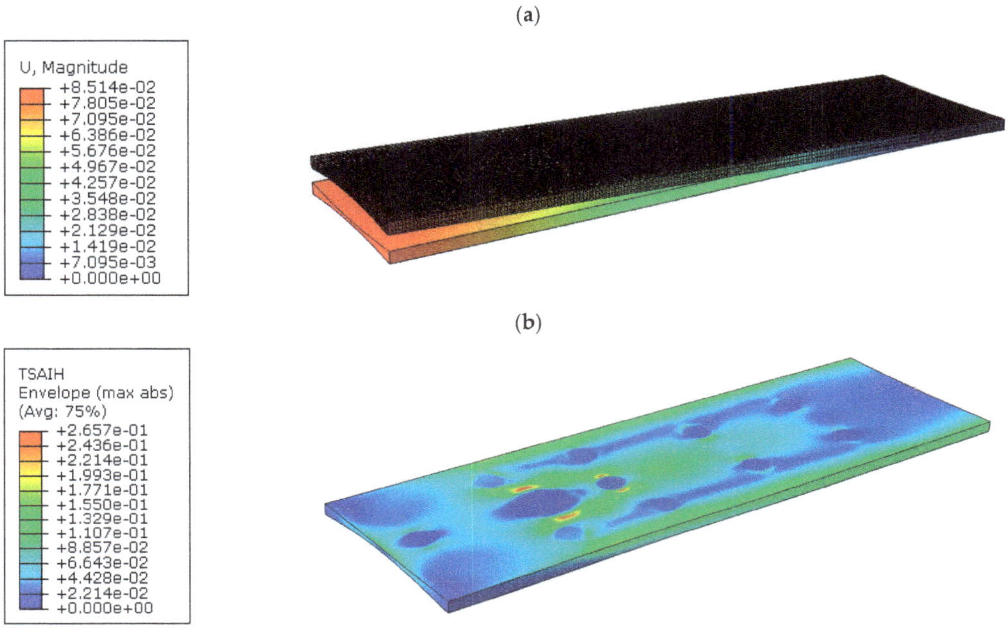

Figure 13. Results for tabletop 2: (**a**) Deflection, (**b**) Tsai–Hill criterion.

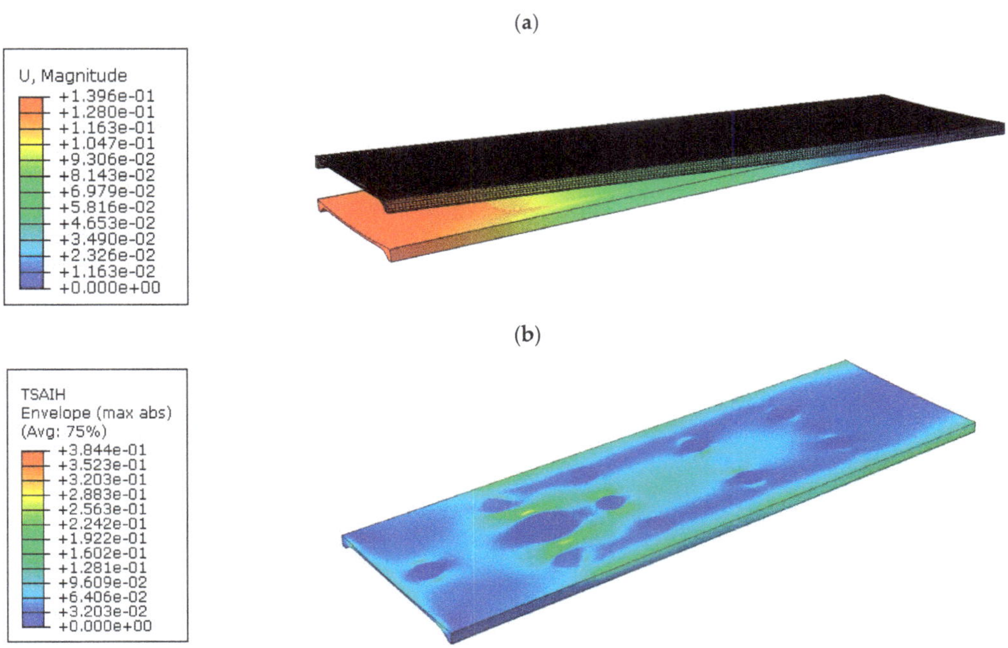

Figure 14. Results for tabletop 3: (**a**) Deflection, (**b**) Tsai–Hill criterion.

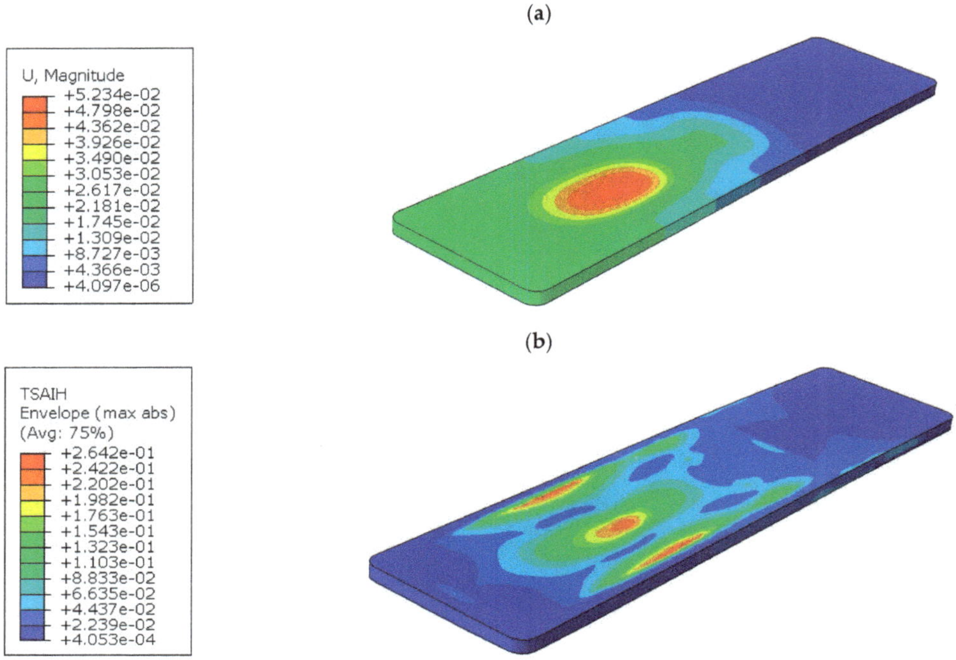

Figure 15. Results for tabletop 4: (**a**) Deflection, (**b**) Tsai–Hill criterion.

The part of the tabletop on the side of the mounting to the column is unloaded, as indicated by the blue area. The maximum displacement of 52 mm occurs where the torso of the body is located. In real conditions, this value will be lower because the weight from the body is distributed more uniformly than shown in Figure 2. The tabletop, due to the maintenance of high translucency, has foam stiffeners only on the perimeter and in the area of attachment to the column, while the inside is hollow. The free end of the tabletop experiences a displacement of 26 mm.

The material effort for tabletop 4 is shown in Figure 15b, with a maximum value of 26.4%, so the safety factor is almost 4. The maximum values are also concentrated in the torso area and in the zones near the edge where the laminate joins the foam filling.

5. Technological Test of Tabletop Bending

To be sure that the simulations were carried out correctly, both additional simulations and a laboratory test were performed. Because it is difficult to reproduce identical conditions as in Figure 1 with gradually increasing load at all points, it was decided to apply the load only at one point 30 cm away from the face of the tabletop for values from 5 to 50 kg with increments of 5 kg. The separated partition for the load in the numerical model can be seen in Figure 16a. Figure 16b shows a section of the tabletop with the sensor attached to a rigid aluminum frame. An intermediate element was placed between the sensor and the composite table in order to not induce stress concentrations [25].

In this way, the tabletop could be subjected to a deflection test precisely and without introducing vibrations or asymmetrical loads. A measuring cell was also attached to the pneumatic actuator so that the load could be precisely controlled. The load cell was attached to the actuator on one side, while the other side was pressed against a rigid plate, which in turn had a soft filling on the contact side of the tabletop so as not to cause significant pressure concentrations that could cause local damage to the laminate. The deflection of the tabletop was measured at its end using an electronic sensor.

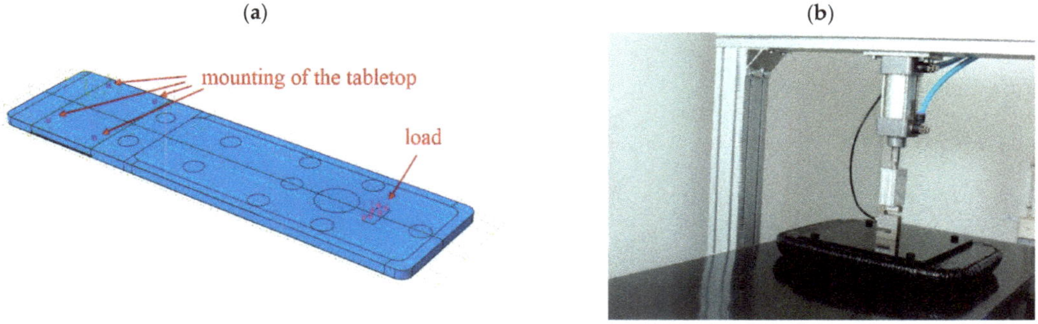

Figure 16. Testing of tabletop 4: (**a**) Mounting and loading method, (**b**) Laboratory test.

After the laboratory test, numerical calculations were made by modifying only the load location and its value in model 4.

Table 2 summarizes the results of deflection at the end of the tabletop with increments of 5 kg. The percentage differences between the laboratory test and numerical simulation are also shown.

Table 2. Results from the laboratory and numerical tests.

Deflection	Load [kg]									
	5	10	15	20	25	30	35	40	45	50
Lab. test [mm]	3.71	7.67	11.17	15.88	19.06	24.37	28.54	33.22	37.90	42.56
Abaqus [mm]	3.68	7.35	11.05	14.71	18.5	22.07	25.77	29.44	33.11	36.8
Difference [%]	0.81	4.17	1.07	7.37	2.94	9.44	9.71	11.38	12.64	13.53

Percentage differences in results in the range of loads up to 25 kg range from 0.81–7.37%, while in the range from 25 kg to 50 kg they reach higher values in the range of 9.44–13.53% and tend to increase. This shows that the elastic response of the tabletop is not linear, which is also visible in Figure 17.

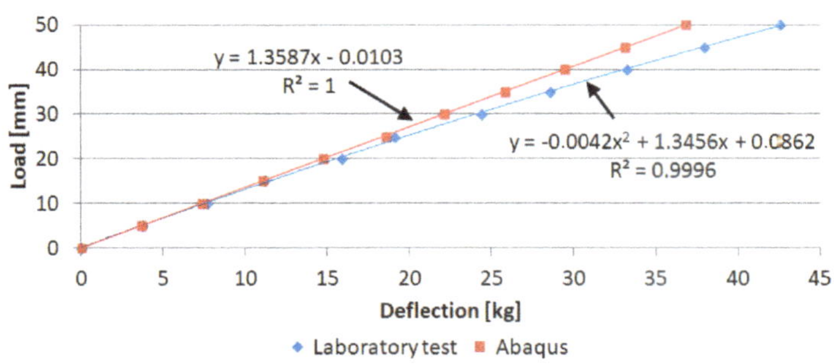

Figure 17. Comparison of results from laboratory test and numerical simulation for tabletop 4.

The laboratory test resulted in a value of 42.56 mm, while the numerical simulation resulted in a value of 36.8 mm. Therefore, the percentage difference concerning the laboratory test is:

$$\Delta = \frac{42.56 - 36.8}{42.56} \cdot 100\% = 13.53\%$$

This is lower than the 20% that was used to determine the milestone in the project from which the research was funded.

Figure 18a shows the displacement map for a maximum load of 50 kg.

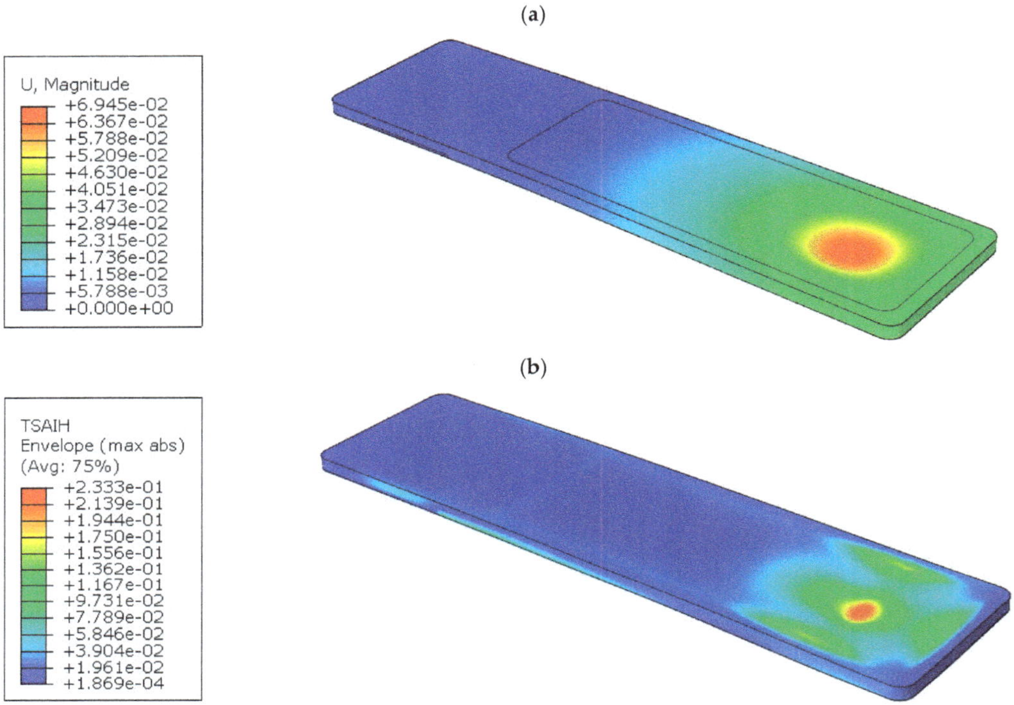

Figure 18. Results for tabletop 4 for point loading (50 kg): (**a**) Deflection, (**b**) Tsai-Hill criterion.

In addition, the effort of the material was determined (Figure 18b). According to the Tsai–Hill hypothesis, it is at the level of 23.3%. Thus, even with a point load application of 50 kg, the safety factor is still high and exceeds the value of 4.

6. Discussion

An analysis of the literature has shown that there are currently no suitable procedures or examples for both designing medical tabletop structures and performing their simulation. As a guideline for design, it is possible to take tabletop loads considering the patient's body and the corresponding safety factor. However, it should also be taken into account that mechanical loads are not the only ones that can affect the durability and strength of the entire structure. Environmental loads [26,27], such as during thermal or chemical disinfection or the effect of UV radiation, would also have to be taken into consideration. In addition, the tabletop structure is often not only made of composite but also of other materials, as shown for model 4, in which the aluminum plate plays an intermediary role and is bonded to the bottom by default. In this case, there is an adhesive joint [28], which should also be properly designed and tested.

Moving on to the analysis of the results, it is important to note their scatter when testing elastic and strength properties. For example, for unidirectional fibers, the value of the relative standard deviation was 6.5% for Young's modulus, so it should be expected that the comparison of simulation results with the laboratory test will also be subject to error. In the analyzed case for tabletop 4, this difference was 13.53%, which can be considered a satisfactory result because in such a complex construction, taking into account all the details of the structure could cause a significant increase in modeling time and numerical calculations. However, a thoughtful analysis can be made of which factors may affect the differences in results for tabletop 4. The tabletop is mounted to the column by a set of four screws. The column can deform when the bolts are not perfectly rigid, as in the numerical model. In the numerical model, the foam infill and the connection to the aluminum plate were made using "tie" constraints; in reality, there are adhesive joints, which can also deform additionally. There may be structural defects that, for example, result in the tabletop not deforming linearly [29]. It has been observed that as the load increases, the respective differences in displacement increments become larger. Therefore, to be able to make more accurate comparisons and detect structural defects, measurements using digital image correlation DIC [30] for the entire structure are necessary. In general, the deflection results for the demonstrator are greater than in the numerical simulation because the virtual model has no defects and is geometrically perfect.

Ergonomic tabletops will be the next stage in the development of this type of design as they allow the patient's body to be aligned with the surface, distribute pressure more evenly, and increase comfort. The surgical procedure itself, while in the hospital and on the medical table, is often a stressful situation, hence the aim should be to improve the ergonomics of equipment to make the patient feel comfortable.

We should also work toward the best possible translucency of the design in terms of X-rays. On the one hand, the use of foam fillings or the use of additional ribs increases stiffness, but on the other, the translucency deteriorates or line disturbances develop where the vertical walls of the laminate are present, as shown in Figure 19 (red arrows).

(a)

(b)

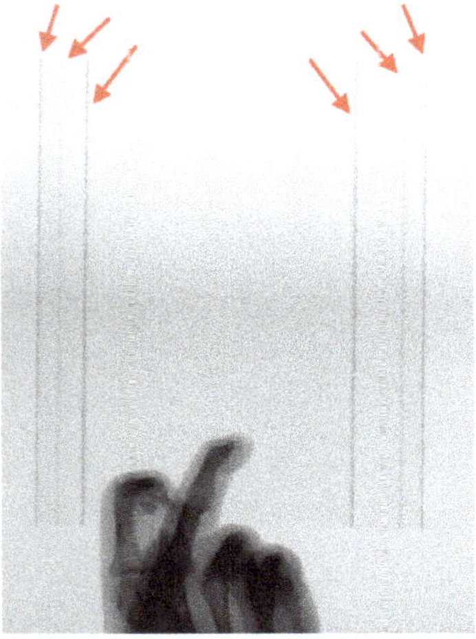

Figure 19. Fragment of the demonstrator with fabricated rail: (**a**) View of a composite part, (**b**) X-ray view (red arrows—vertical walls).

Such walls can occur, for example, close to the edge, where the rail for attaching accessories is located. However, the topics of accessory mounting and transparency testing will be the subject of future articles.

7. Conclusions

In this work, a complete approach to designing, and performing numerical simulations and laboratory tests for medical operating tables was presented. This is the first work of its kind in the world literature on medical tabletops. The following conclusions are drawn from the research.

- The compressive strength of laminates both made of unidirectional fibers and fabrics is more than twice the tensile strength. Compressive strength should be taken into account in research and numerical testing because the tabletop works like a cantilever beam, and thus the lower part of it is compressed.

- The occurring scatter of results in elastic and strength properties leads to a situation in which the result of the numerical simulation will also be subject to error compared to the laboratory test. It is therefore necessary to maintain an appropriate safety factor when designing the structure.
- Three types of ergonomic tops were presented, whose change in cross-section significantly affected the deflection, but the effort of the material remained at a similar level.
- A satisfactory difference (13.53%) was obtained in the results for tabletop 4 between the deflection obtained for the demonstrator and in the numerical simulation.

Author Contributions: Conceptualization, P.G. and A.M.W.-R.; Methodology, P.G. and D.P.; Software, D.P. and P.G.; Validation, P.G. and T.S.; Formal Analysis, P.G. and T.S.; Investigation, D.P.; Resources, T.S. and A.M.W.-R.; Data Curation, D.P. and P.G.; Writing—Original Draft Preparation, P.G.; Writing—Review and Editing, P.G. and T.S.; Visualization, D.P. and P.G.; Supervision, T.S. and A.M.W.-R.; Project Administration, A.M.W.-R.; Funding Acquisition, A.M.W.-R. All authors have read and agreed to the published version of the manuscript.

Funding: Financial support of the European Regional Development Fund under the Operational Program Smart Growth 2014-2020 Project "Composite system for fast mechanical connections for the medical industry" implemented by WIT-Composites, contact number POIR.01.01.01-00-0944/17-00. T.S. was funded under the grant "Subvention for Science" (MEiN), project No. FD-20/IL-4/046.

Institutional Review Board Statement: Not applicable.

Informed Consent Statement: Not applicable.

Data Availability Statement: The data presented in this study are available on reasonable request from the corresponding author.

Conflicts of Interest: The authors declare no conflict of interest.

Appendix A

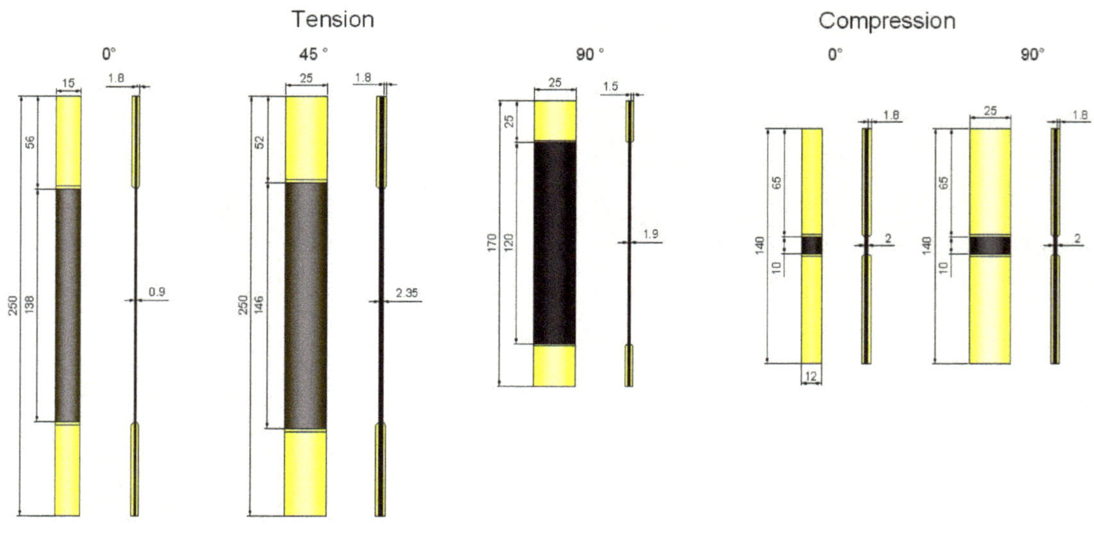

Figure A1. Shapes and dimensions of samples made of KORDCARBON-CPREG-200-UD-3K-EP-120-38 (UD).

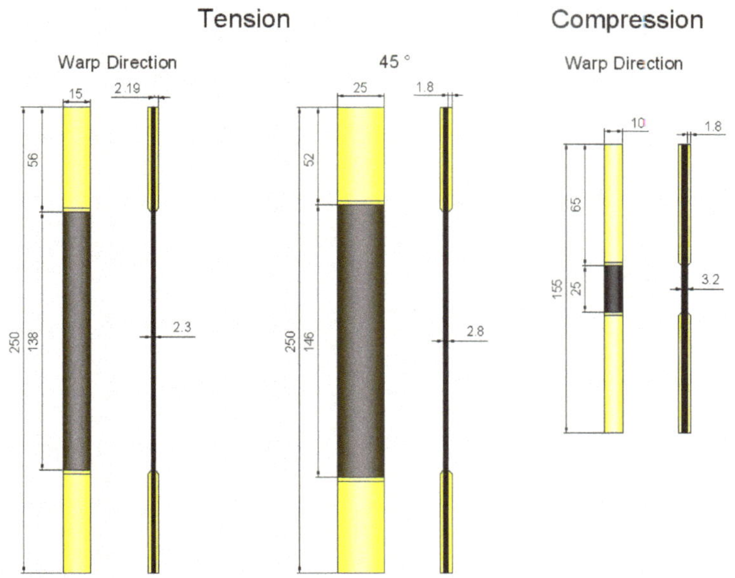

Figure A2. Shapes and dimensions of samples made of KORDCARBON-CPREG-245-T-3K-EP1.5-42-T (TW).

Figure A3. Test stand: (**a**) Tensile test, (**b**) Compression test.

Appendix B

Summary of the results obtained for the materials used in the study:
(1) KORDCARBON-CPREG-200-UD-3K-EP-120-38 (UD),
(2) KORDCARBON-CPREG-245-T-3K-EP1.5-42-T (TW).

Property	Symbol	Unit	Value	
			(1) UD	(2) TW
Young's modulus in the direction of fiber orientation (warp direction for TW)	E_{t0}	GPa	121.965	48.069
Young's modulus transverse to the direction of fiber orientation	E_{t90}	GPa	7.817	-
Kirchhoff's modulus	G_{12}	GPa	3.434	2.652
Poisson's ratio	ν_{12}	-	0.298	0.064
Tensile strength in the fiber direction (warp direction for TW)	X_t	MPa	2003.119	453.385
Tensile strength in the direction perpendicular to the fibers	Y_t	MPa	45.12	-
Shear strength	S	MPa	58.749	58.73
Young's modulus in the direction of fiber orientation (warp direction for TW)	E_{c0}	GPa	111.564	43.4
Young's modulus transverse to the direction of fiber orientation	E_{c90}	GPa	8.094	-
Compressive strength in the fiber direction (warp direction for TW)	X_c	MPa	793.504	208.99
Compressive strength in the direction perpendicular to the fibers	Y_c	MPa	131.592	-

References

1. Campbell, M.H. What is the Basic Furniture in the Operating Theatre? In *Theatre Routine*; Elsevier: Amsterdam, The Netherlands, 1971; pp. 9–12. [CrossRef]
2. Campbell, M.H. How does the Operating Table Function? In *Theatre Routine*; Elsevier: Amsterdam, The Netherlands, 1971; pp. 53–54. [CrossRef]
3. Pisklakov, S.; Houten, J.K.; Echt, M.; Kosse, A.; Abramowicz, A.E. O-arm failure traps patient prone on a Jackson table. *J. Clin. Neurosci.* **2015**, *22*, 1365–1366. [CrossRef] [PubMed]
4. Tan, Y.; Tanaka, M.; Fujiwara, Y.; Uotani, K.; Yamauchi, T.; Yorimitsu, M.; Yokoyama, Y.; Sonawane, S. Effect of an Adjustable Hinged Carbon Fiber Operating Table on the Coronal Alignment of the Lumbar Spine During Oblique Lateral Interbody Fusion. *World Neurosurg.* **2021**, *149*, e958–e962. [CrossRef] [PubMed]
5. Tanaka, M.; Desai, D.; Fujiwara, Y.; Arataki, S.; Latka, K.; Sake, N.; Liang, W.; Kodama, Y.; Miyamoto, Y.; Yamauchi, T. Effect of an Adjustable Hinged Carbon Fiber Operating Table on Sagittal Alignment of the Lumbar Spine. *Appl. Sci.* **2023**, *13*, 138. [CrossRef]
6. Al-Qaisi, S.K.; El Tannir, A.; Younan, L.A.; Kaddoum, R.N. An ergonomic assessment of using laterally-tilting operating room tables and friction reducing devices for patient lateral transfers. *Appl. Ergon.* **2020**, *87*, 103122. [CrossRef]
7. McAllister, R.K.; Booth, R.T.; Bittenbinder, T.M. Two loose screws: Near-miss fall of a morbidly obese patient after an operating room table failure. *J. Clin. Anesth.* **2016**, *33*, 47–50. [CrossRef] [PubMed]
8. Demelio, G.; Genovese, K.; Pappalettere, C. An experimental investigation of static and fatigue behaviour of sandwich composite panels joined by fasteners. *Compos. Part B Eng.* **2001**, *32*, 299–308. [CrossRef]
9. Han, M.J.; Ko, S. Comparison of interface pressures and subjective comfort of pressure-relieving overlays on the operating table for healthy volunteers. *Int. J. Environ. Res. Public Health* **2017**, *18*, 2640. [CrossRef] [PubMed]
10. Basli, A.A.; Van Giersbergen, M.Y. Comparison of interface pressures on three operating table support surfaces during surgery. *J. Tissue Viability* **2021**, *30*, 410–417. [CrossRef] [PubMed]
11. Modjarrad, K. Introduction. In *Handbook of Polymer Applications in Medicine and Medical Devices*; Elsevier: Amsterdam, The Netherlands, 2014; pp. 1–7. [CrossRef]
12. Ashter, S.A. Introduction to polymers and plastics for medical devices. In *Applications of Polymers and Plastics in Medical Devices*; Elsevier: Amsterdam, The Netherlands, 2022; pp. 1–26. [CrossRef]
13. Ashter, S.A. Selection of materials for construction of medical devices. In *Applications of Polymers and Plastics in Medical Devices*; Elsevier: Amsterdam, The Netherlands, 2022; pp. 45–64. [CrossRef]
14. McKeen, L.W. Plastics Used in Medical Devices. In *Handbook of Polymer Applications in Medicine and Medical Devices*; Elsevier: Amsterdam, The Netherlands, 2014; pp. 21–53. [CrossRef]
15. Ashter, S.A. Classification of plastics and elastomers used in medical devices. In *Applications of Polymers and Plastics in Medical Devices*; Elsevier: Amsterdam, The Netherlands, 2022; pp. 65–78. [CrossRef]

16. Tang, W.; Xu, Y.; Hui, X.; Zhang, W. Multi-Objective Optimization of Curing Profile for Autoclave Processed Composites: Simultaneous Control of Curing Time and Process-Induced Defects. *Polymers* **2022**, *14*, 2815. [CrossRef] [PubMed]
17. Kojnoková, T.; Nový, F.; Markovičová, L. Evaluation of tensile properties of carbon fiber reinforced polymers produced from commercial prepregs. *Mater. Today Proc.* **2022**, *62*, 2663–2668. [CrossRef]
18. *D 3518*; Standard Test Method for In-Plane Shear Response of Polymer Matrix Composite Materials by Tensile Test of a ±45° Laminate. ASTM International: West Conshohocken, PA, USA, 2018. [CrossRef]
19. *ASTM D3039/D3039M*; Standard Test Method for Tensile Properties of Polymer Matrix Composite Materials. ASTM International: West Conshohocken, PA, USA, 2014. [CrossRef]
20. *ASTM D6641*; Standard Test Method for Compressive Properties of Polymer Matrix Composite Materials Using a Combined Loading Compression (CLC) Test Fixture. ASTM International: West Conshohocken, PA, USA, 2014.
21. Li, S.; Zhang, Y.; Dou, X.; Zuo, P.; Liu, J. Hard to be killed: Load-bearing capacity of the leech Hirudo nipponia. *J. Mech. Behav. Biomed. Mater.* **2018**, *86*, 345–351. [CrossRef] [PubMed]
22. Fuchs, C.; Bhattacharyya, D.; Fakirov, S. Microfibril reinforced polymer-polymer composites: Application of Tsai-Hill equation to PP/PET composites. *Compos. Sci. Technol.* **2006**, *66*, 3161–3171. [CrossRef]
23. Suchý, T.; Balík, K.; Sedláček, R.; Sucharda, Z.; Sochor, M.; Prokop, J.; Beneš, J.; Křena, J. Radiolucent composites providing high resistance against sterilization decomposition. *Ceramics-Silikaty* **2011**, *55*, 401–409.
24. Bier, B.; Ravikumar, N.; Unberath, M.; Levenston, M.; Gold, G.; Fahrig, R.; Maier, A. Range imaging for motion compensation in C-arm cone-beam CT of knees under weight-bearing conditions. *J. Imaging* **2018**, *4*, 13. [CrossRef]
25. Sburlati, R. The contact behavior between a foam core sandwich plate and a rigid indentor. *Compos. Part B Eng.* **2002**, *33*, 325–332. [CrossRef]
26. Golewski, P.; Sadowski, T.; Pietras, D.; Dudzik, S. The influence of aging in salt chamber on strength of aluminum—CFRP single lap joints. *Mater. Today Proc.* **2021**, *45*, 4264–4267. [CrossRef]
27. Golewski, P.; Sadowski, T. An Influence of UV Ageing Process on Tensile Strength and Young's Modulus of Polymeric Fiber Composite Materials. *IOP Conf. Ser. Mater. Sci. Eng.* **2018**, *416*, 012057. [CrossRef]
28. Golewski, P. Tensile Behaviour of Double- and Triple-Adhesive Single Lap Joints Made with Spot Epoxy and Double-Sided Adhesive Tape. *Materials* **2022**, *15*, 7855. [CrossRef] [PubMed]
29. Kim, J.S.; Oh, J.; Cho, M. Efficient analysis of laminated composite and sandwich plates with interfacial imperfections. *Compos. Part B Eng.* **2011**, *42*, 1066–1075. [CrossRef]
30. Golewski, P.; Sadowski, T.; Kneć, M.; Budka, M. The effect of thermal aging degradation of CFRP composite on its mechanical properties using destructive and non-destructive methods and the DIC system. *Polym. Test.* **2023**, *118*, 107902 [CrossRef]

materials

MDPI

Article

Assessing Effects of van der Waals Corrections on Elasticity of $Mg_3Bi_{2-x}Sb_x$ in DFT Calculations

Qing Peng [1,2,3,*], Xinjie Ma [2,4], Xiaoyu Yang [2,5], Shuai Zhao [2,6], Xiaoze Yuan [2] and Xiaojia Chen [1,*]

1 School of Science, Harbin Institute of Technology, Shenzhen 518055, China
2 The State Key Laboratory of Nonlinear Mechanics, Institute of Mechanics, Chinese Academy of Sciences, Beijing 100190, China; yuanxz@imech.ac.cn (X.Y.)
3 Guangdong Aerospace Research Academy, Guangzhou 511458, China
4 Beijing MaiGao MatCloud Technology Co., Ltd., Beijing 100190, China
5 Computer Network Information Center, Chinese Academy of Sciences, Beijing 100190, China
6 Department of Modern Mechanics, University of Science and Technology of China, Hefei 230026, China
* Correspondence: pengqing@imech.ac.cn (Q.P.); xjchen@hit.edu.cn (X.C.); Tel.: +86-(10)-82544239 (Q.P.)

Abstract: As a promising room-temperature thermoelectric material, the elastic properties of $Mg_3Bi_{2-x}Sb_x$ ($0 \leq x \leq 2$), in which the role of van der Waals interactions is still elusive, were herein investigated. We assessed the effects of two typical van der Waals corrections on the elasticity of $Mg_3Bi_{2-x}Sb_x$ nanocomposites using first-principles calculations within the frame of density functional theory. The two van der Waals correction methods, PBE-D3 and vdW-DFq, were examined and compared to PBE functionals without van der Waals correction. Interestingly, our findings reveal that the lattice constant of the system shrinks by approximately 1% when the PBE-D3 interaction is included. This leads to significant changes in certain mechanical properties. We conducted a comprehensive assessment of the elastic performance of $Mg_3Bi_{2-x}Sb_x$, including Young's modulus, Poisson's ratio, bulk modulus, etc., for different concentration of Sb in a 40-atom simulation box. The presence or absence of van der Waals corrections does not change the trend of elasticity with respect to the concentration of Sb; instead, it affects the absolute values. Our investigation not only clarifies the influence of van der Waals correction methods on the elasticity of $Mg_3Bi_{2-x}Sb_x$, but could also help inform the material design of room-temperature thermoelectric devices, as well as the development of vdW corrections in DFT calculations.

Keywords: thermoelectric materials; PBE-D3; vdW-DFq; first-principles calculation; $Mg_3Bi_{2-x}Sb_x$

1. Introduction

The thermoelectric effect is a phenomenon in which a material generates an electric current when exposed to a temperature gradient. Materials with exceptional thermoelectric properties have significant applications in energy conversion processes, such as refrigeration and waste heat harvesting [1,2]. In recent years, there has been a growing interest in investigating the thermoelectric properties of $Mg_3Bi_{2-x}Sb_x$ [3–11], a promising room-temperature thermoelectric material. Notably, $Mg_3Bi_{2-x}Sb_x$ exhibits a remarkable thermoelectric figure of merit, reaching approximately 1.5 when x = 0.5 [12]. This places it in a competitive position when compared to more traditional thermoelectric materials like Bi_2Te_3 [13] and Ag_2Se [14], offering advantages such as cost-effectiveness. Researchers have delved into various aspects of $Mg_3Bi_{2-x}Sb_x$ over the past few years. These investigations have covered various areas, including band topology [15–18], phonon dynamics [19–23], and the topological thermoelectric properties of nodal-line semimetals [22–27]. Notable contributions, such as the work of Kanno [28], have shed light on specific characteristics of $Mg_3Bi_{2-x}Sb_x$. For instance, it has been demonstrated that this material exhibits higher charge carrier mobility and lower thermal conductivity, which can be attributed to its disordered structure.

As a prospective room-temperature high-performance thermoelectric material, it is crucial to investigate its mechanical properties. Despite prior computational investigations into the elastic characteristics of $Mg_3Bi_{2-x}Sb_x$ using first principles, as seen in works like those by Peng et al. [29], the consideration of van der Waals (vdW) correction is noticeably absent although desirable. The interactions due to the fluctuation of nonlocal charge densities are hard to model but crucial in first-principles calculations. To improve the accuracy of our calculations and understand the impact of vdW correction within the system, we conducted a comprehensive comparative study employing both the vdW-DFq [30] and PBE-D3 [31] methods with reference to the impact of non-vdW correction (and thus, PBE). Our analysis revealed intriguing findings: the inclusion of vdW interactions resulted in a reduction in lattice constants, particularly noticeable using the PBE-D3 method, which displayed a more significant decrease of approximately 1%. Importantly, the PBE-D3 calculations demonstrated a closer alignment with experimental data, thereby augmenting their persuasiveness. Our investigation encompassed a thorough exploration of 40-atom systems with varying Sb content. Remarkably, except for the specific point at x = 0.375, neither the PBE-D3 nor the vdW-DFq methods altered the trend of elastic properties in $Mg_3Bi_{2-x}Sb_x$ as the Sb content varied. The changes were solely confined to absolute values, usually within a range of about 10%. This comprehensive analysis not only enhances our understanding of the mechanical behavior of $Mg_3Bi_{2-x}Sb_x$, but also underscores the nuanced yet pivotal role of vdW correction.

2. Materials and Methods

Both Mg_3Bi_2 and Mg_3Sb_2 belong to the trigonal crystal system, with a symmetry that belongs to the space group $P\bar{3}m1$. We have plotted all the structures in Figure 1. The primitive unit cell consists of five atoms, as shown in Figure 1a,b for Mg_3Bi_2 and Mg_3Sb_2, respectively. In the case of $Mg_3Bi_{2-x}Sb_x$, the objective is to maintain a constant number of Mg atoms while gradually substituting Bi atoms with Sb atoms. To cover a wide range of concentrations, we expanded the original unit cell to a 2 × 2 × 2 supercell, resulting in a total of 40 atoms. Among these atoms, there are 16 atoms that are either Bi or Sb, with concentrations ranging from 0 to 16 Sb atoms, as shown in Figure 1c–s. The initial structures of these 17 configurations were obtained from LAsou [32,33]. LAsou is an active learning approach that overcomes the exponential-wall problem for the effective structure prediction of chemical-disordered materials. The lowest-energy configuration was chosen as the representative structure for that particular concentration. For the calculation of elastic constants, each representative structure was further optimized to a higher precision. We applied both positive and negative strains to the six independent degrees of freedom in the system, resulting in a total of 12 different configurations. Through first-principles calculations, we determined the internal pressures for these 12 configurations, enabling us to obtain the elastic constants for the entire system.

To obtain the energies and stresses of configurations, we utilized the Vienna Ab initio Simulation Package (VASP) [34–37], a piece of software based on first-principles density functional theory. The Perdew–Burke–Ernzerhof (PBE) [37] exchange-correlation functional was chosen for the calculations. For the 40-atom system calculations, we utilized the Methfessel–Paxton [38] method to generate k-points centered at the gamma point, with a grid setting of 4 × 4 × 2. The energy cutoff for the plane wave basis used in the calculations was 450 eV. During structural optimization, a force convergence criterion of 10^{-3} eV/Å was used, while the electronic convergence criterion was 10^{-6} eV. We adopted Gaussian smearing and the smearing width was 0.02 eV. We used VASPKIT [39] to process the obtained elastic constants tensor, allowing us to extract Young's moduli and Poisson's ratios for different directions [40]. Given the significant number of configurations, we integrated the calculation workflow onto the MatCloud [41] platform for improved efficiency. This cloud-based approach allowed us to efficiently distribute multiple jobs to high-performance computing clusters, optimizing resource utilization for extensive system calculations.

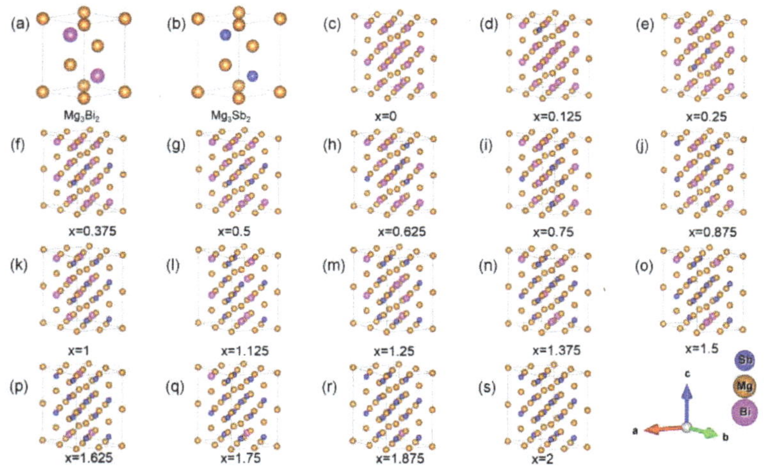

Figure 1. The atomistic structures of $Mg_3Bi_{2-x}Sb_x$: the 5-atom unit cell of (**a**) Mg_3Bi_2 and (**b**) Mg_3Sb_2, and (**c–s**) the 40-atom cells of $Mg_3Bi_{2-x}Sb_x$ for x ranging from 0 to 2 with increments of 0.125.

For vdW corrections, two distinct approaches were utilized: PBE-D3 and vdW-DFq. The PBE-D3 method, which was developed by Grimme et al. [30], is designed to accurately model the vdW dispersion energy-correction term. The incorporation of empirical parameters in density functional theory has been proposed as a means to address its limitations. This approach has shown promising results, especially in the case of systems that involve heavy elements. On the contrary, vdW-DFq correction, as proposed by Peng et al. [31], presents a novel approach for precise density and geometry calculations, with a specific focus on semihard materials. Both of these methods have been effectively integrated into the VASP-5.4.4 software. By manipulating various parameters, we successfully performed a range of calculations using these methodologies. PBE-D3 addresses the limitations of density functional theory by integrating empirical parameters, whereas vdW-DFq prioritizes the precise computation of the density and geometric characteristics of semihard materials.

3. Results

Prior to examining the vdW corrections, we conducted a reference study [29] to compare the effects of precision enhancement on various system parameters without considering vdW correction. This comparison encompasses lattice constants "*a*" (in Å) and "*c*" (in Å), elastic constants C_{11}, C_{33}, C_{44}, and C_{12}, as well as bulk and shear moduli (in GPa). The results of our precision enhancement are shown in red, while the results from previous work are shown in blue, in Figure 2.

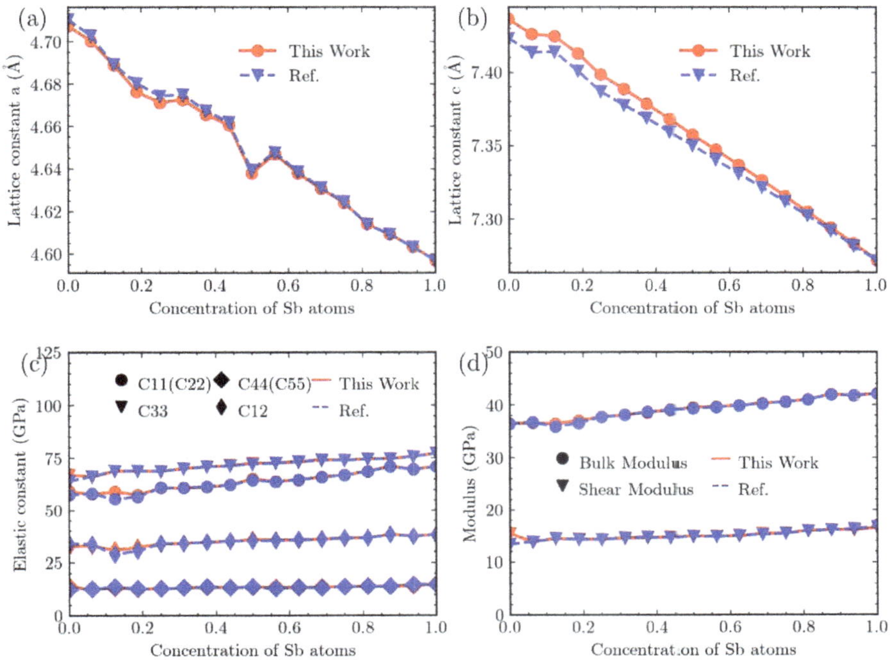

Figure 2. Lattice constants *a* (**a**) and *c* (**b**), elastic constants (**c**), and elastic modulus (**d**) of $Mg_3Bi_{2-x}Sb_x$ as a function of the concentration of Sb atoms, compared with ref. [29]. The red lines denote values from this work and the blue ones represent values from ref. [29].

Upon comparison, it becomes evident that the impact of precision enhancement on lattice constant "*a*" is not particularly significant. However, for lattice constant "*c*", especially at lower values of Sb content, the impact is relatively more significant. In contrast, for higher concentrations of Sb, the precision used in previous calculations appears to be sufficient. Regarding elastic constants and moduli, the impact of enhancing precision is usually negligible, except for the Mg_3Bi_2 system, which exhibits more noticeable variations. These comparisons suggest that when performing calculations on $Mg_3Bi_{2-x}Sb_x$, it may be possible to decrease precision to some extent in systems with higher Sb content in order to improve computational efficiency, while still maintaining acceptable accuracy. Conversely, for systems with lower Sb content, it is advisable to increase precision.

The comparison with previous work [29] indicates that precision adjustments in $Mg_3Bi_{2-x}Sb_x$ calculations can be customized based on the Sb concentration in the system. It is recommended to use higher-precision for systems with lower Sb content to ensure accuracy, while systems with higher Sb content can benefit from reduced precision to improve computational efficiency. Additionally, this study adds credibility to our current efforts involving the inclusion of vdW corrections.

All the calculated configurations of $Mg_3Bi_{2-x}Sb_x$ belong to the trigonal crystal system. Using first-principles calculations, we obtained the optimal structures for each configuration. For the trigonal crystal system, our focus was on lattice constants "*a*" and "*c*". We plotted the lattice constants that vary with different Sb atom content in Figure 3. It is clear from Figure 3 that, regardless of the calculation method used, the lattice constants decrease as the Sb atom content increases. Considering the impact of different functionals, it can be observed that the influence of vdW-DFq on the lattice constants is minimal. For lattice constant "*a*", there is a decrease of approximately 0.4%. The effect on lattice constant "*c*" is negligible. However, PBE-D3 correction has a relatively more noticeable impact. Taking the

Mg_3Bi_2 system as an example, lattice constant "*a*" decreases by 1.3%, while lattice constant "*c*" is affected by 1.1%.

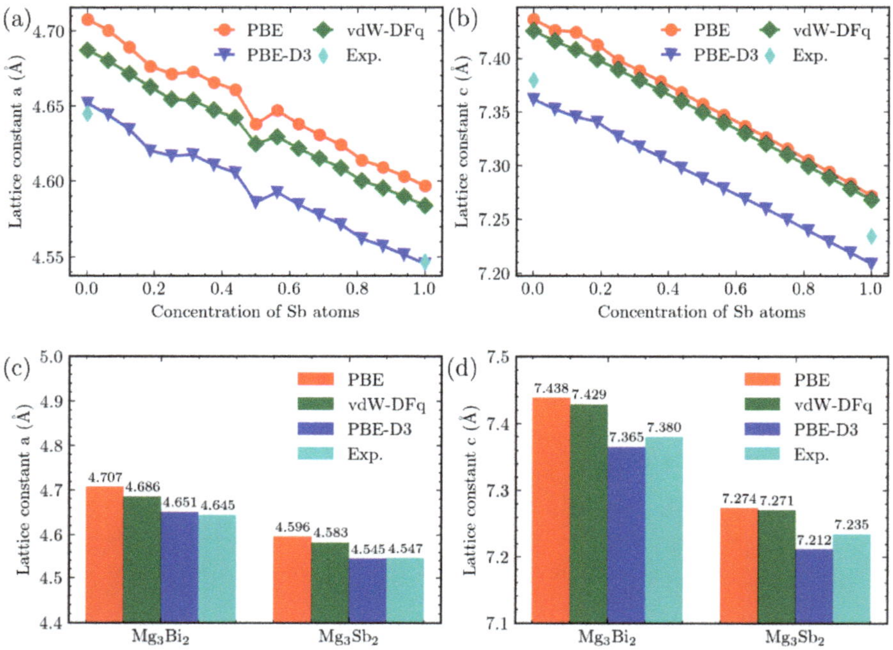

Figure 3. Lattice constants of $Mg_3Bi_{2-x}Sb_x$. The four methods are color-coded: red (PBE), green (vdW-DFq), blue (PBE-D3), and cyan (experimental data [42]). (**a**) The trend of lattice constant "*a*" with respect to Sb content. (**b**) Lattice constant "*c*" as a function of Sb content. (**c**) Comparison of lattice constant "*a*" among different Mg_3Bi_2 and Mg_3Sb_2 structures using various methods. (**d**) Comparison of lattice constant "*c*" among different Mg_3Bi_2 and Mg_3Sb_2 structures using various methods.

We conducted a comparison between the experimental lattice constants of Mg_3Bi_2 and Mg_3Sb_2, as illustrated in Figure 3 c,d, respectively. The figure clearly demonstrates a close alignment between the experimental lattice constants and those obtained through the application of the PBE-D3 method. Therefore, this comparison provides robust evidence in favor of employing PBE-D3 correction in calculations related to $Mg_3Bi_{2-x}Sb_x$.

Upon subjecting the system to deformation along its six independent degrees of freedom, we obtained a 6×6 elastic constant tensor. For the trigonal crystal system, the elastic tensor possesses eight independent components, namely, C_{11}, C_{33}, C_{44}, C_{12}, C_{13}, C_{14}, C_{15}, and C_{45}. We have illustrated the variations in Figure 4a,b, depicting the changes in Sb content. Regardless of the adoption of vdW functionals, the values of these components exhibit a linear increase as x changes, except for the points at Sb concentrations of 0.1875 and 0.25 (corresponding to x = 0.375 and x = 0.5). Specifically, for the C_{44} component, the PBE-D3 calculations generally yield higher results than those without vdW corrections, showing an increase of approximately 16%, except for the pure Mg_3Bi_2 state. Comparatively, the results of vdW-DFq exhibit some variation, with values oscillating and differing by approximately 8% from the PBE calculations.

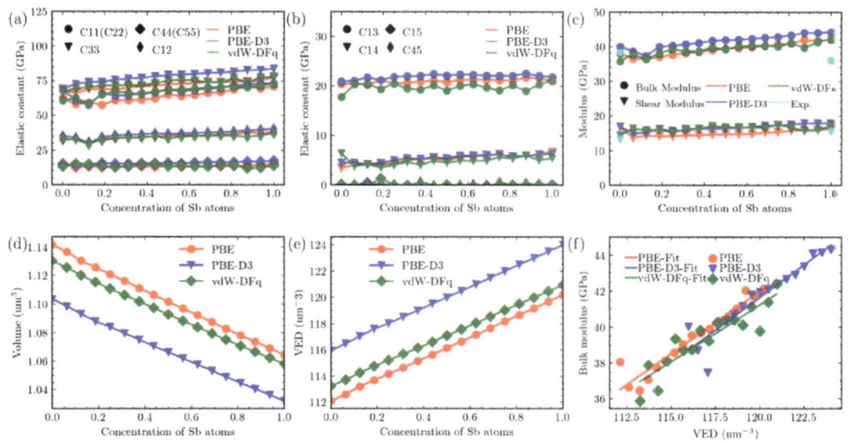

Figure 4. Comparison of elastic constants from different calculation methods: red (PBE), green (vdW-DFq), blue (PBE-D3), cyan (Exp.). (**a**,**b**) Elastic constants, (**c**) bulk modulus and shear modulus, (**d**) volume, (**e**) valence electron density (VED), (**f**) bulk modulus with changing average VED.

For the C_{12} component, it is noticeable that the results from all three methods are quite similar. Comparing the results obtained from vdW-DFq with those obtained using only PBE without considering vdW corrections, there are some variations in the values for different structures within $Mg_3Bi_{2-x}Sb_x$, with an average change of approximately 2.5%. For PBE-D3, the results for all structures, except for the one at an Sb concentration of 0.1875, are generally larger than those obtained with PBE. The structure at x = 0.375 is about 2.3% smaller than PBE. Overall, there is a numerical difference of about 5.5%.

Concerning the C_{33} component, once again, the results from vdW-DFq and PBE are relatively close, with differences of approximately 2.4%. However, the results obtained with PBE-D3 are, on average, about 8.8% larger than those obtained with PBE. As for the C_{44} component, all methods show a decrease at an Sb concentration of 0.1875, while at other points, they increase with increasing Sb content. In the calculation of this component, the influence of vdW is more significant. Compared to PBE calculations, the results from vdW-DFq are, on average, 8.3% higher, while the results from PBE-D3 are, on average, 16.8% higher. As for the components in Figure 4b, C_{15} and C_{45} are both close to 0 overall, while both C_{13} and C_{14} show an increase with Sb content. The results obtained from various vdW methods differ from those of the PBE method by approximately 6%.

Typically, the bulk modulus is used to describe a material's resistance to compression and expansion in volume, while the shear modulus characterizes its resistance to shearing forces. In the context of crystalline systems, both the bulk and shear moduli exhibit varying degrees of anisotropy. To facilitate the comparative analysis of the variation in bulk and shear moduli for different Sb contents, we adopted the Voigt-Reuss-Hill approximation [43–45] to describe the changes in these moduli. We utilized the following formulas to calculate the bulk modulus and shear modulus:

$$\begin{aligned} B &= \tfrac{1}{2}(B_V + B_R) \\ G &= \tfrac{1}{2}(G_V + G_R) \end{aligned} \tag{1}$$

where B_V, B_R, G_V, and G_R can be obtained from the following expressions based on the elastic constant and its compliance S, which is often used to describe the compliance of materials and can be calculated by taking the inverse of the elastic tensor:

$$\begin{aligned} B_V &= \tfrac{1}{9}(C_{11} + C_{22} + C_{33}) + \tfrac{2}{9}(C_{23} + C_{13} + C_{12}) \\ B_R &= \tfrac{1}{3(S_{11}+S_{22}+S_{33})+6(S_{23}+S_{13}+S_{12})} \end{aligned} \tag{2}$$

$$G_V = \frac{1}{5}[(C_{11} + C_{22} + C_{33}) - (C_{23} + C_{13} + C_{12})] + \frac{3}{5}(C_{44} + C_{55} + C_{66})$$
$$G_R = \frac{1}{12[(S_{11}+S_{22}+S_{33})-(S_{23}+S_{13}+S_{12})]+9((S_{44}+S_{55}+S_{66}))} \quad (3)$$

This relationship is depicted in Figure 4c. It is evident that with increasing Sb content, except for a few points, both the shear and bulk moduli show an increase. Regarding the influence of vdW corrections on the bulk modulus, the results obtained from the PBE-D3 calculations are generally higher compared to those from vdW-DFq and PBE. Taking Mg_3Sb_2 as an example, the bulk moduli calculated using the three methods are 44.34 GPa (PBE-D3), 42.41 GPa (vdW-DFq), and 42.14 GPa (PBE), respectively. The PBE-D3 results are approximately 5.2% higher than the PBE results. Overall, the bulk moduli obtained using PBE-D3 are approximately 5.5% higher than those obtained without vdW corrections. In comparison, the impact of vdW-DFq is relatively smaller, with numerical differences of approximately 2%.

Concerning the shear modulus, both the PBE-D3 and vdW-DFq results are higher by about 10% compared to the PBE results. However, there are some minor differences in the details. For instance, in the case of PBE-D3, the result at the point where the Sb concentration is 0.1875 is similar to those obtained without vdW corrections. On the other hand, for vdW-DFq, the points at Sb concentrations of 0.9375 and 0.875 (corresponding x = 1.875 and x = 1.75) closely align with the PBE results. This alignment is understandable because the corresponding moduli are derived from the elements of the elastic constants' matrix, and their trends also align with the trends of the elastic constants. For comparison, we have indicated the experimental [46] values for the bulk and shear moduli of Mg_3Bi_2 and Mg_3Sb_2 as cyan dashed points in Figure 4c. It can be observed that, except for the bulk modulus of Mg_3Sb_2, the calculated moduli closely match the experimental values.

We have simultaneously illustrated the trends in volume, bulk modulus, and average valence electron density (VED). VED is defined as the number of valence electrons in a unit volume. These trends are compared for the three methods with respect to the concentration of Sb atoms, as shown in Figure 4. Since the numbers of valence electrons for the elements Mg, Bi, and Sb are 2, 5, and 5, respectively, all 40-atom cell structures of $Mg_3Bi_{2-x}Sb_x$ have the same number of total valence electrons, which is 128. Due to the disparities in lattice constants obtained from different calculation methods, these differences are further magnified at the volume level. The lattice volume decreases as the content of Sb increases. Additionally, when employing the PBE-D3 method for vdW simulations, the volume is approximately 5% smaller than when not using vdW corrections. It is worth noting that VED is different from valence electron concentration (VEC), which is defined as the number of valence electrons in a formula unit [47].

According to the findings of a previous study [29], in the $Mg_3Bi_{2-x}Sb_x$ system, except for the special point at an Sb concentration of 0.25, the bulk modulus demonstrates a linear relationship with the VED. This conclusion is supported by the observations in Figure 4f. However, when improving computational precision, the linear relationship becomes less noticeable when using the vdW-DFq results. In contrast, the results from PBE-D3 still exhibit a linearly increasing trend, although with some deviations at several points for $x < 0.5$. Such a trend can be understood as follows: a higher valence electron density implies stronger atomic interactions, and consequently, a higher bulk modulus. This ascending trend indicates that we can intentionally adjust the relevant elastic properties of the material by consciously manipulating VED, such as through doping with other elements like Co and Te [28,48].

Furthermore, we employed Jiang's model [49] to investigate the Vickers hardness of the $Mg_3Bi_{2-x}Sb_x$ system. This empirical model establishes a linear relationship between the Vickers hardness and Young's modulus. Utilizing the previously calculated bulk and shear moduli, we can derive the Young's modulus (E) using the following formula. Subsequently, we can deduce the Vickers hardness (V):

$$\begin{aligned} E &= \frac{9GB}{3B+G} \\ V &= 0.0608E \end{aligned} \tag{4}$$

We calculated the Young's Modulus and the corresponding Vickers hardness (in HV) for various Sb concentrations in $Mg_3Bi_{2-x}Sb_x$ and graphed the hardness values in Figure 5a. For the Mg_3Sb_2 system, we obtained Young's Modulus values of 43.69 GPa, 47.58 GPa, and 45.82 GPa, using the PBE, PBE-D3, and vdW-DFq methods, respectively. These values are in close agreement with the experimental Young's Modulus of approximately 41 GPa for Mg_3Sb_2, indicating good consistency between our calculations and experimental data [46].

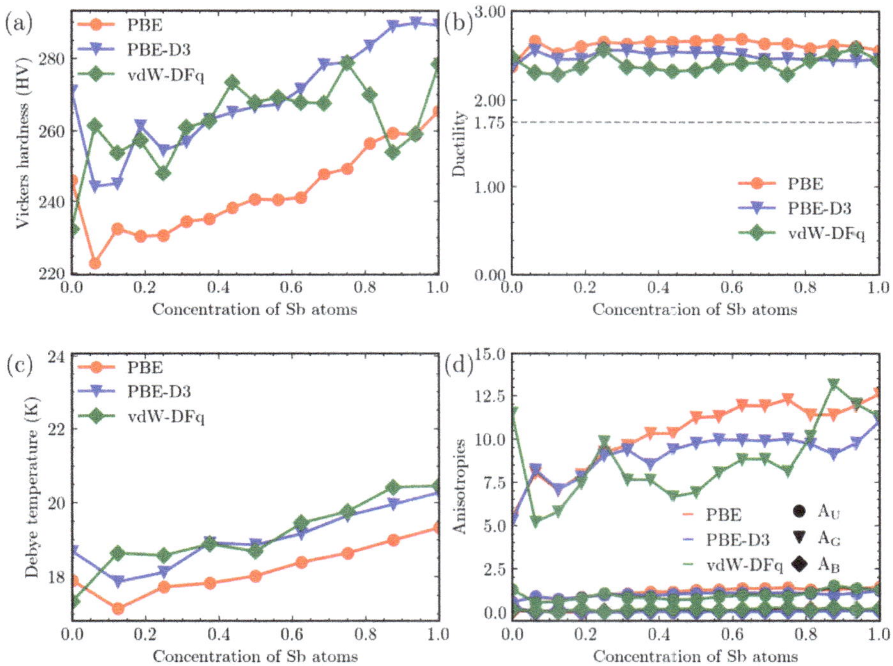

Figure 5. Comparison of (**a**) Vickers hardness, (**b**) ductility, (**c**) Debye temperature and (**d**) anisotropies from the three methods. Red (PBE), green (vdW-DFq), blue (PBE-D3).

As the Sb content increases, the hardness also shows an upward trend. The three calculation methods exhibit slight discrepancies in detail. The overall hardness calculated using the PBE-D3 and vdW-DFq methods is higher than that obtained without considering vdW corrections. However, at the point of Mg_3Bi_2, the results obtained using vdW-DFq are lower than those from PBE, by approximately 5.5%. At Sb concentrations of 0.9375 and 0.875, the vdW-DFq result closely resembles that without vdW, while the PBE-D3 result is significantly higher.

Similarly, the Pugh ratio [50] was employed to evaluate the ductility of the system. The Pugh ratio is derived using a specific formula (B/G), where the symbols B and G in the equation represent the bulk modulus and shear modulus, respectively. The outcomes of this comparison are graphically represented in Figure 5b. Regardless of the computational approach employed, the outcomes exhibit a range between 2 and 3. According to empirical observations, materials that have a Pugh ratio greater than 1.75 are regarded as exhibiting toughness. Consequently, all concentrations within the system under study can be considered as demonstrating resilience. In contrast, the widely recognized thermoelectric material Bi_2Te_3 exhibits a Pugh ratio of 1.62, suggesting that the $Mg_3Bi_{2-x}Sb_x$ system possesses relatively lower brittleness in comparison.

The Debye temperature [51] is a parameter that represents the highest-frequency lattice vibrations and is commonly employed as an indicator of the strength of interatomic bonding. It can be determined by employing the following formula:

$$\Theta = \frac{h}{k_B}\left(\frac{3q}{4\pi}\frac{N\rho}{M}\right)^{\frac{1}{3}} v_m \quad (5)$$

where h, k_B, and N are the Planck constant, Boltzmann constant, and Avogadro constant, respectively; ρ is the mass density; M is the molecular weight of all the atoms in the supercell; and q is the number of atoms in the unit cell. The average sound velocity (v_m) is defined as a combination of the shear sound velocity (v_s) and the longitudinal sound velocity (v_l):

$$v_m = \left[\frac{1}{3}\left(\frac{2}{v_s^3} + \frac{1}{v_l^3}\right)\right]^{-\frac{1}{3}} \quad (6)$$

where $v_s = \sqrt{G/\rho}$ and $v_l = \sqrt{(B + \frac{3}{4}G)/\rho}$, respectively.

The results of the Debye temperature calculations are presented in Figure 5c. Here, it is apparent that the inclusion of PBE-D3 and vdW-DFq increases the Debye temperature compared to cases where vdW is not considered, and the value is roughly 7% higher when PBE-D3 is used. As the content of Sb increases, there is a corresponding increase in the Debye temperature, indicating strengthening of the interatomic bonding within the system. The results obtained using the vdW-DFq method generally align with the trend observed when vdW correction is not considered.

We describe the anisotropy of the system through three indices, A_U, A_G, and A_B, which are obtained using the following formula [52,53]:

$$\begin{aligned} A_U &= \tfrac{5G_V}{G_R} + \tfrac{B_V}{B_R} - 6 \\ A_G &= \tfrac{G_V - G_R}{G_V + G_R} \times 100 \\ A_B &= \tfrac{B_V - B_R}{B_V + B_R} \times 100 \end{aligned} \quad (7)$$

We have presented the results in Figure 5d, where A_B is observed to be close to 0. This suggests that the system exhibits relatively consistent properties in terms of bulk modulus across different crystallographic directions. This observation is further confirmed by the 3D plot of the bulk modulus. The maximum and minimum values of the bulk modulus in different directions for all systems are within 7 GPa. In contrast, when compared to Bi_2Te_3, which has an A_B value of 1.58, indicating strong anisotropy as it exceeds 1, $Mg_3Bi_{2-x}Sb_x$ exhibits significantly lower anisotropy. In the case of the shear modulus, the systems at various concentrations exhibit noticeable anisotropy due to the relatively large values of A_G. This anisotropy increases with higher Sb content. From the perspective of the A_U parameter, $Mg_3Bi_{2-x}Sb_x$ exhibits significant directionality when compared to other alloys such as Zr_5Sn_3X (X = B, Nb, and Sn), where their A_U values are all within 0.5. Except for Mg_3Bi_2, all other Sb concentrations have A_U values above 0.7, and these values increase with higher Sb content. When considering the A_U parameter, the use of vdW introduces some bias in the results, but it does not affect the overall trend with increasing Sb content.

To enhance our understanding of the mechanical anisotropy of $Mg_3Bi_{2-x}Sb_x$, we employed VASPKIT to extract Young's moduli for different crystallographic orientations. We then visualized these data as 3D surface plots, where the distance from the origin to each point on the surface represents the Young's modulus in the corresponding direction. Furthermore, to visualize the magnitude of Young's modulus more conveniently, we used different colors on the surface to represent different values of Young's modulus. In Figure 6, we present the Young's modulus for different Sb concentrations, obtained via the vdW-DFq calculation method.

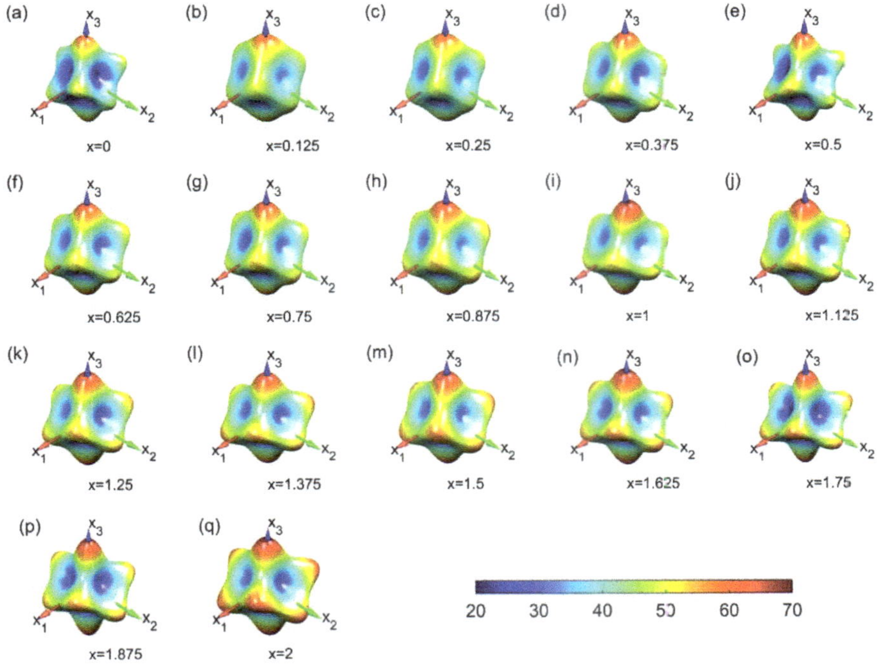

Figure 6. Three-dimensional Young's modulus distribution of $Mg_3Bi_{2-x}Sb_x$ ($0 \leq x \leq 2$) at various x of Sb contents (**a**–**q**) from the vdW-DFq calculations. Direction-dependent Young's modulus for every potency of Sb at an interval of 0.125, ranging from 0 to 2. The color bar shows the rage of Young's modulus with unit of GPa.

It is evident that, for all concentrations, the maximum Young's modulus is obtained along the x_3 direction, which corresponds to the lattice vector "*c*" within the crystal cell. All surfaces for the systems take the form of a hexahedron (a six-faced figure), with the degree of concavity varying at different concentrations, reflecting differences in anisotropy. Within the x_1–x_2 plane, the edges of these hexahedra are visible. In these materials, the Young's modulus is relatively high. The manifestation of this characteristic is primarily due to the occupation of Mg^{2+} ions at the symmetric positions 3 m and −3 m within the unit cell. As the Sb content increases, it can be observed that the surface expands significantly. This also implies that the Young's modulus is increasing in all directions. As mentioned earlier, this is related to the enhanced atomic interactions resulting from the addition of Sb atoms.

We examined the Young's modulus in the x_3 direction using the three calculation methods. The results are displayed in Figure 7a. In all other cases, the Young's modulus in the x_3 direction ranges from 55 GPa to 75 GPa. When vdW corrections are taken into account, the Young's modulus in the x_3 direction shows a linear increase with increasing Sb content. However, for the vdW-DFq calculations, this linear trend deviates somewhat. Nevertheless, the values obtained with vdW-DFq are relatively close to those obtained without vdW corrections.

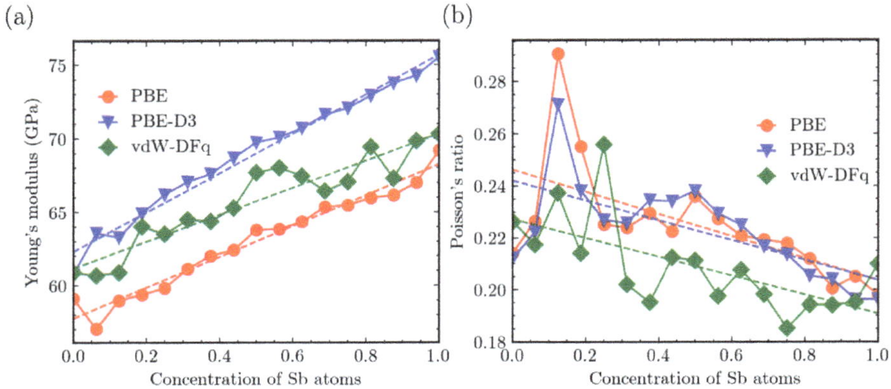

Figure 7. Young's modulus and Poisson's ratio from the three methods: red (PBE), green (vdW-DFq), blue (PBE-D3). (**a**) Trends in Young's Modulus in x_3 direction as function of Sb atom concentration. (**b**) Variation in Poisson's ratio in the x_3 direction with changing Sb concentration.

Considering the influence of vdW corrections, the Young's modulus was generally higher than when these vdW corrections were not considered. This aligns with expectations because both an increase in Sb content and the inclusion of vdW corrections lead to a reduction in lattice constants. This reduction results in closer atomic distances and stronger interatomic interactions, which consequently yield higher Young's moduli.

When a material is subjected to deformation in one direction, it also undergoes strain in the transverse directions in general. This phenomenon is known as the Poisson effect. Poisson's ratio is used to measure this effect and is defined as the ratio of transverse strain to longitudinal strain. Poisson's ratio reflects a material's hardness to some extent. Similar to Young's modulus, Poisson's ratio of $Mg_3Bi_{2-x}Sb_x$ also exhibits significant anisotropy.

To better visualize this anisotropy, we created 3D plots of the Poisson's ratio for various Sb concentrations, as shown in Figure 8. These results are based on calculations with vdW-DFq for the vdW corrections. Unlike Young's modulus, Poisson's ratio displays irregularities, with values ranging from −0.1 to 0.8 in different directions. In certain directions, Poisson's ratio approaches zero or even becomes negative. This indicates that stretching in these directions does not result in transverse contraction, which is a phenomenon that is less commonly observed in traditional alloys.

Similarly, to compare the influence of vdW corrections, we examined Poisson's ratio in the x_3 direction, as depicted in Figure 7b. With few exceptions, the Poisson's ratio in the x_3 direction of $Mg_3Bi_{2-x}Sb_x$ decreases linearly with increasing Sb content. With the inclusion of vdW effects, for the PBE-D3 method, the Poisson's ratio results are relatively close to those without vdW. In contrast, for the vdW-DFq method, the results are slightly lower than those without vdW, but its linear trend of decreasing with increasing Sb content is more pronounced.

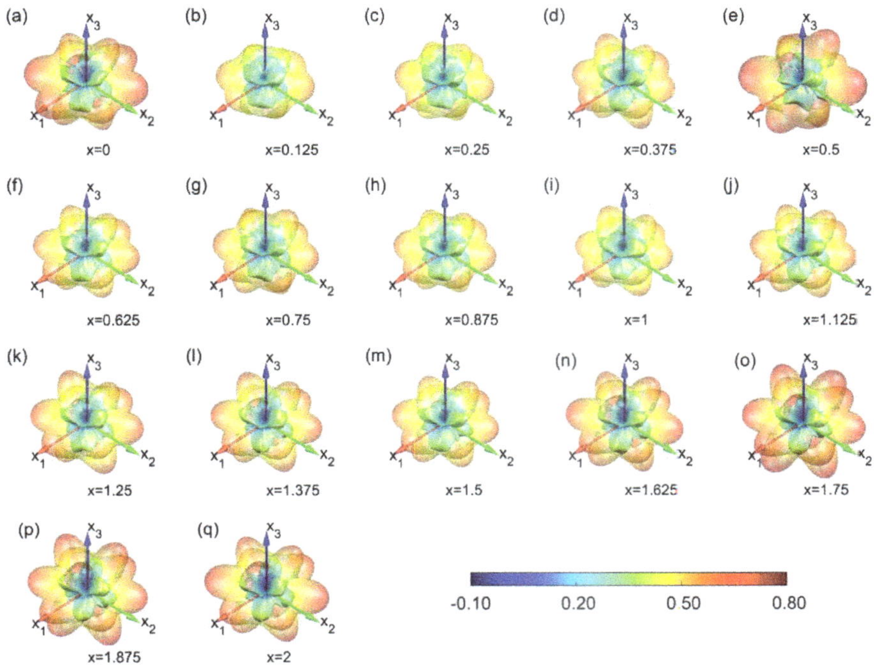

Figure 8. Poisson's ratio distribution of $Mg_3Bi_{2-x}Sb_x$ ($0 \leq x \leq 2$) at various Sb contents. (**a–q**) Direction-dependent Poisson's ratio at different Sb contents of x, ranging from 0 to 2, calculated via vdW-DFq. Outer surface denotes the maximum Poisson's ratio and inner surface is the contrary.

4. Conclusions

A systematic investigation was conducted into the mechanical properties of the room-temperature thermoelectric material $Mg_3Bi_{2-x}Sb_x$ ($0 \leq x \leq 2$) through the utilization of first-principles calculations within the frame of density functional theory. By conducting a comparative analysis of the impacts of distinct vdW corrections, we have provided a comprehensive understanding of the computational performance of these three methods and their outcomes. The influence of precision is contingent upon the concentration of Sb atoms. Higher-precision calculations have a negligible impact on the final optimized lattice constants, and this impact diminishes as the concentration of Sb increases. Additionally, in the context of elastic constants and elasticity, it was found that higher precision has minimal impact, except for the Mg_3Bi_2 system.

We conducted a comprehensive assessment of the influence of vdW corrections on the elasticity of the $Mg_3Bi_{2-x}Sb_x$ ($0 \leq x \leq 2$) nanocomposites. We employed both the PBE-D3 and vdW-DFq methodologies for nonlocal vdW corrections. The PBE-D3 method has a discernible effect of approximately 1% on lattice constants. In terms of other properties, such as elastic constant and bulk modulus, the influence of this method is generally within a range of 10%, except for a few outliers. The vdW-DFq method has a limited influence. Furthermore, the inclusion of vdW interactions does not significantly alter the patterns observed in various properties as the concentration of Sb vary, except for a few anomalous data points. In the case of lattice constants, the results obtained using the PBE-D3 method are closer to the experimental values.

Our results suggest the necessity of employing the PBE-D3 method to consider vdW corrections for more accurate prediction of the properties of $Mg_3Bi_{2-x}Sb_x$ by means of density functional theory calculations. This assessment might be helpful in the further development of vdW corrections and the material design of room-temperature thermoelectric materials.

Author Contributions: Q.P. conceived the project; X.M., S.Z. and Q.P. performed the first-principle-based calculation and analysis. All authors analyzed the data, discussed the results, wrote the draft, and reviewed and edited the manuscript. All authors have read and agreed to the published version of the manuscript.

Funding: The authors are grateful for the financial support from the Shenzhen Science and Technology Program (Grant No. KQTD20200820113045081), High-level Innovation Research Institute Program of Guangdong Province (Grant No. 2020B0909010003), and the Institute of Mechanics, Chinese Academy of Sciences (Grant No. E1Z1011001).

Institutional Review Board Statement: Not applicable.

Informed Consent Statement: Not applicable.

Data Availability Statement: Not applicable.

Acknowledgments: The main numerical calculations in this study were carried out using the MatCloud computational infrastructure for high-throughput material simulation.

Conflicts of Interest: The authors declare no conflict of interest.

References

1. Disalvo, F.J. Thermoelectric cooling and power generation. *Science* **1999**, *285*, 703–706. [CrossRef] [PubMed]
2. Bell, L.E. Cooling, heating, generating power, and recovering waste heat with thermoelectric systems. *Science* **2008**, *321*, 1457–1461. [CrossRef] [PubMed]
3. Zhang, F.; Chen, C.; Yao, H.; Bai, F.; Yin, L.; Li, X.; Li, S.; Xue, W.; Wang, Y.; Cao, F.; et al. High-Performance N-type Mg_3Sb_2 towards Thermoelectric Application near Room Temperature. *Adv. Funct. Mater.* **2020**, *30*, 1906143. [CrossRef]
4. Imasato, K.; Kang, S.D.; Ohno, S.; Snyder, G.J. Band engineering in Mg_3Sb_2 by alloying with Mg_3Bi_2 for enhanced thermoelectric performance. *Mater. Horiz.* **2018**, *5*, 59–64. [CrossRef]
5. Shi, X.; Sun, C.; Bu, Z.; Zhang, X.; Wu, Y.; Lin, S.; Li, W.; Faghaninia, A.; Jain, A.; Pei, Y. Revelation of Inherently High Mobility Enables Mg_3Sb_2 as a Sustainable Alternative to n-Bi_2Te_3 Thermoelectrics. *Adv. Sci.* **2019**, *6*, 1802286. [CrossRef]
6. Zhang, J.; Song, L.; Pedersen, S.H.; Yin, H.; Hung, L.T.; Iversen, B.B. Discovery of high-performance low-cost n-type Mg_3Sb_2-based thermoelectric materials with multi-valley conduction bands. *Nat. Commun.* **2017**, *8*, 13901. [CrossRef]
7. Imasato, K.; Kang, S.D.; Snyder, G.J. Exceptional thermoelectric performance in $Mg_3Sb_{0.6}Bi_{1.4}$ for low-grade waste heat recovery. *Energy Environ. Sci.* **2019**, *12*, 965–971. [CrossRef]
8. Imasato, K.; Wood, M.; Kuo, J.J.; Snyder, G.J. Improved stability and high thermoelectric performance through cation site doping in n-type La-doped $Mg_3Sb_{1.5}Bi_{0.5}$. *J. Mater. Chem. A* **2018**, *6*, 19941–19946. [CrossRef]
9. Wood, M.; Kuo, J.J.; Imasato, K.; Snyder, G.J. Improvement of Low-Temperature zT in a Mg_3Sb_2–Mg_3Bi_2 Solid Solution via Mg-Vapor Annealing. *Adv. Mater.* **2019**, *31*, 1902337. [CrossRef]
10. Zhang, J.; Song, L.; Iversen, B.B. Insights into the design of thermoelectric Mg_3Sb_2 and its analogs by combining theory and experiment. *Npj Comput. Mater.* **2019**, *5*, 76. [CrossRef]
11. Pan, Y.; Yao, M.; Hong, X.; Zhu, Y.; Zhu, Y.; Fan, F.; Imasato, K.; He, Y.; Hess, C.; Fink, J.; et al. $Mg_3(Bi,Sb)_2$single crystals towards high thermoelectric performance. *Energy Environ. Sci.* **2020**, *13*, 1717–1724. [CrossRef]
12. Tamaki, H.; Sato, H.K.; Kanno, T. Isotropic Conduction Network and Defect Chemistry in $Mg_{3}+\delta Sb_2$-Based Layered Zintl Compounds with High Thermoelectric Performance. *Adv. Mater.* **2016**, *28*, 10182–10187. [CrossRef] [PubMed]
13. Ao, D.W.; Liu, W.; di Chen, Y.X.; Wei, M.; Jabar, B.; Li, F.; Shi, X.L.; Zheng, Z.H.; Liang, G.X.; Zhang, X.H.; et al. Novel Thermal Diffusion Temperature Engineering Leading to High Thermoelectric Performance in Bi_2Te_3-Based Flexible Thin-Films. *Adv. Sci.* **2022**, *9*, 2103547. [CrossRef]
14. Yang, Q.; Yang, S.; Qiu, P.; Peng, L.; Wei, T.-R.; Zhang, Z.; Shi, X.; Chen, L. Flexible thermoelectrics based on ductile semiconductors. *Science* **2022**, *377*, 854–858. [CrossRef] [PubMed]
15. Yu, R.; Fang, Z.; Dai, X.; Weng, H. Topological nodal line semimetals predicted from first-principles calculations. *Front. Phys.* **2017**, *12*, 1–14. [CrossRef]
16. Teshome, T.; Datta, A. Topological Phase Transition in Mg_3Sb_2 Assisted by Strain. *ACS Omega* **2019**, *4*, 8701–8706. [CrossRef]
17. Li, S.; Yu, Z.M.; Liu, Y.; Guan, S.; Wang, S.S.; Zhang, X.; Yao, Y.; Yang, S.A. Type-II nodal loops: Theory and material realization. *Phys. Rev. B* **2017**, *96*, 081106. [CrossRef]
18. Chen, W.; Lu, H.Z.; Zilberberg, O. Weak localization and antilocalization in nodal-line semimetals: Dimensionality and topological effects. *Phys. Rev. Lett.* **2019**, *122*, 196603. [CrossRef]
19. Zhu, Y.; Liu, J.; Wei, B.; Xu, S.; Song, Y.; Wang, X.; Xia, T.-L.; Chen, J.; Snyder, G.J.; Hong, J. Giant phonon anharmonicity driven by the asymmetric lone pairs in Mg_3Bi_2. *Mater. Today Phys.* **2022**, *27*, 100791. [CrossRef]

20. Xin, J.; Li, G.; Auffermann, G.; Borrmann, H.; Schnelle, W.; Gooth, J.; Zhao, X.; Zhu, T.; Felser, C.; Fu, C. Growth and transport properties of Mg_3X_2 (X = Sb, Bi) single crystals. *Mater. Today Phys.* **2018**, *7*, 61–68. [CrossRef]
21. Agne, M.T.; Imasato, K.; Anand, S.; Lee, K.; Bux, S.K.; Zevalkink, A.; ERettie, A.J.; Young Chung, D.; Kanatzidis, M.G.; Jeffrey Snyder, G. Heat capacity of Mg_3Sb_2, Mg_3Bi_2, and their alloys at high temperature. *Mater. Today Phys.* **2018**, *6*. 83–88. [CrossRef]
22. Ding, J.; Lanigan-Atkins, T.; Calderón-Cueva, M.; Banerjee, A.; Abernathy, D.L.; Said, A.; Zevalkink, A.; Delaire, O. Soft anharmonic phonons and ultralow thermal conductivity in Mg_3 (Sb, Bi)$_2$ thermoelectrics. *Sci. Adv.* **2021**, *7*, eabg1449. [CrossRef] [PubMed]
23. Peng, W.; Petretto, G.; Rignanese, G.M.; Hautier, G.; Zevalkink, A. An Unlikely Route to Low Lattice Thermal Conductivity: Small Atoms in a Simple Layered Structure. *Joule* **2018**, *2*, 1879–1893. [CrossRef]
24. Gooth, J.; Schierning, G.; Felser, C.; Nielsch, K. Quantum materials for thermoelectricity. *MRS Bull.* **2018**, *43*, 187–192. [CrossRef]
25. Li, M.; Chen, G. Thermal transport for probing quantum materials. *MRS Bull.* **2020**, *45*, 348–356. [CrossRef]
26. Barati, S.; Abedinpour, S.H. Thermoelectric response of nodal-line semimetals: Probing the Fermi surface topology. *Phys. Rev. B* **2020**, *102*, 125139. [CrossRef]
27. Hosoi, M.; Tateishi, I.; Matsuura, H.; Ogata, M. Thin Films of Topological Nodal Line Semimetals as a Candidate for Efficient Thermoelectric Converters. *Phys. Rev. B* **2022**, *105*, 085406. [CrossRef]
28. Kanno, T.; Tamaki, H.; Yoshiya, M.; Uchiyama, H.; Maki, S.; Takata, M.; Miyazaki, Y. High-Density Frenkel Defects as Origin of N-Type Thermoelectric Performance and Low Thermal Conductivity in Mg3Sb2-Based Materials. *Adv. Funct. Mater.* **2021**, *31*, 2008469. [CrossRef]
29. Peng, Q.; Zhao, S.; Yuan, X.; Chen, X.J. Elasticity of Mg_3Bi_2-xSbx. *Materials* **2022**, *15*, 7161. [CrossRef]
30. Peng, Q.; Wang, G.; Liu, G.R.; De, S. Van der Waals density functional theory vdW-DFq for semihard materials. *Crystals* **2019**, *9*, 243. [CrossRef]
31. Grimme, S.; Antony, J.; Ehrlich, S.; Krieg, H. A consistent and accurate ab initio parametrization of density functional dispersion correction (DFT-D) for the 94 elements H-Pu. *J. Chem. Phys.* **2010**, *132*, 154104. [CrossRef]
32. Yuan, X.; Zhou, Y.; Peng, Q.; Yang, Y.; Li, Y.; Wen, X. Active learning to overcome exponential-wall problem for effective structure prediction of chemical-disordered materials. *NPJ Comput. Mater.* **2023**, *9*, 12. [CrossRef]
33. Peng, Q.; Yuan, X.; Zhao, S.; Zhou, Y.; Wen, X.; Chen, X.J. Active-learning search for unitcell structures: A case study on Mg_3Bi_2-xSbx. *Comput. Mater. Sci.* **2023**, *226*, 112260. [CrossRef]
34. Kohn, W.; Sham, L.J. Self-Consistent Equations Including Exchange and Correlation Effects*. *Phys. Rev.* **1965**, *140*, A1133. [CrossRef]
35. Kresse, G.; Furthmiiller, B.J. Efficiency of ab-initio total energy calculations for metals and semiconductors using a plane-wave basis set. *Comput. Mater. Sci.* **1996**, *6*, 15–50. [CrossRef]
36. Blochl, P.E. Projector augmented-wave method. *Phys. Rev. B* **1994**, *50*, 24. [CrossRef]
37. Perdew, J.P.; Burke, K.; Ernzerhof, M. Generalized Gradient Approximation Made Simple. *Phys. Rev. Lett.* **1996**, *77*, 3865. [CrossRef]
38. Monkhorst, H.J.; Pack, J.D. Special points for Brillonin-zone integrations. *Phys. Rev. B* **1976**, *13*, 5188. [CrossRef]
39. Wang, V.; Xu, N.; Liu, J.C.; Tang, G.; Geng, W.T. VASPKIT: A user-friendly interface facilitating high-throughput computing and analysis using VASP code. *Comput. Phys. Commun.* **2021**, *267*, 108033. [CrossRef]
40. Marmier, A.; Lethbridge, Z.A.D.; Walton, R.I.; Smith, C.W.; Parker, S.C.; Evans, K.E. ElAM: A computer program for the analysis and representation of anisotropic elastic properties. *Comput. Phys. Commun.* **2010**, *181*, 2102–2115. [CrossRef]
41. Yang, X.; Wang, Z.; Zhao, X.; Song, J.; Zhang, M.; Liu, H. MatCloud: A high-throughput computational infrastructure for integrated management of materials simulation, data and resources. *Comput. Mater. Sci.* **2018**, *146*, 319–333. [CrossRef]
42. Zhou, D.W.; Liu, J.S.; Xu, S.H.; Peng, P. Thermal stability and elastic properties of Mg_3Sb_2 and Mg_3Bi_2 phases from first-principles calculations. *Phys. B Condens. Matter* **2010**, *405*, 2863–2868. [CrossRef]
43. Reuß, A. Berechnung der fließgrenze von mischkristallen auf grund der plastizitätsbedingung für einkristalle. *ZAMM-J. Appl. Math. Mech./Z. Für Angew. Math. Mech.* **1929**, *9*, 49–58. [CrossRef]
44. Hill, R. The elastic behaviour of a crystalline aggregate. *Proc. Phys. Soc. Sect. A* **1952**, *65*, 349. [CrossRef]
45. Voigt, W. *Lehrbuch der Kristallphysik*; Teubner: Leipzig, Germany, 1908.
46. Isotta, E.; Peng, W.; Balodhi, A.; Zevalkink, A. Elastic Moduli: A Tool for Understanding Chemical Bonding and Thermal Transport in Thermoelectric Materials. *Angew. Chem. Int. Ed.* **2023**, *62*, e202213649. [CrossRef]
47. Balasubramanian, K.; Khare, S.V.; Gall, D. Valence electron concentration as an indicator for mechanical properties in rocksalt structure nitrides, carbides and carbonitrides. *Acta Mater.* **2018**, *152*, 175–185. [CrossRef]
48. Chen, Y.; Wang, C.; Ma, Z.; Li, L.; Li, S.; Wang, J. Improved thermoelectric performance of n-type Mg_3Sb_2–Mg_3Bi_2 alloy with Co element doping. *Curr. Appl. Phys.* **2021**, *21*, 25–30. [CrossRef]
49. Jiang, X.; Zhao, J.; Wu, A.; Bai, Y.; Jiang, X. Mechanical and electronic properties of B12-based ternary crystals of orthorhombic phase. *J. Phys. Condens. Matter* **2010**, *22*, 315503. [CrossRef]
50. Pugh, S. Relations between the elastic moduli and the plastic properties of polycrystalline pure metals. *Lond. Edinb. Dublin Philos. Mag. J. Sci.* **1954**, *45*, 823–843. [CrossRef]
51. Anderson, O.L. A simplified method for calculating the Debye temperature from elastic constants. *J. Phys. Chem. Solids Pergamon Press.* **1963**, *24*, 909–917. [CrossRef]

52. Ravindran, P.; Fast, L.; Korzhavyi, P.A.; Johansson, B.; Wills, J.; Eriksson, O. Density functional theory for calculation of elastic properties of orthorhombic crystals: Application to $TiSi_2$. *J. Appl. Phys.* **1998**, *84*, 4891–4904. [CrossRef]
53. Chen, H.; Cao, Y.; Liu, K.; Tao, X.; Zhou, Y.; Ouyang, Y.; Gao, F.; Du, Y.; Peng, Q. Stability and physical properties tuning via interstitials chemical engineering of Zr_5Sn_3: A first-principles study. *J. Mater. Sci.* **2019**, *54*, 10284–10296. [CrossRef]

materials MDPI

Article

Crash Performance of Inward-Inverting Composite Tubes Filled with Foam: Experimentation and Simulation

Pu Yu [1,2], Zhefeng Yu [2,*], Xiang Zhou [2] and Wu Xu [2]

1 Shanghai Aircraft Manufacturing Co., Ltd., Shanghai 201324, China; yupu@comac.cc
2 Aerospace Structure Research Center, School of Aeronautics and Astronautics, Shanghai Jiao Tong University, Shanghai 200240, China; xiangzhou@sjtu.edu.cn (X.Z.); xuwu@sjtu.edu.cn (W.X.)
* Correspondence: yuzf@sjtu.edu.cn; Tel./Fax: +86-21-34206387

Abstract: This study presents a novel shock absorber with an inward-inverting composite foam-filled tube. Under the compression of a pressing cap and the action of an internal inversion cap, the composite tube inverted inward. During the crushing, the fronds of the composite tube compacted the foam, thereby enhancing the energy absorption. Three types of foams were applied to the absorber, and a drop-weight impact test was performed to obtain the assessment parameters. The foam increased the specific energy absorption (SEA) of the composite tube to 103 kJ/kg. Finite-element simulation based on the user-defined material subroutine was performed for the initial failure and stable stages of the crushing, and a foam model was identified through the experimental data. The mean crush force from the simulation agreed with the experimental data, and the SEA maximum error was <7%, thus validating the crush simulation of the proposed shock absorber. The development of the damage modes of the plies was analyzed based on the simulation results, showing a good energy absorption mechanism of this shock absorber.

Keywords: shock absorber; inward-inverting composite tube; foam-filled; specific energy absorption

1. Introduction

Crashworthiness has become a basic requirement for transportation vehicles, such as helicopters, airplanes, and automobiles [1–3]. For structures made of aluminum and other metals, the plastic deformation of the structural members absorbs the kinetic energy transferred during the impact. Composite materials have been widely used to fabricate lighter structures; some of them, such as carbon-fiber-reinforced plastic (CFRP), are brittle; thus, their crashworthiness is a key issue. The kinetic energy of the CFRP structure dissipates through material failures, including fiber breakage, crushing of the matrix, and delamination. For automobiles, crush boxes are typically employed [4], whereas for aircraft, struts in the fuselage serve as energy absorbers [5–7]. Composite tubes are widely used in structures such as struts [8–10] and bumpers. Some composite tubes are subjected to transverse loads [11,12], whereas other loads are axial. The energy absorption of the former is lower; therefore, most research has focused on the axial crush problem.

The techniques used to trigger stable crushing and enhance the specific energy absorption (SEA) present two key issues. The triggering mechanism at the end of a tube that initiates axial crushing may be "a chamfer" or "steeple" [13,14]. The fiber layout and triggering technique of composite tubes have been studied [15–17]. The SEA of CFRP tubes generally ranges from 50 to 80 kJ/kg [16,18,19].

The failure-triggering method can be employed for a composite tube to achieve a higher SEA and a more-stable crushing procedure. A rigid die is typically used to trigger the failure of the tube [20]. Heimbs et al. [21] presented a method wherein a composite column was cut into strips by a special machine under an axial load. Similarly, Ueda et al. [22] proposed a trigger with a double-sided plug, and Tong et al. [23] proposed a chamfer trigger, which guides an external flap of the composite tube. In the abovementioned studies, the crushed debris was spread out across the tube. By contrast, Siromani et al. [24] studied the performance of CFRP tubes under the action of an inward-folding cap under quasi-static loading. Yu et al. [25,26] proposed a shock absorber based on inward-folding composite laminate tubes and sandwich tubes [27]. The debris filled the hollow cavity and further increased the SEA or passed through the hole at the other end. The SEA during the steady-state procedure in the impact test exceeded 80 kJ/kg. The SEA could be further improved by filling the hollow with other substances.

Foam is lightweight; consequently, it is used in some structures to increase energy absorption. Foam can be made from polymers or aluminum. In one study, it was applied to an internally stiffened tube subject to a transverse load [28], where the SEA was below 8 kJ/kg. Yang et al. applied aluminum foam to a skeleton-filled tube subjected to an axial load [29] and achieved an SEA of 72 kJ/kg, which was higher than that when no foam was used (approximately 63 kJ/kg). Alia and Zhou et al. studied a structure consisting of foam reinforced by small carbon fiber tubes; the SEA values in the quasi-static test were close to 93 kJ/kg [30,31]. Therefore, a foam-filled tube is a prospective energy absorber.

Finite-element (FE) analysis of the crushing of composite laminates is also a key issue with many challenges. Although laminates have a complex structure, their failure during crushing can be simulated with shell elements. Feraboli et al. [6] simulated the crushing of a composite specimen with LS-DYNA (R6.0), where the "Enhanced Composite Damage" material model MAT54 was applied to the shell elements, which represented the entirety of the laminates. This model requires experimentally determined parameters, rather than those based on the damage mechanics. Siromani et al. [32] also simulated the crushing of a composite tube using the material model MAT54, where one layer of the shell elements represented several plies. Inward-folding crushing was also simulated. A more-popular method is to establish a model with a stacked shell or brick element, and a cohesive element is typically used to connect different layers to simulate interlaminar damage, even when simulating the crushing procedure of a complex strut in a fuselage [33]. The detailed intralaminar damage, including matrix cracks and fiber failures, can be predicted using the 3D Hashin criteria, which are generally applied to the material through the coded user-defined material subroutine (VUMAT) in ABAQUS/Explicit [34]. The interlaminar damage (primarily delamination) is simulated using cohesive zone elements [35–37].

This study proposes a shock absorber with an internal inversion cap [25–27], for which the SEA exceeded 80 kJ/kg under dynamic testing. Herein, the tube was filled with polymer foam to reinforce it. During crushing, the folded fronds of the composite tube in also compacted the foam, thereby enhancing the energy absorption. The remainder of this paper is organized as follows. Section 2 presents the shock absorber with the three types of foam used and the experimental method employed. Section 3 describes the FE simulation of specimen crushing. In Section 4, the results of the experiment and simulation are given and discussed. Finally, Section 5 provides the conclusions of the study.

2. Experimental Setup

2.1. Configuration of the Energy Absorber

A schematic of the energy absorber is shown in Figure 1a. It consisted of a composite tube filled with polymer foam, an internal inversion cap, and a pressing cap. A composite tube with a length of L_{t0}, a diameter of D, and a thickness of t_w was connected to the two caps. The internal inversion cap destructed the composite tube through a triggering fillet with a radius R_t and folded the destructed tube wall through the inverted surface with a radius R_f. Herein, the inverted composite fronds continued to compress the foam (Figure 1b). When the folded tube reached the internal inversion cap (Figure 1c), composite debris began filling the cavity, thereby increasing the reaction load. The metal inversion cap and composite tube filled with foam are shown in Figure 2.

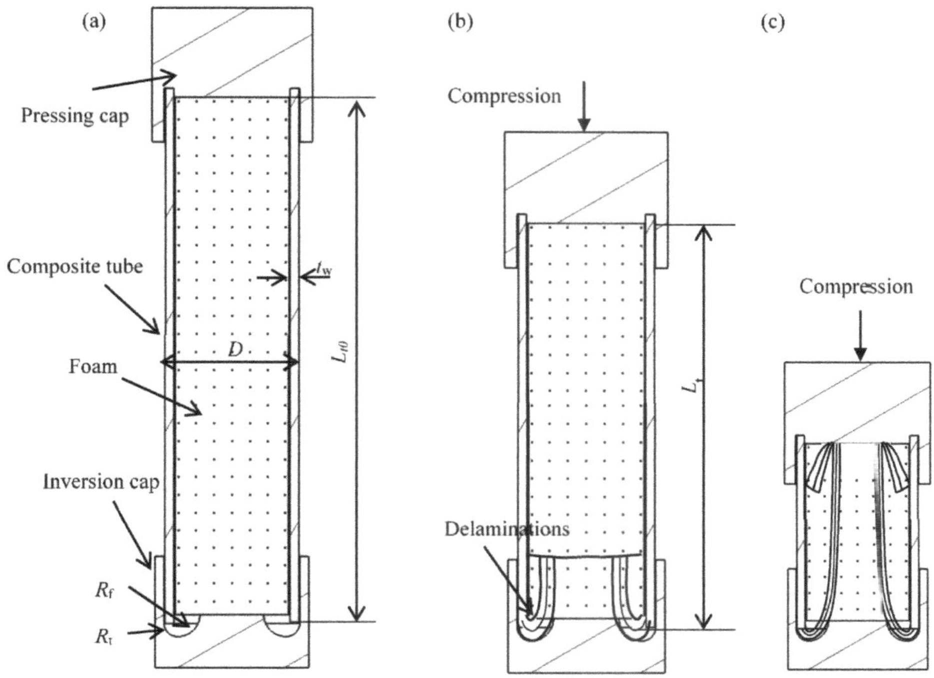

Figure 1. Concept of the shock absorber: (**a**) configuration of the structure, (**b**) stable crushing stage, and (**c**) the compression of composite debris after it reaches the pressing cap.

Figure 2. Inversion cap (**left**) and foam-filled composite tube (**right**).

2.2. Parameters of Shock Absorber

The behavior of a shock absorber without foam reinforcement was demonstrated previously in [25,26], where the internal inversion cap with R_t = 3 mm and R_f = 5 mm provided a steady crush triggering and high SEA; therefore, the same internal inversion cap was adopted in this study. The composite tube was mainly made of carbon fiber/epoxy prepreg unidirectional tape, and a layer of plain woven fabric was coated on the outside of the tube, which was not considered in the numerical simulation in this study. Two thicknesses of 2 and 1.5 mm with ply setups of $[90/0]_5$ and $[(0/90)_3 0]$ were considered, where the tape at 0° coincides with the axis of the tube. The composite tubes had a length of

100 mm and a diameter D of 30 mm. The other mechanical parameters of the unidirectional tape are listed in Table 1.

Table 1. Mechanical properties of and symbols for carbon-fiber/epoxy prepreg unidirectional tape.

Properties	Value
Density	1529 kg/m^3
Young's modulus E_1	115 GPa
Young's modulus $E_2 = E_3$	8 GPa
Poisson's ratio v_{12}	0.3
Poisson's ratio v_{13}	0.3
Poisson's ratio v_{23}	0.35
Shear modulus $G_{12} = G_{13}$	4.9 GPa
Shear modulus G_{23}	3 GPa
Tensile fracture toughness G_{1C}^{t}	133 N/mm
Compressive fracture toughness G_{1C}^{c}	40 N/mm
Tensile fracture toughness $G_{2C}^{t} = G_{3C}^{t}$	0.6 N/mm
Compressive fracture toughness $G_{2C}^{c} = G_{3C}^{c}$	2.1 N/mm
Tensile strength X_t	1600 MPa
Compressive strength X_c	1000 MPa
Tensile strength Y_t	40 MPa
Compressive strength Y_c	135 MPa
Tensile strength Z_t	40 MPa
Compressive strength Z_c	135 MPa
Shear strength $S_{12} = S_{13} = S_{23}$	89 MPa

The mechanical properties of the three types of polymethacrylimide (PMI) foam provided by the supplier are listed in Table 2. The parameters needed for the FE simulation through ABAQUS were calculated by the embedded code based on the input stress–strain data, whereas the stress–strain curves were obtained from static tests conducted on the MTS servo-hydraulic system, as shown in Figure 3.

Table 2. Mechanical properties of PMI foam.

Type	Density (kg/m^3)	Compressive Strength (MPa)	Tensile Strength (MPa)	Young's Modulus (MPa)	Fracture Elongation (%)	Shear Strength (MPa)	Shear Modulus (MPa)
50 RS	50	0.85	1.68	83	2.6	0.85	30
75 RS	70	1.70	2.30	108	2.8	1.25	48
110 RS	110	3.60	3.70	197	2.8	2.38	80

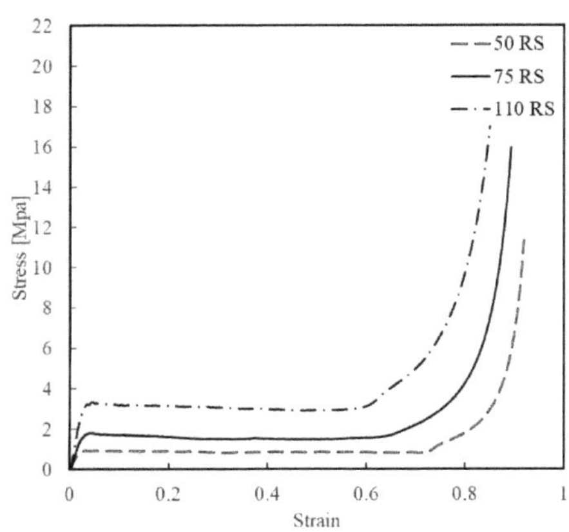

Figure 3. Stress–strain curves of PMI foam samples.

Eight types of absorbers were fabricated by combining the abovementioned composite tube and foam, as listed in Table 3. Two or more samples of each type were prepared and tested.

Table 3. Parameters of tested specimens and characteristics of energy absorption.

Type of Absorber	t_w (mm)	Foam	MICF (kN)	MCF in Stage 1 (kN)	CFE	SEA (kJ/kg)	SEA from Simulation (kJ/kg)	Error of SEA Simulation (%)
S1	2	None	21.5	18.8	0.87	70.1	73.5	4.59
S2	2	50 RS	25.9	22.9	0.88	77.6	82.2	5.68
S3	2	75 RS	31.5	28.2	0.90	91.8	91.3	−0.56
S4	2	110 RS	38.3	33.5	0.87	102.8	106.3	3.32
S5	1.5	None	14.2	9.4	0.66	45.6	42.7	−6.81
S6	1.5	50 RS	15.8	13.5	0.866	58.0	58.3	0.51
S7	1.5	75 RS	22.2	17.4	0.78	70.4	68.1	−3.38
S8	1.5	110 RS	22.8	22.2	0.98	83.4	78.7	−5.97

2.3. Equipment for Impact Tests

Impact tests on the shock absorbers were conducted using a drop-weight tower, as introduced in [26]; see Figure 4. The weight of the dropping hammer ranged from 30 to 90 kg, and the lift height ranged from 1 to 1.5 m, depending on the energy necessary to test the different absorbers. To evaluate the energy absorbed, crush distance and other characteristics of the absorber, the impact velocity v_i was measured using an optical sensor; the impact load $F(t)$ was measured using a dynamic force sensor, a PCB 200C20 mounted at the bottom of absorber.

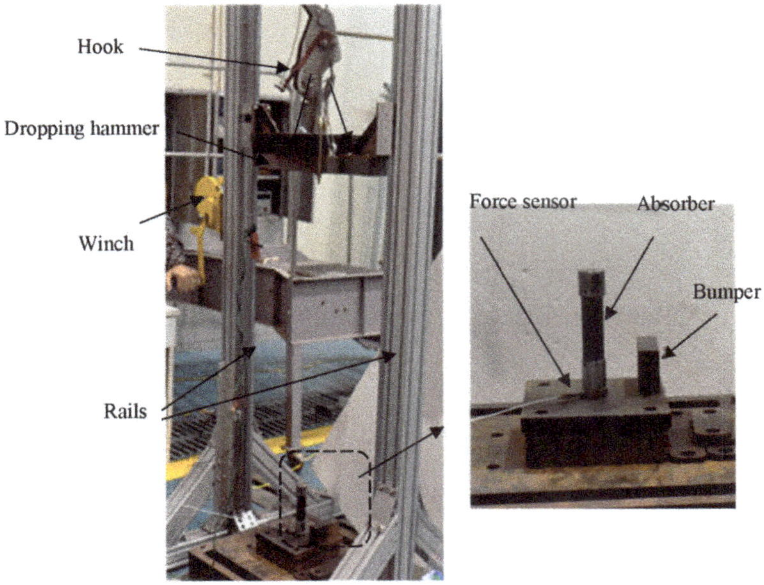

Figure 4. Setup of drop-weight tower for impact test.

A DH-5922 digital data acquisition system was used to record the impact force $F(t)$ and the impact velocity at a sampling rate of 100 kHz. The velocity of dropping hammer at time t was obtained by [38]

$$v(t) = v_i + gt - \int_0^t \frac{F(t)}{m} dt, \tag{1}$$

where g denotes the gravitational acceleration; m denotes the hammer mass. The crushing length at time t is expressed by:

$$\delta_t = v_t + \frac{gt^2}{2} - \int_0^t \left(\int_0^t \frac{F(t)}{m} dt \right) dt. \tag{2}$$

In order to compare the behavior of the absorbers with different lengths, the dimensionless distance is defined as:

$$\overline{\delta_{\rm t}} = \frac{\delta_{\rm t}}{L_{\rm t0}} = \frac{L_{\rm t0} - L_{\rm t}}{L_{\rm t0}}, \tag{3}$$

where $L_{\rm t0}$ and $L_{\rm t}$ are the original and real-time tube length, respectively, as shown in Figure 1.

Based on the impact load and crushing length, several criteria used to assess the energy absorption abilities of thin-walled columns can be calculated. The SEA is the value of the energy absorbed per unit mass; for a composite tube, it can be expressed as:

$$\text{SEA} = \frac{\int_0^{\delta_{\rm C}} F(\delta) d\delta}{\overline{m}\delta_{\rm C}}, \tag{4}$$

where $\delta_{\rm C}$ denotes the concerned crushing length; $\overline{m}$ denotes the mass per unit length; and $F(\delta)$ is the load at crush distance δ. The mean crush force (MCF) over the crush distance is given as:

$$\text{MCF} = \frac{1}{\delta_{\rm C}} \int_0^{\delta_{\rm C}} F(\delta) d\delta. \tag{5}$$

The crushing force efficiency (CFE) is the ratio of the MCF to the maximum initial collapse force (MICF).

$$\text{CFE} = \frac{\text{MCF}}{\text{MICF}}. \tag{6}$$

The following load should not be higher after the MICF. A high CFE implies that the energy absorber can absorb more energy with a relatively lower reaction load to the occupants, which further suggests that the material efficiency is high.

3. Numerical Simulation

3.1. Model of Intralaminar Damage in Composite Tube

The plies of the composite tube were meshed using eight-node C3D8R reduced integration. The intralaminar damage was simulated using the ABAQUS/Explicit solver through a user material VUMAT subroutine developed in the FORTRAN language. Failure initiation was determined based on the Chang–Lessard failure criteria [39,40], which include four failure modes: matrix crushing, matrix cracking, fiber–matrix shearing failure, and fiber failure. In this study, fiber compressive failure was also considered. The failure criteria are listed in Table 4, and their strength parameters are listed in Table 1.

Herein, a linear stress–strain behavior was assumed for the composite laminar damage, demonstrated with the tensile stress shown as Figure 5. Once the damage was initiated, the stress began to reduce to zero linearly as the tensile strain increased. The failure initiation strain in tension $\varepsilon_{0,1}^{\rm t}$ is

$$\varepsilon_{0,1}^{\rm t} = \frac{X_{\rm T}}{E_1}. \tag{7}$$

Table 4. Failure criteria for the unidirectional ply.

Failure Modes	Damage Factors
Fiber tensile failure criterion ($\sigma_1 \geq 0$)	$F_{\mathrm{ft}} = \sqrt{\left(\frac{\sigma_1}{X_T}\right)^2 + \left(\frac{\tau_{12}}{S_{12}}\right)^2 + \left(\frac{\tau_{13}}{S_{13}}\right)^2} \geq 1$
Fiber compressive failure ($\sigma_1 < 0$)	$F_{\mathrm{fc}} = \sqrt{\left(\frac{\sigma_1}{X_C}\right)^2} \geq 1$
Matrix tensile failure in 2 directions ($\sigma_2 \geq 0$)	$F_{\mathrm{mt}} = \sqrt{\left(\frac{\sigma_2}{Y_T}\right)^2 + \left(\frac{\tau_{12}}{S_{12}}\right)^2 + \left(\frac{\tau_{23}}{S_{23}}\right)^2} \geq 1$
Matrix crushing failure criterion for in-plane compression ($\sigma_2 < 0$)	$F_{\mathrm{mc}} = \sqrt{\left(\frac{\sigma_2}{Y_C}\right)^2 + \left(\frac{\tau_{12}}{S_{12}}\right)^2 + \left(\frac{\tau_{23}}{S_{23}}\right)^2} \geq 1$
Matrix crushing failure criterion for out-of-plane tension ($\sigma_3 \geq 0$)	$F_{\mathrm{dt}} = \sqrt{\left(\frac{\sigma_3}{Z_T}\right)^2 + \left(\frac{\tau_{13}}{S_{13}}\right)^2 + \left(\frac{\tau_{23}}{S_{23}}\right)^2} \geq 1$
Matrix crushing failure criterion for out-of-plane compression ($\sigma_3 < 0$)	$F_{\mathrm{dc}} = \sqrt{\left(\frac{\sigma_3}{Z_C}\right)^2 + \left(\frac{\tau_{13}}{S_{13}}\right)^2 + \left(\frac{\tau_{23}}{S_{23}}\right)^2} \geq 1$

$\sigma_1, \sigma_2, \sigma_3$—normal stress;
$\tau_{12}, \tau_{13}, \tau_{23}$—shearing stress;
F_k (k = ft, fc, mt, mc, dt, dc)—damage factor.

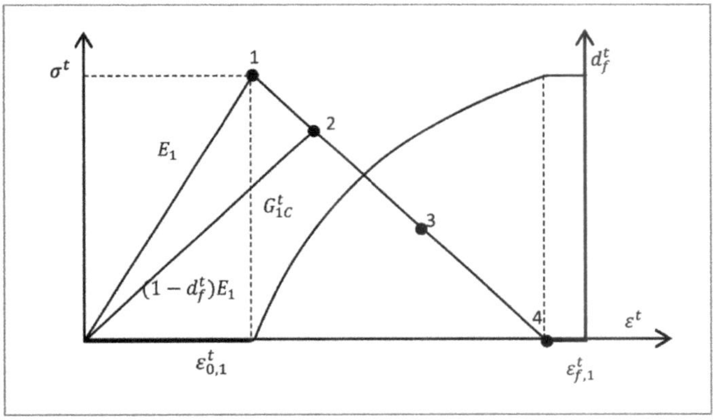

Figure 5. Intralaminar damage behavior model for tensile failure mode.

The maximum strain $\varepsilon_{\mathrm{f},1}^{\mathrm{t}}$ is given as

$$\varepsilon_{\mathrm{f},1}^{\mathrm{t}} = \frac{2G_{1\mathrm{C}}^{\mathrm{t}}}{\sigma_{\mathrm{t}} l}, \tag{8}$$

where G_{1C}^t is the fracture toughness, which equals the area below the strain–stress curve, including the linear and failure procedure; σ^t is the ultimate tensile strength of the material; and l is the characteristic length of the crack growth, which is relative to the element volume, as the fracture energy is distributed over the volume of the represented element. Herein, in the VUMAT subroutine, the cube root of the element volume was transferred through the variable *charLength*. The element had the same width and length, and the thickness t was 0.2 mm; thus, $l = \sqrt{charLength^3/t}$.

The damage variable in the tension of the fiber also includes the effect of shear strain, which is expressed as follows.

$$d_{\mathrm{f}}^{\mathrm{t}} = \frac{\varepsilon_{\mathrm{f},1}^{\mathrm{t}}}{\varepsilon_{\mathrm{f},1}^{\mathrm{t}} - \varepsilon_{0,1}^{\mathrm{t}}}\left(1 - \frac{1}{r_{\mathrm{ft}}}\right), \tag{9}$$

where

$$r_{\mathrm{ft}} = \sqrt{\left(\frac{S_{11}}{\varepsilon_{0,1}^{\mathrm{t}} E_1 \left(1 - d_{\mathrm{f,old}}^{\mathrm{t}}\right)}\right)^2 + \left(\frac{\varepsilon_{12}}{\varepsilon_{12}^{\mathrm{c}}}\right)^2 + \left(\frac{\varepsilon_{13}}{\varepsilon_{13}^{\mathrm{c}}}\right)^2}, \tag{10}$$

where $d_{\mathrm{f,old}}^{\mathrm{t}}$ is the damage coefficient of the last step; and $\varepsilon_{12}^{\mathrm{c}}$ and $\varepsilon_{13}^{\mathrm{c}}$ are the strains at damage initiation. The same damage evolution was applied to the damage of the unidirectional tape in 2 and 3 directions. Then, the material stiffness matrix could be obtained as follows.

$$C_{\mathrm{d}} = \begin{bmatrix} \alpha C_{11} & \alpha\beta C_{12} & \alpha\psi C_{13} & 0 & 0 & 0 \\ \alpha\beta C_{12} & \beta C_{22} & \beta\psi C_{23} & 0 & 0 & 0 \\ \alpha\psi C_{13} & \beta\psi C_{23} & \psi C_{33} & 0 & 0 & 0 \\ 0 & 0 & 0 & \alpha\beta C_{44} & 0 & 0 \\ 0 & 0 & 0 & 0 & \alpha\psi C_{55} & 0 \\ 0 & 0 & 0 & 0 & 0 & \beta\psi C_{66} \end{bmatrix}, \tag{11}$$

where $\alpha = \left(1 - d_{\mathrm{f}}^{\mathrm{t}}\right)\left(1 - d_{\mathrm{f}}^{\mathrm{c}}\right)$; $\beta = \left(1 - d_{\mathrm{m}}^{\mathrm{t}}\right)\left(1 - d_{\mathrm{m}}^{\mathrm{c}}\right)$; $\psi = \left(1 - d_{\mathrm{d}}^{\mathrm{t}}\right)\left(1 - d_{\mathrm{d}}^{\mathrm{c}}\right)$; superscripts t and c denote the tension and compression, respectively; subscripts f, m, and d denote the 1, 2, and 3 directions of the unidirectional tape. In fact, the maximum of every damage variable was limited to approximately 0.8 to avoid the distortion of elements and maintain enough stiffness for simulating the supporting effect of composite debris. The element was removed when the strain was greater than 0.8 or less than −0.6, which was calculated using the tensor matrix transferred by variable stretchNew in the VUMAT subroutine.

3.2. Model of Interlaminar Damage in Composite Tube

To simulate the separation of the plies, the cohesive elements (COH3D8) were established between adjacent layers to simulate their interaction and the interlaminar damage. The delamination onset was determined by the traction–separation law shown in Figure 6, where δ_{m} is the equivalent displacement, given by [41]

$$\delta_{\mathrm{m}} = \sqrt{\langle\delta_{\mathrm{n}}\rangle^2 + \delta_{\mathrm{s}}^2 + \delta_{\mathrm{t}}^2}, \tag{12}$$

where δ_{n} is the normal displacement, with $\langle\delta_{\mathrm{n}}\rangle = (\delta_{\mathrm{n}} + |\delta_{\mathrm{n}}|)/2$; δ_{s} and δ_{t} are the in-plane shear displacements. The cohesive stresses t_{n}, t_{s}, and t_{t} were obtained as follows.

$$\begin{Bmatrix} t_{\mathrm{n}} \\ t_{\mathrm{s}} \\ t_{\mathrm{t}} \end{Bmatrix} = \frac{1}{T_0} \begin{bmatrix} E_{\mathrm{nn}} & & \\ & E_{\mathrm{ss}} & \\ & & E_{\mathrm{tt}} \end{bmatrix} \begin{Bmatrix} \delta_{\mathrm{n}} \\ \delta_{\mathrm{s}} \\ \delta_{\mathrm{t}} \end{Bmatrix}, \tag{13}$$

where $E_{ii}(i = \mathrm{n, s, t})$ is the elasticity coefficient, as listed in Table 5, and T_0 is the thickness of the cohesive element, which was set as 0.01 mm in this study.

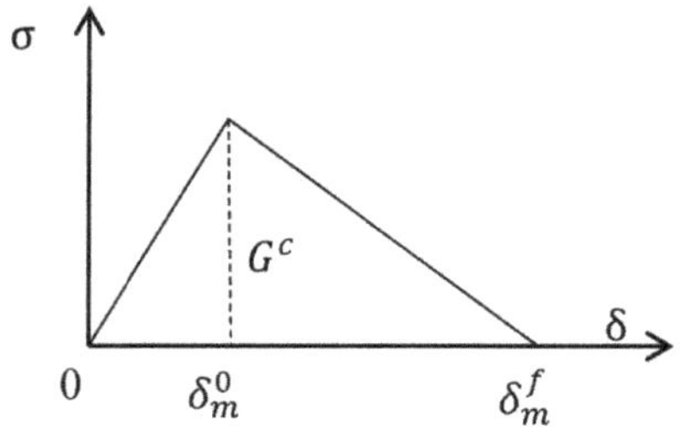

Figure 6. Law of the traction–separation for the cohesive zone.

Table 5. Properties of cohesive model.

Properties	Values
E_{nn}	3000 MPa
$E_{ss} = E_{tt}$	1154 MPa
t_n^0	38 MPa
$t_s^0 = t_t^0$	55 MPa
G_n^C	0.533 N/mm
$G_s^C = G_t^C$	1.1 N/mm

The crack occurred when the quadratic interaction criterion on stress was satisfied:

$$\left\{\frac{\langle t_n\rangle}{t_n^0}\right\}^2+\left\{\frac{t_s}{t_s^0}\right\}^2+\left\{\frac{t_t}{t_t^0}\right\}^2\geq 1;\ \langle t_n\rangle=\frac{t_n+|t_n|}{2}, \tag{14}$$

where t_n^0, t_s^0, and t_t^0 denote the peak values of the nominal direct, first shear, and second shear stress on the interface, respectively.

The Benzeggagh–Kenane formulation based on the fracture energy [42] was used to describe the damage evolution, as follows.

$$G^C=G_n^C+\left(G_s^C-G_n^C\right)\left(\frac{G_s+G_t}{G_n+G_s+G_t}\right)^{\eta}, \tag{15}$$

where G_n^C, G_s^C, and G_t^C are the values of the critical fracture energy per unit area necessary to initiate failure with the normal fracture modes and the two in-plane modes; G_n, G_s, and G_t are the values of fracture energy; and η ranges from 0.5 to 2.0 and was determined to be 2.0 in this study.

Linear damage softening behavior was adopted; the damage evolution variable is expressed as:

$$D=\frac{\delta_m^f\left(\delta_m^{max}-\delta_m^0\right)}{\delta_m^{max}\left(\delta_m^f-\delta_m^0\right)}, \tag{16}$$

where δ_m^0 is the effective displacement at damage initiation; δ_m^{max} is the maximum value of the effective displacement in the loading procedure; and δ_m^f is the effective displacement at complete failure, with $\delta_m^f=2G^C/T_{eff}^0$, where T_{eff}^0 is the effective stress at damage initiation. Once the overall damage variable reached the specified value (0.85 in this study), the element was removed from the model. The stress components are affected by damage as follows.

$$\begin{gathered} t_n=\begin{cases}(1-D)\bar{t}_n, \bar{t}_n\geq 0\\ \bar{t}_n, \text{otherwise}\end{cases}\\ t_s=(1-D)\bar{t}_s\\ t_t=(1-D)\bar{t}_t\end{gathered}, \tag{17}$$

where $\bar{t}_n$, $\bar{t}_s$, and $\bar{t}_t$ are the stress components calculated according to the traction–separation law for the current strains without damage.

3.3. Model of Foam Behaviors

A hyperfoam material was used to simulate the behavior of the PMI foam. The elastic behavior of the foam is based on the strain energy function:

$$U=\sum_{i=1}^{N}\frac{2\mu_i}{\alpha_i^2}\left[\hat{\lambda}_1^{\alpha_i}+\hat{\lambda}_2^{\alpha_i}+\hat{\lambda}_3^{\alpha_i}-3+\frac{1}{\beta_i}\left(\left(J^{el}\right)^{-\alpha_i\beta_i}-1\right)\right], \tag{18}$$

where N is a material parameter less than 6; and μ_i, α_i, and β_i are temperature-dependent material parameters.

$$\hat{\lambda}_i=\left(J^{th}\right)^{-\frac{1}{3}}\lambda_i, \tag{19}$$

where λ_i is the principal stretches and J^{el} is the elastic volume ratio, expressed as follows.

$$J^{el} = \frac{J}{J^{th}} = \hat{\lambda}_1\hat{\lambda}_2\hat{\lambda}_3, \tag{20}$$

where J is the volume ratio of the current volume to the reference volume and J^{th} is the thermal volume ratio, expressed as follows.

$$J^{th} = \left(1+\varepsilon^{th}\right)^3, \tag{21}$$

For each term in the energy function, the coefficient β_i determines the degree of compressibility, which is related to Poisson's ratio, ν_i, as follows.

$$\beta_i = \frac{\nu_i}{1-2\nu_i}, \tag{22}$$

$$\nu_i = \frac{\beta_i}{1+2\beta_i}. \tag{23}$$

In this study, N was determined to be three, and the abovementioned material parameters were calculated in ABAQUS when the test data shown in Table 6 were specified.

Table 6. Parameters of foam for FE simulation.

Type of Foam	50 RS	75 RS	110 RS
μ_1	−24.59	−39.53	−71.04
μ_2	45.99	80.15	146.54
μ_3	−1.17	−2.18	−4.00
α_1	16.74	15.40	15.26
α_2	25.00	25.00	25.00
α_3	−1.25	−1.24	−1.24
ν_1	0.045	0.045	0.045
ν_2	0.045	0.045	0.045
ν_3	0.045	0.045	0.045

4. Results and Discussion

4.1. Experimental Results

The time histories of the impact load recorded in the impact test (Figures 7a and 8a) were used to calculate the crush distance, and then, the absorbed energy was calculated. As the impact velocity and impact energy were not the same in different tests, the time histories of the impact load were not synchronous. Further, the crush load was plotted with the crush distance and dimensionless load (Figures 7b,c and 8b,c). Evidently, two main stages were observed after the initial failure. First was the stable stage corresponding to when the destructed tube was pushed into the tube hollow; essentially, the folded composite frond squeezed the foam, which, in turn, blocked the movement of the composite tube and increased the blocking force of the absorber. Second is the load-increasing stage after the crush distance was greater than half of the tube; essentially, the fronds reached one pressing cap and began to be compressed further.

The initial peak load of the absorber filled with the 70 RS and 110 RS foams was lower than that of the MCF, and the CFE was greater than 1 according to Equation (6). Therefore, in this study, the MICF was determined as the maximum load in the steady stage. The MICF, MCF, CFE, and SEA values in the stable stage of each type of absorber are listed in Table 3. Evidently, the CFE of the foam-filled absorber was greater than 0.86.

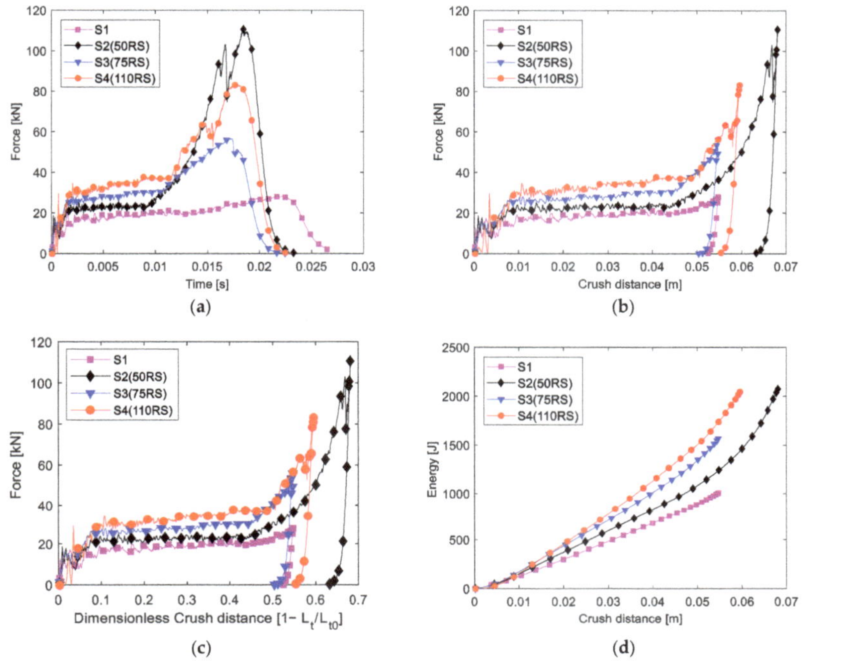

Figure 7. Experimental results of the 2.0 mm-thick samples: (**a**) load and time, (**b**) load and distance, (**c**) load and dimensionless crush distance, and (**d**) energy absorbed and crush distance.

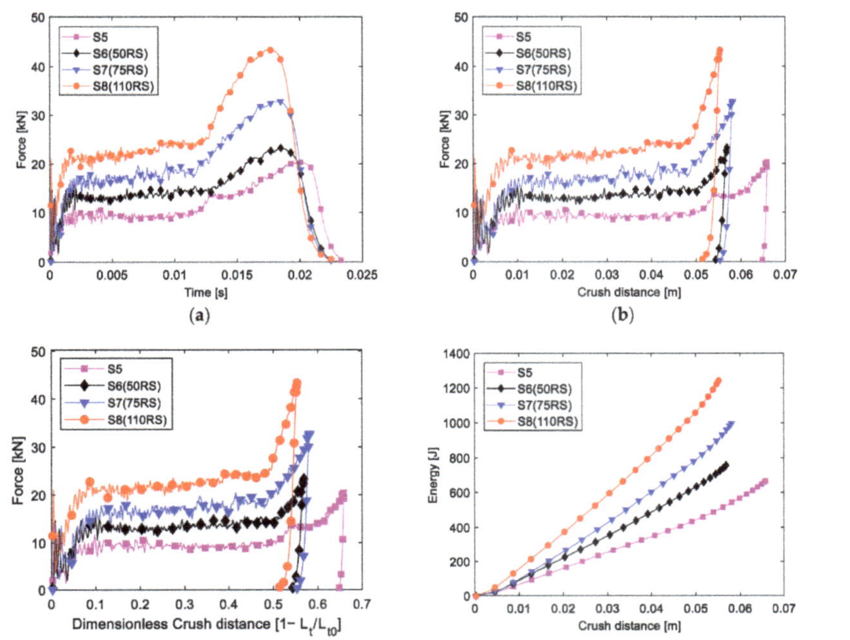

Figure 8. Experimental results of the 1.5 mm-thick samples: (**a**) load and time, (**b**) load and distance, (**c**) load and dimensionless crush distance, and (**d**) energy absorbed and crush distance.

It should be mentioned that the maximum force is not discussed, which is located in the last stage in the curve dominated by the reaction of the compacted debris at the final crush distance. Some changes of the impact energy or absorber parameters affected the final crush distance and the maximum force, so the maximum force in the last stage was an incredible factor for the presented tests.

The SEA values of different specimens were also compared, as shown in Figure 9. The increment in the SEA from foam ranged from 10 to 46% for the 2 mm-thick tube and from 27 to 85% for the 1.5 mm-thick tube. The 2.0 mm-thick tube without foam had a higher SEA. Note that the higher the elastic modulus of the foam, the higher the energy absorption achieved was. Herein, the best SEA was achieved by the 2.0 mm-thick tube filled with the 110 RS foam (SEA, 103 kJ/kg), 47% higher than that of the 2.0 mm-thick tube without foam.

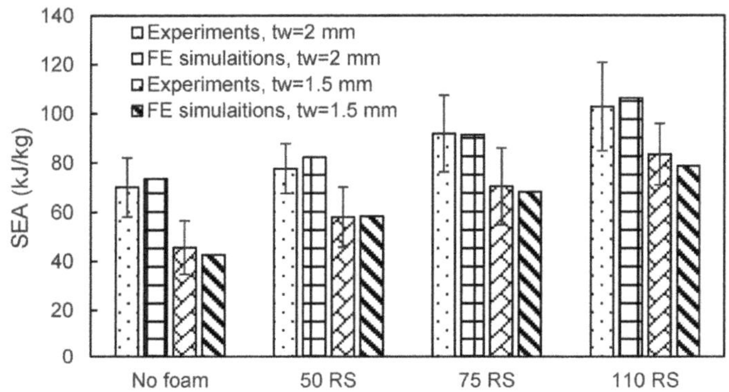

Figure 9. SEA comparison of absorber with different wall thicknesses and foams.

The energy absorbed was plotted against the crush distance, as shown in Figures 7d and 8d, where the slope of the curve reflects the SEA; evidently, the slope increased after the dimensionless crush distance was greater than 0.5.

4.2. Simulation Results

Numerical models were established using ABAQUS; the parts of the FE model are shown in Figure 10. The pressing and inversion caps were modelled using a rigid material. The length of the composite tube was 60 mm in the FE model to reduce the computational time. General contact was applied to the model with a friction factor of 0.27 between the composites and 0.12 between the composites and inversion cap. The inversion cap was fixed at the reference point. An initial velocity of 2.5 m/s was applied to the pressing cap.

As the tube length in the simulation was different from that in the experiment, the load curves were plotted with the dimensionless crush distance and for comparison, as shown in Figure 11. Evidently, the simulation curve agreed well with the experimental results in the stable stage. It should be noted that the simulated load in the initial dimensionless crush distance (within 0.15) was lower than the experimental one, due to the tube length (60 mm) in the simulation being different from that of the experiment (100 mm). The behavior of the sample in the load-increasing stage was complicated to simulate using the presented FE model; therefore, the simulation was stopped when the crush distance exceeded half of the tube length. The SEA of the stable stage from the simulation was compared with that from the experiment, as listed in Table 3, where the maximum error was less than 7%.

To further investigate the failure of the composite tube, the damage morphologies of Sample S2 at different crush distances obtained from the quasi-static compressive test are given in Table 7, and the cross-section of Sample S2 after the impact test is also given. Correspondingly, the damage factors of the four failure modes shown in Table 4 are also given for each crush stage, respectively.

When the crush distance was equal to 3 mm, the tube end began to be compressed in the radial direction under the effect of the inverted surface. There were fiber breakages and matrix cracks along the periphery. Delaminations also occurred at the tip. According to the simulation results, all damage modes began to emerge, except the fiber tensile failure.

The plies were compressed further and began to intersect when the crush distance was equal to 4.5 mm, and fiber tensile failure occurred due to fiber bending. Other types of damage grew, especially the matrix crushing due to in-plane compression.

The tip of composite tube was completely inverted after the crush distance became greater than 10 mm, while most plies were bent, then fiber tensile failure was presented in the simulation results. Other failures developed, in which the quantity of fiber compressive failure was relatively less due to the buckling of the thin ply.

From the cross-section of the sample after the impact test, the composite wall was destructed by the internal inversion cap; the fronds were inverted and compacted; the foam was also compressed. The collapse and buckling of the composite fronds in the simulation were similar to those in the experimentation. The damage factors showed that the plies were destructed seriously, which is beneficial for energy absorption.

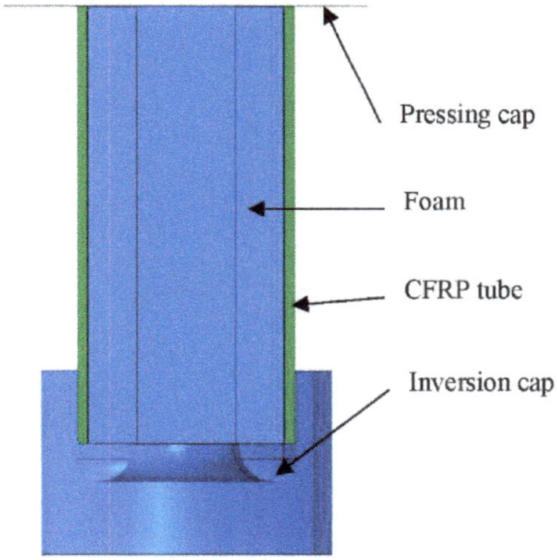

Figure 10. Components of FE model.

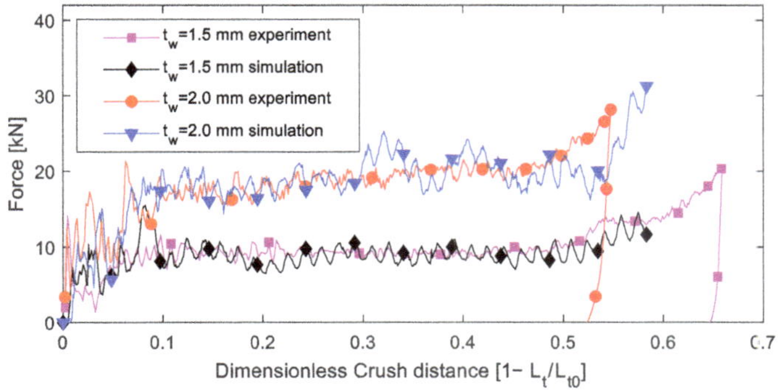

(**a**)

Figure 11. *Cont.*

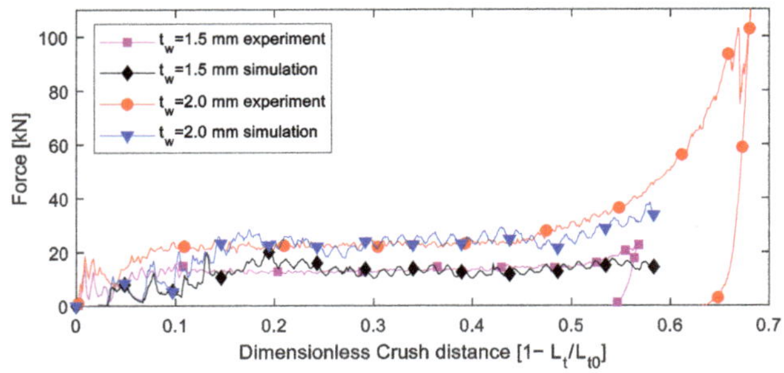

(**b**)

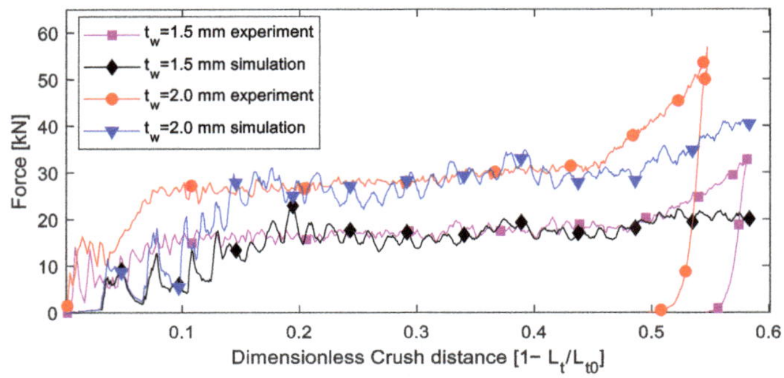

(**c**)

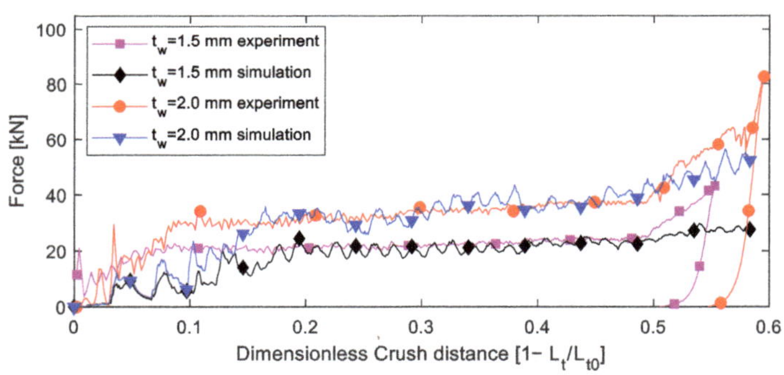

(**d**)

Figure 11. Comparison of impact load from the experimentation and simulation: (**a**) without foam, (**b**) with 50 RS foam, (**c**) with 75 RS foam, and (**d**) with 110 RS foam.

Table 7. Morphologies of the specimen damage at different crush distances and the corresponding damage factors of FE simulation.

Crush Stage	$\delta = 3$ mm	$\delta = 4.5$ mm	$\delta = 10$ mm	Cross-Section of Impacted Test Sample
Damage morphology	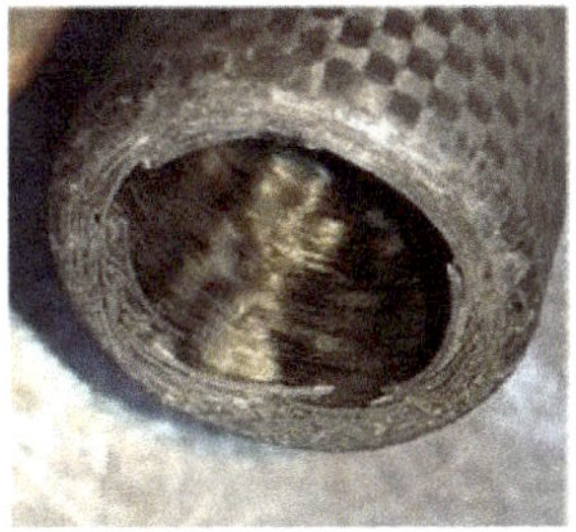			
F_{ft}	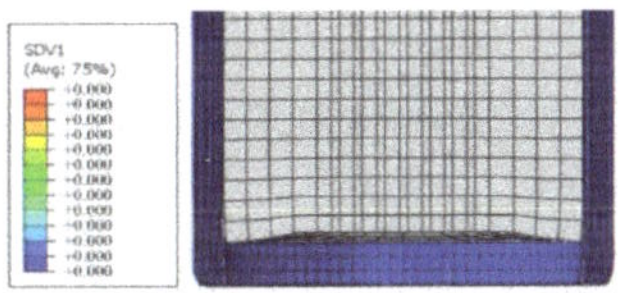	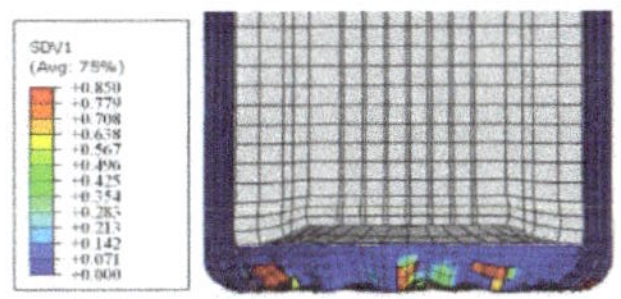	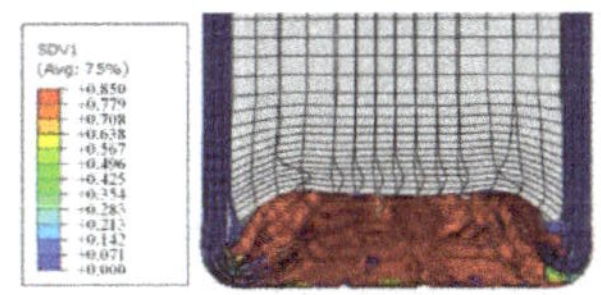	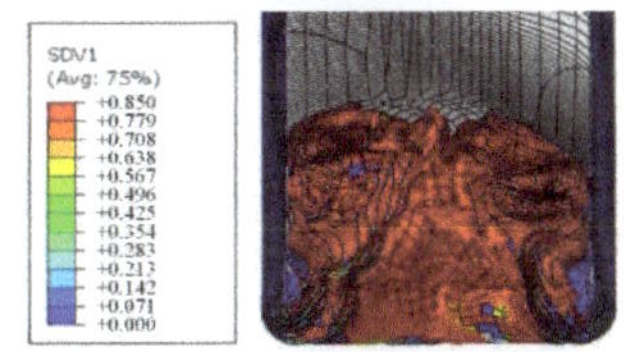
F_{fc}	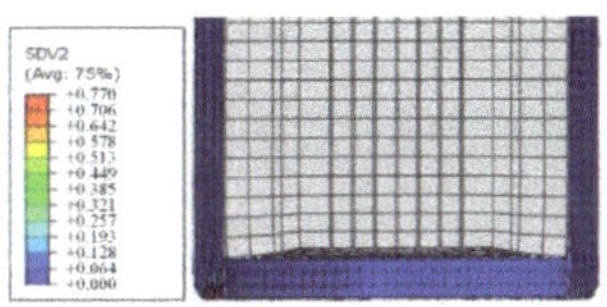	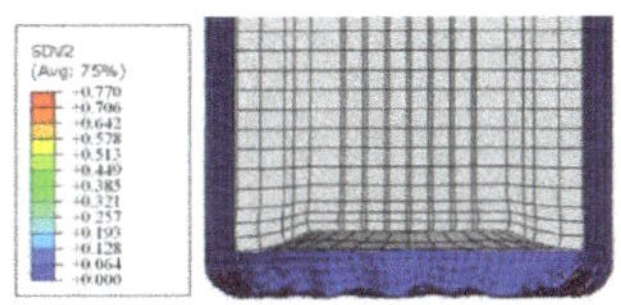	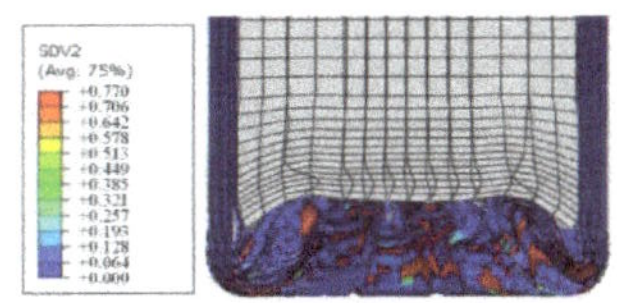	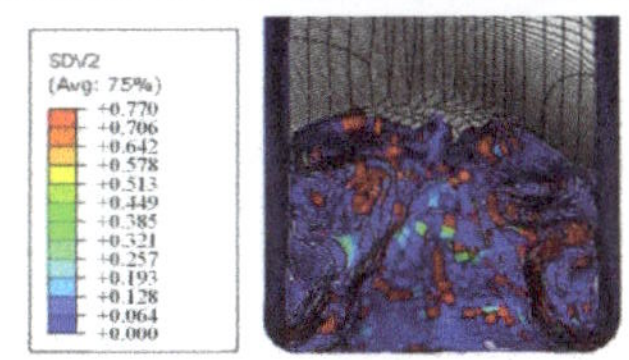

Table 7. *Cont.*

Crush Stage	δ = 3 mm	δ = 4.5 mm	δ = 10 mm	Cross-Section of Impacted Test Sample
F_{mt}	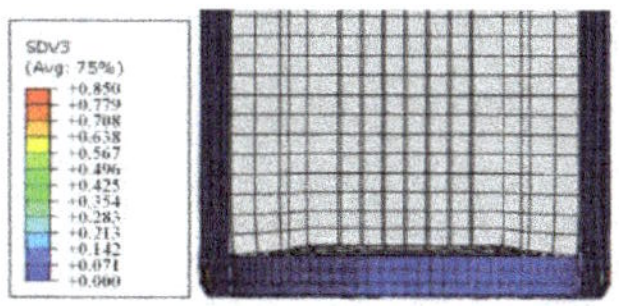	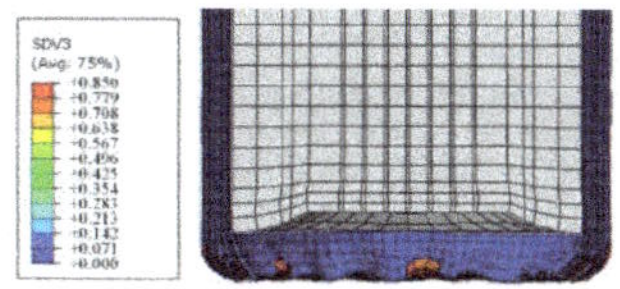	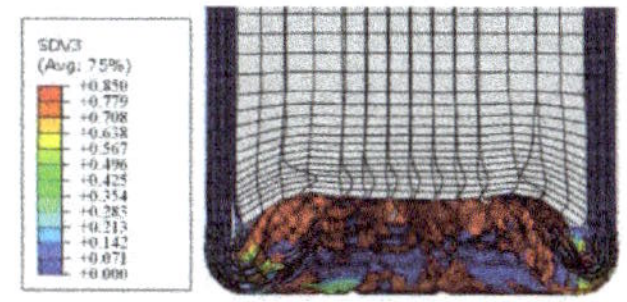	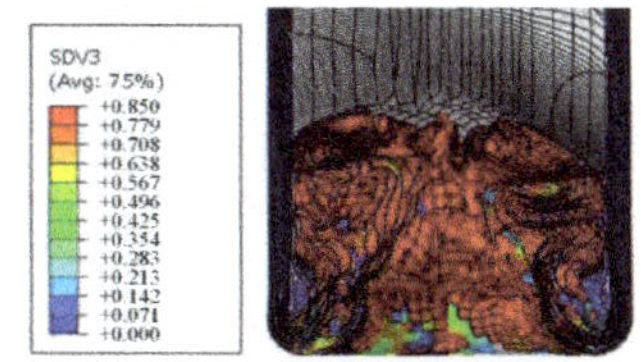
F_{mc}	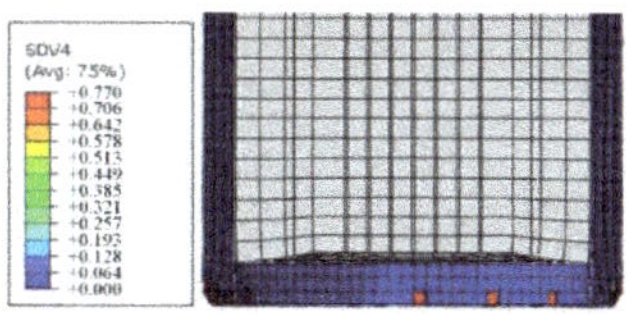	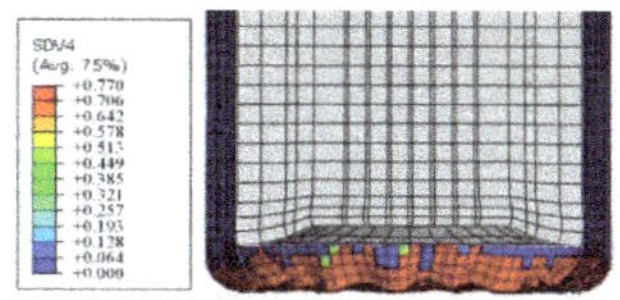	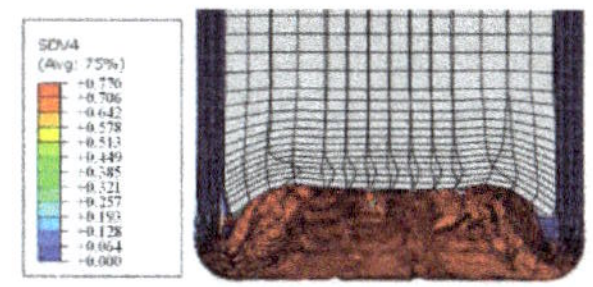	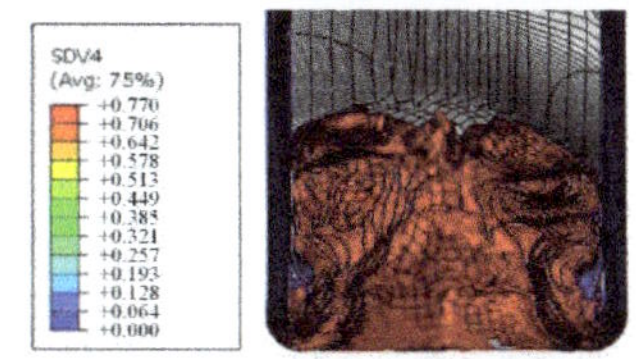
F_{dt}	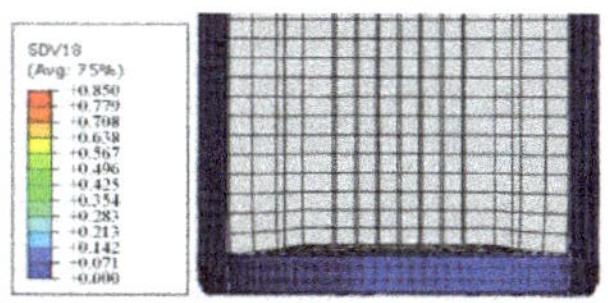	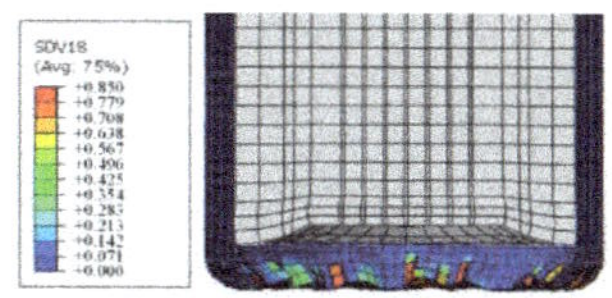	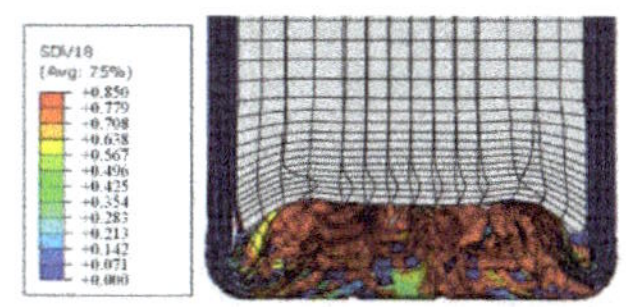	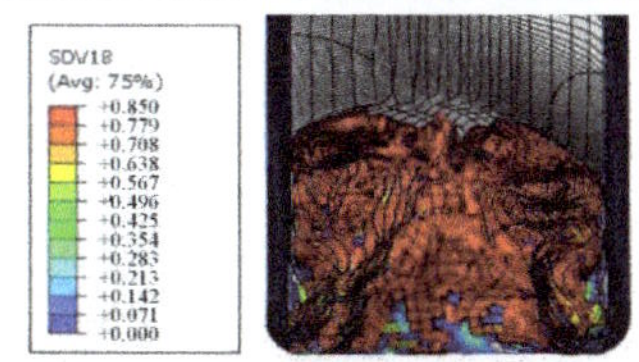
F_{dc}	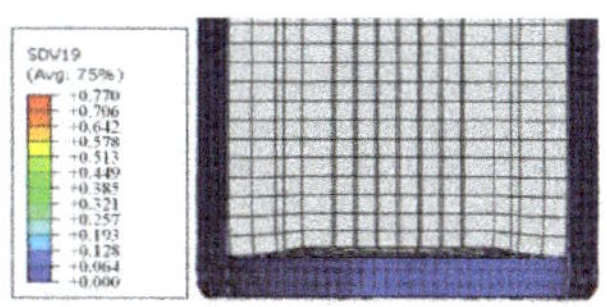	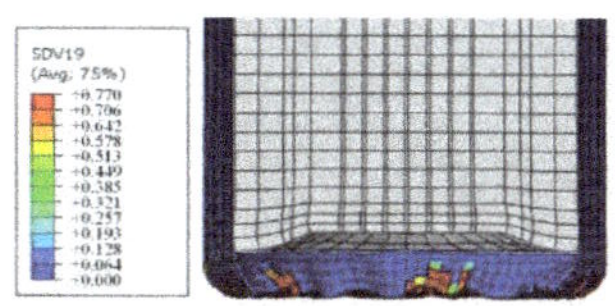	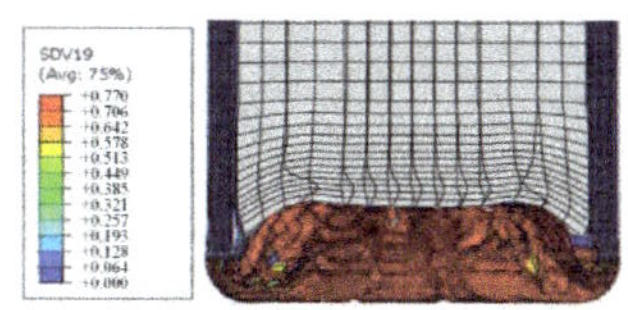	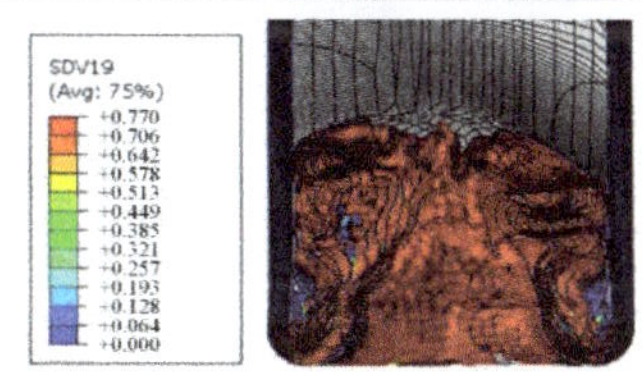

5. Conclusions

The crush performance under the internal inversion cap of composite tubes with 1.5 mm and 2 mm thicknesses filled with different PMI foams was investigated. The experimental results revealed that the 2 mm-thick tube had a higher SEA than the 1.5 mm-thick tube. Evidently, the higher the elastic modulus of the filled foam, the higher the SEA achieved was. The best SEA was achieved by the 2.0 mm-thick tube filled with the 110 RS foam, with a value of 103 kJ/kg, which was 47% higher than that of the 2.0 mm-thick tube without foam.

A numerical simulation of the collapse of the foam-filled tubes based on ABAQUS/Explicit was conducted. The material behaviors of the composite tube were defined using the VUMAT user subroutine. The parameters of the foam model were determined through experimentation on the foam. The improved Chang–Lessard failure criteria were used to judge the damage initiation; subsequently, the stiffness of the element was reduced through the VUMAT user subroutine. The initial failure and stable stage of the crushing were simulated. The results revealed that the mean crush force from the simulation was in good agreement with the experimental data with errors of the SEA less than 7%, thus indicating that the method is appropriate for the crush simulation of the proposed shock absorber. The development of six damage modes to the plies were analyzed, five of which were developed thoroughly, except the fiber compressive failure, showing a good energy absorption mechanism under the action of the internal inversion cap.

The shock absorber proposed in this study has a long stroke with a steady reaction load and a high SEA, and therefore, it has good prospects for crashworthiness engineering structures. For cars, it can be used in the crash box behind the front bumper. A shock absorber with a similar configuration as that in this study, but not filled with foam, was utilized in the legged landing gear of a drone [26]. According to the existing research, the shock absorber could be used in the subfloor stanchion of passenger airplanes [9,25]. Further studies will focus on the application to engineering structures and the improvement of the SEA through ply sequence optimization.

Author Contributions: Conceptualization, Z.Y. and X.Z.; validation, P.Y. and W.X. All authors have read and agreed to the published version of the manuscript.

Funding: This research was funded by the National Natural Science Foundation of China Grant Number 12072199.

Data Availability Statement: The data presented in this study are available upon request from the corresponding author.

Conflicts of Interest: The authors declare no conflict of interest.

References

1. Mou, H.; Xie, J.; Feng, Z. Research status and future development of crashworthiness of civil aircraft fuselage structures: An overview. *Prog. Aerosp. Sci.* **2020**, *119*, 100644. [CrossRef]
2. Guida, M.; Marulo, F.; Abrate, S. Advances in crash dynamics for aircraft safety. *Prog. Aerosp. Sci.* **2018**, *98*, 106–123. [CrossRef]
3. Zhang, J.; Zheng, D.; Lu, B.; Zhang, T. Energy absorption performance of hybrid cross section tubes under oblique loads. *Thin-Walled Struct.* **2021**, *159*, 107133. [CrossRef]
4. Isaac, C.W.; Ezekwem, C. A review of the crashworthiness performance of energy absorbing composite structure within the context of materials, manufacturing and maintenance for sustainability. *Compos. Struct.* **2021**, *257*, 113081.
5. Collins, J.S.; Johnson, E.R. Static and dynamic response of graphite-epoxy curved frames. *J. Compos. Mater.* **1992**, *26*, 792–803. [CrossRef]
6. Feraboli, P.; Wade, B.; Deleo, F.; Rassaian, M.; Higgins, M.; Byar, A. LS-DYNA mat54 modeling of the axial crushing of a composite tape sinusoidal specimen. *Compos. Part A* **2011**, *42*, 1809–1825. [CrossRef]
7. Ren, Y.; Jiang, H.; Gao, B.; Xiang, J. A progressive intraply material deterioration and delamination based failure model for the crashworthiness of fabric composite corrugated beam: Parameter sensitivity analysis. *Compos. Part B Eng* **2018**, *135*, 49–71. [CrossRef]
8. Di Palma, L.; Di Caprio, F.; Chiariello, A.; Ignarra, M. Vertical Drop Test of Composite Fuselage Section of a Regional Aircraft. *AIAA J.* **2020**, *58*, 474–487. [CrossRef]

9. Riccio, A.; Saputo, S.; Sellitto, A.; Di Caprio, F.; Di Palma, L. A numerical-experimental assessment on a composite fuselage barrel vertical drop test: Induced damage onset and evolution. *Compos. Struct.* **2020**, *248*, 112519. [CrossRef]
10. Yang, H.; Lei, H.; Lu, G.; Zhang, Z.; Li, X.; Liu, Y. Energy absorption and failure pattern of hybrid composite tubes under quasi-static axial compression. *Compos. Part B Eng.* **2020**, *198*, 108127. [CrossRef]
11. Chen, D.; Sun, G.; Jin, X.; Li, Q. Quasi-static bending and transverse crushing behaviors for hat-shaped composite tubes made of CFRP, GFRP and their hybrid structures. *Compos. Struct.* **2020**, *239*, 111842.
12. Li, S.; Guo, X.; Liao, J.; Li, Q.; Sun, G. Crushing analysis and design optimization for foam-filled aluminum/CFRP hybrid tube against transverse impact. *Compos. Part B Eng.* **2020**, *196*, 108029. [CrossRef]
13. Jimenez, M.A.; Miravete, A.; Larrode, E.; Revuelta, D. Effect of trigger geometry on energy absorption in composite profiles. *Compos. Struct.* **2000**, *48*, 107–111. [CrossRef]
14. Palanivelu, S.; Paepegem, W.V.; Degrieck, J.; Kakoglannis, D.; Van Ackeren, J.; Van Melrijck, D.; Wastiels, J.; Vantomme, J. Parametric study of crushing parameters and failure patterns of pultruded composite tubes using cohesive elements and seam, Part I: Central delamination and triggering modelling. *Polym. Test.* **2010**, *29*, 729–741. [CrossRef]
15. Palanivelu, S.; Paepegem, W.; Degrieck, J.; Kakogiannis, D.; Ackeren, J.; Hemelrijck, D.; Wastiels, J. Crushing and energy absorption performance of different geometrical shapes of small-scale glass/polyester composite tubes under quasi-static loading conditions. *Compos. Struct.* **2011**, *93*, 992–1007. [CrossRef]
16. Mamalis, A.G.; Robinson, M.; Manolakos, D.E.; Demosthenous, G.A.; Ioannidis, M.B.; Carruthers, J.J. Crashworthy capability of composite material structures. *Compos. Struct.* **1997**, *37*, 109–134. [CrossRef]
17. Garner, D.M.; Adams, D. Test methods for composites crashworthiness: A review. *J. Adv. Mater.* **2008**, *40*, 5–26.
18. Jacob, G.C.; Fellers, J.F.; Simunovic, S.; Starbuck, J.M. Energy absorption in polymer composites for automotive crashworthiness. *J. Compos. Mater.* **2002**, *36*, 813–849. [CrossRef]
19. Asl, O.N.; Pol, M.H.; Golshan, N.R. Experimental investigation of parameters affected on behavior of composite tubes under quasi static and dynamic axial loading. *Compos. Part B Eng.* **2019**, *163*, 471–486.
20. Li, Y.; You, Z. External inversion of thin-walled corrugated tubes. *Int. J. Mech. Sci.* **2018**, *144*, 54–66. [CrossRef]
21. Heimbs, S.; Strobl, F.; Middendorf, P.; Guimard, J.M. Composite Crash Absorber for Aircraft Fuselage Applications. In Proceedings of the International Conference on Structures under Shock and Impact, Tallinn, Estonia, 28–30 July 2010; Volume 113.
22. Ueda, M.; Tsuji, T.; Jeong, T.K. Controlling the Energy Absorption Capability of a Unidirectional Carbon Fiber Reinforced Plastic Tube Using a Double-Sided Plug. *Open J. Compos. Mater.* **2015**, *5*, 30–40. [CrossRef]
23. Tong, Y.; Xu, Y. Improvement of crash energy absorption of 2D braided composite tubes through an innovative chamfer external triggers. *Int. J. Impact Eng.* **2018**, *111*, 11–20. [CrossRef]
24. Siromani, D.; Henderson, G.; Mikita, D.; Mirarchi, K.; Park, R.; Smolko, J.; Awerbuch, J.; Tan, T.M. An experimental study on the effect of failure trigger mechanisms on the energy absorption capability of CFRP tubes under axial compression. *Compos. Part A Appl. Sci. Manuf.* **2014**, *64*, 25–35. [CrossRef]
25. Yu, Z.; Zhou, X.; Zhou, X.; Zhang, Y.; Zhu, Q. Crashworthy subfloor structure of civil aircraft via inclined inward-folding composite tubes. *Compos. Part B Eng.* **2020**, *189*, 107887. [CrossRef]
26. Yu, Z.; Liu, X.; Wang, J.; You, Q.; Fu, Y. Shock absorber with inward-folding composite tube and its application in legged landing gear. *Int. J. Crashworthiness* **2021**, *26*, 388–403. [CrossRef]
27. Huang, K.; Ye, H.; Yu, Z.; Zhou, X. Energy absorption properties of composite sandwich tubes with pre-folded cores. *Compos. Struct.* **2022**, *294*, 115737. [CrossRef]
28. Sebaey, T.A.; Rajak, D.K.; Mehboob, H. Internally stiffened foam-filled carbon fiber reinforced composite tubes under impact loading for energy absorption applications. *Compos. Struct.* **2021**, *255*, 112910. [CrossRef]
29. Yang, H.; Lei, H.; Lu, G. Crashworthiness of circular fiber reinforced plastic tubes filled with composite skeletons/aluminum foam under drop-weight impact loading. *Compos Part B Eng.* **2021**, *160*, 10380. [CrossRef]
30. Alia, R.A.; Cantwell, W.J.; Langdon, G.S.; Yuen, S.C.K.; Nurick, G.N. The energy-absorbing characteristics of composite tube-reinforced foam structures. *Compos. Part B Eng.* **2014**, *61*, 127–135. [CrossRef]
31. Zhou, J.; Guan, Z.; Cantwell, W.J. The energy-absorbing behavior of composite tube reinforced foams. *Compos. Part B Eng.* **2018**, *139*, 227–237. [CrossRef]
32. Siromani, D.; Awerbuch, J.; Tan, T.M. Finite element modeling of the crushing behavior of thin-walled CFRP tubes under axial compression. *Compos. Part B Eng.* **2014**, *64*, 50–58. [CrossRef]
33. Waimer, M.; Siemann, M.H.; Feser, T. Simulation of CFRP components subjected to dynamic crash loads. *Int. J. Impact Eng.* **2017**, *101*, 115–131. [CrossRef]
34. Hashin, Z. Failure criteria for unidirectional fiber composites. *J. Appl. Mech.* **1980**, *47*, 329–334. [CrossRef]
35. Li, X.; Ma, D.; Liu, H.; Tan, W.; Gong, X.; Zhang, C.; Li, Y. Assessment of failure criteria and damage evolution methods for composite laminates under low-velocity impact. *Compos. Struct.* **2018**, *207*, 727–739. [CrossRef]
36. Maziz, A.; Tarfaoui, M.; Gemi, L.; Rechak, S.; Nachtane, M. A progressive damage model for pressurized filament-wound hybrid composite pipe under low-velocity impact. *Compos. Struct.* **2021**, *276*, 114520. [CrossRef]
37. Camanho, P.P.; Davila, C.G. *Mixed-Mode Decohesion Finite Elements for the Simulation of Delamination in Composite Materials*; NASA/TM-2002–211737; NASA: Washington, DC, USA, 2002; pp. 1–37.

38. *ASTM D7136/D7136-15*; Standard Test Method for Measuring the Damage Resistance of a Fiber Reinforced Polymer Matrix Composite to a Drop Weight Impact Event. ASTM International: West Conshohocken, PA, USA, 2015.
39. Chang, F.K.; Lessard, L.B. Damage tolerance of laminated composites containing an open hole and subject to compressive loadings: Part I—Analysis. *J. Compos. Mater.* **1991**, *25*, 2–43. [CrossRef]
40. Olmedo, A.; Santiuste, C. On the prediction of bolted single-lap composite joints. *Compos. Struct.* **2012**, *94*, 2110–2117. [CrossRef]
41. *DS.SIMULIA.ABAQUS User Manual*; Version 6.10; ABAQUS Inc.: Providence, RI, USA, 2010.
42. Benzeggagh, M.L.; Kenane, M. Measurement of mixed-mode delamination fracture toughness of unidirectional glass/epoxy composites with mixed-mode bending apparatus. *Compos. Sci. Technol.* **1996**, *56*, 439–449. [CrossRef]

materials

MDPI

Article

Filament-Reinforced 3D Printing of Clay

Julian Jauk [1,*], Lukas Gosch [1], Hana Vašatko [1], Markus Königsberger [2], Johannes Schlusche [3] and Milena Stavric [1]

1 Institute of Architecture and Media, Graz University of Technology, 8010 Graz, Austria
2 Institute of Mechanics of Materials and Structures, TU Wien, 1040 Vienna, Austria
3 Institute of Structure and Design, University of Innsbruck, 6020 Innsbruck, Austria
* Correspondence: julian.jauk@tugraz.at

Abstract: This research resulted in the development of a method that can be used for filament-reinforced 3D printing of clay. Currently, clay-based elements are mixed with randomly dispersed fibrous materials in order to increase their tensile strength. The advantages of taking this new approach to create filament-reinforced prints are the increased bridging ability while printing, the increased tensile strength of the dried elements, and the achievement of non-catastrophic failure behavior. The research methodology used involves the following steps: (1) evaluating properties of various filament materials with respect to multiple criteria, (2) designing a filament guiding nozzle for co-extrusion, and (3) conducting a comprehensive testing phase for the composite material. This phase involves comparisons of bridging ability, tensile strength evaluations for un-reinforced clay prints and filament-reinforced prints, as well as the successful production of an architectural brick prototype. (4) Finally, the gathered results are subjected to thorough analysis. Compared to conventional 3D printing of clay, the developed method enables a substantial increase in bridging distance during printing by a factor of 460%. This capability facilitates the design of objects characterized by reduced solidity and the attainment of a more open, lightweight, and net-like structure. Further, results show that the average tensile strength of the reinforced sample in a dry state exhibited an enhancement of approximately 15%. The combination of clay's ability to resist compression and the filament's capacity to withstand tension has led to the development of a structural concept in this composite material akin to that of reinforced concrete. This suggests its potential application within the construction industry. Producing the prototype presented in this research would not have been possible with existing 3D printing methods of clay.

Keywords: clay; reinforcement; 3D printing; composite material

1. Introduction

Surpassing traditional methods of clay processing [1], 3D printing of ceramics started to gain ground in building technology research a few years ago [2,3], yet digital fabrication, allowing for material efficiency [4] as well as unique aesthetics [5], primarily focuses on non-sustainable materials such as concrete [6]. This study aims to address the inquiry of which structural design [7,8], material compositions, and processing techniques [9] must be devised and what basic material properties [10] must be acquired to enhance the utilization of 3D printed clay in architectural designs. Building upon research on fiber reinforcement of clay [11], interest has recently emerged in the possible industrial applications of fibrous reinforcement in the extrusion of 3D printing viscous materials, which is further discussed in this section.

So far, the addition of fibers prior to printing has been used to efficiently reinforce a material mixture on a small scale (diameter of 0.08 to 1.5 mm and a length of 2 to 20 mm) [12]. This research, however, focuses on using large-scale reinforcements (endless filament with a diameter from 0.5 to 2 mm). The following references are organized according to the used material: concrete, for which numerous examples can be found, and clay, for which fewer scientific publications exist. Furthermore, the references will be categorized by using temporal processes: pre-printing reinforcement, while-printing reinforcement, and post-printing reinforcement [13].

An example of concrete and pre-printing reinforcement is Mesh Mould, a project using a digital fabrication technique developed at the Swiss Federal Institute of Technology (ETH) in Zürich that involves an in situ welded steel reinforcement and later applied concrete [14]. This allows for the reinforcing mesh to be produced using a specific and fully automated method. The even distribution of concrete was applied manually, however, as the accessibility of the cavities within the highly complex reinforcement is low.

Pre-printing linear reinforcement of clay can be found in a project at IAAC Barcelona using tense ropes and other kinds of fiber reinforcement to strengthen 3D-printed clay. Ropes were arranged in two-dimensional layers and covered with clay. The main advantage of using this method was that it enabled the user to apply pre-tension to the reinforcement. However, this approach requires the use of formwork to tighten the ropes and manual effort [15]. Another student's project at IAAC presents a method that can be used for 3D printing around or on top of Polylactide (PLA) structures so that they can be used as initial support structures. It was envisioned that two robots work in parallel: one 3D printing a PLA reinforcing structure and the other one printing clay around it [16]. In this way, the geometry of the reinforcement can be designed to have a specific complex and optimized shape.

The use of concrete and post-printing reinforcement is commonly utilized in commercially available methods to reinforce 3D-printed concrete. This typically involves the initial printing of outer wall segments, the placement of steel grids, and the subsequent casting of concrete to fill all recesses [17]. This method is already practically applied, as this process does not include any other digital fabrication processes than 3D printing concrete. This method of fabrication, however, also involves manual intervention.

An exemplary use of clay and post-printing was not found in ongoing research, but such examples can be found when examining the existing production techniques, e.g., filling hollow chambers of conventional masonry bonds with further structural elements or tie rods.

Incorporating concrete and while-printing techniques, an active reinforcement entrainment device called RED was developed at the Eindhoven University of Technology to enhance steel cable reinforcement in printed concrete [18]. This makes using a single automated fabrication process possible. However, the authors also reported substantial defects in the concrete matrix, particularly underneath the reinforcement cable, and efforts that need to be made to improve the bond between the two material components.

Depending on the material and the nature of the extrusion tool, a specialized technical solution is required to achieve reinforcement during the printing process [19,20]. Concurrently with the creation of this paper, a similar project was initiated at RWTH Aachen University, which successfully implemented the primary goal of using while-printing reinforcement for clay [21]. This underscores the significance of this topic within the field of architecture. The present work contributes to the existing research by offering a broader spectrum of filament materials used, additional experiments regarding mechanical properties, and the fabrication of a unique prototype.

2. Materials and Methods

Several experiments were carried out to validate the potential of filament-reinforced 3D printed clay and to develop its production and shaping methods. These were divided into two phases. Phase one (described in Section 2.1) consisted of (a) the preparation of a suitable clay mixture, (b) the comparison of different filament materials, and (c) the evaluation of several nozzle designs. Phase two included all results of phase one and experiments carried out to (a) evaluate the improvement of the bridging ability of wet, unfired clay (described in Section 2.3) and (b) to perform material testing of the tensile strength of dried, unfired test specimens (described in Section 2.4) and (c), ultimately, to produce an exemplary 1:1 brick module to demonstrate a potential architectural application (described in Section 2.5) (Figure 1). Methods for statistically analyzing the data (described in Section 3.2) comprise standard methods such as determination of average, standard deviation, and moving averages.

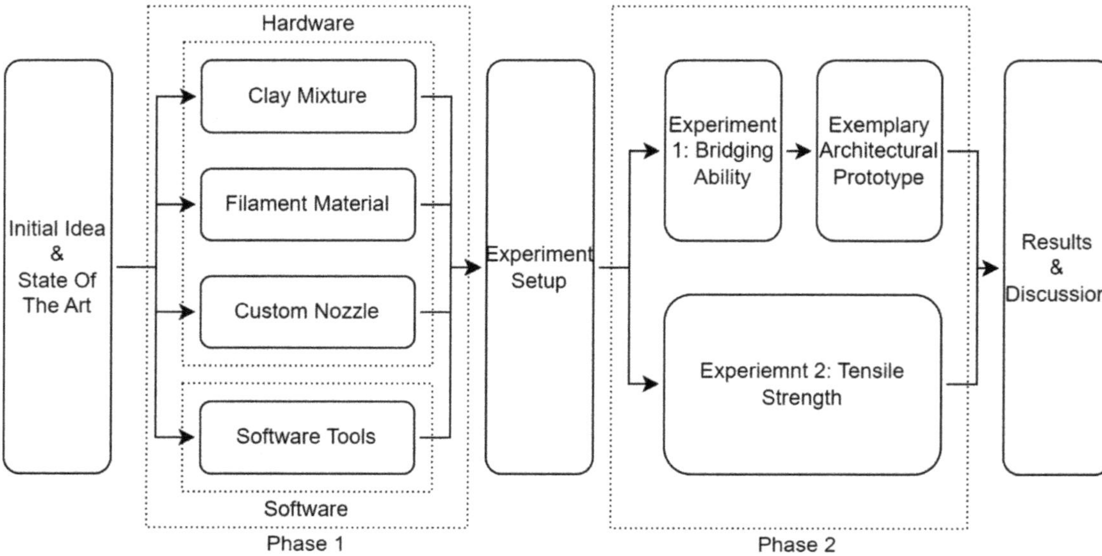

Figure 1. Scheme illustrating the research process.

2.1. Hardware Setup

In order to achieve an even extrusion of clay encasing the filament material, various hardware and software parameters needed to be determined by performing a series of experiments. These parameters included a specific clay mixture, a filament material, a filament guiding nozzle, and a custom-made machine code defining the printing path. These experiments were also necessary to ensure the reproducibility and comparability of the output from both phases.

2.1.1. Clay Mixture

For this study, clay powder type 209 (Goerg & Schneider, Boden, Germany) was used, consisting of 72.2% SiO_2, 22.4% Al_2O_3, 2.8% K_2O, 1.3% TiO_2, 0.9% Fe_2O_3, and 0.3% MgO, as well as amounts of CaO and Na_2O smaller than 0.1%. The clay powder, compared to clay modeling mass, allows clay mixtures to be printed while ensuring an exact moisture content in the material. Furthermore, unlike modeling mass, clay powder takes up less storage space and weighs less, making transportation more viable.

For the printing process, attempts were made to keep the moisture content of the final mixture as low as possible to decrease the chance of negative effects associated with shrinking, such as cracking and uneven alterations in geometry. The need to be able to extrude the material smoothly through the hardware components, as described in the subsequent sections, restricted the amount of water that could be used and, hence, the viscosity of the clay mixture.

The mixture was blended by using an industrial mixing machine containing a kneading arm designed for dough for a minimum of 10 min and allowed to rest for a minimum of 24 h, as an observation was made that the adhesive nature of the mixtures increased significantly after resting. The weight ratio of clay powder to water was 10:3 and was applied in phase two. Preliminary investigations of this clay material regarding viscosity revealed that the yield stress was reached at 3300 pascals. Once this point was reached, the shear stress remained constant while the shear rate increased. For the tests, a Netzsch

Kinexus Prime lab+ rheometer was employed, with samples measuring 1 mm in height and tested at 20 °C.

2.1.2. Filament Material

An initial experiment was carried out to cover filament materials with a clay mixture and to evaluate suitable types of filaments. A set of different filament materials with thicknesses ranging between 0.2 and 2 mm was selected based on existing studies [22,23] and preliminary work [24]. For testing purposes, each filament was positioned at the center of a circular clay extrusion with an 8mm diameter, leading to clay-to-filament cross-sectional area ratios ranging from 100:0.06 to 100:6. The filaments were tested with regard to their stiffness, elasticity, surface texture, accessibility (availability and affordability) and, above all, their ability to compound with wet clay (Table 1). The maximum diameter was limited by the nozzle design, which is explained in Section 2.1.3. After performing these initial experiments, metallic filaments (fine ropes constructed from stainless steel, obtained in 1 × 7, 1 × 19, and 6 × 7 configurations, each with an overall diameter of 0.45 mm and a tensile strength of 1770 MPa) exhibiting a repellent behavior with wet clay were excluded from further research. The high degree of stiffness and low accessibility of certain filaments also influenced the decision to discard them from this research. Viable results were obtained while using polyamide filaments with different surface textures, such as extruded polyamide, as well as with mineral filaments with rough surface textures, such as silicate filaments. Work with these filaments was continued in phase two.

Table 1. Comparison of all filament materials according to the following criteria: adhesion, stiffness, elasticity, texture, and accessibility.

Thread Type	Adhesion	Stiffness	Elasticity	Texture	Accessibility
Synthetic	Low	Low	Low	Medium	High
Cotton	High	Low	Medium	Medium	High
Nylon	Low	Medium	Medium	Plain	High
Carbon	Medium	Low	Low	Rough	Medium
Glass	Medium	Low	Low	Rough	Medium
E-Glass	Medium	Low	Low	Rough	Medium
Basalt	Medium	Low	Low	Rough	Medium
Stainless Steel	Low	High	Low	Medium	Low
Silicate	Medium	Low	Low	Rough	medium

2.1.3. Customized Nozzle

The starting point of the hardware setup was a 3D paste-based extrusion printer for clay, a Delta WASP 40100 Clay model from the year 2020 (Massa Lombarda RA, Italy), with a maximum printable cylindrical region of 40 cm in diameter and f 100 cm in height. The 3D printer's further specifications include a minimum layer height of 0.5 mm and a maximum printing speed of 9000 mm/m. The stepper motor-driven extruder of the printer is built for nozzle diameters between 2 and 8 mm. The nozzles, printing bed, and material tanks were custom-made.

The design of the filament-guiding nozzle was crucial for successfully printing the filament-reinforced clay objects (Figure 2, left). The nozzle consists of (a) a thread used to mount it tightly onto the extruder, (b) a smaller inner nozzle that guides the filament toward the center of the extrusion channel (the filament is to be fed from the outside), and (c) a larger surrounding nozzle that shapes the extruded clay. Furthermore, a spool holder was added to the printer head to ensure that the filament was unrolled easily (Figure 2, right). If unwinding the filament from the spool required too much force, the filament would shorten printing paths and thereby interfere with the geometry of the already extruded clay.

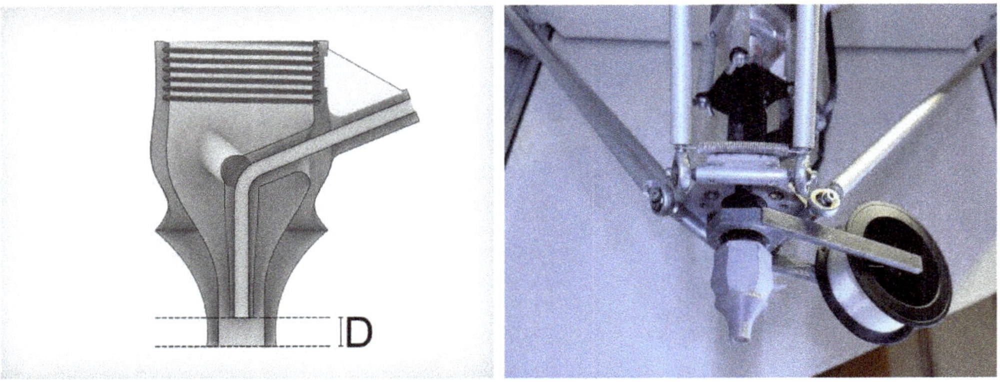

Figure 2. Nozzle design displayed as section drawing showing a distance (D) of 5 mm (**left**) and mounting the nozzle together with the spool holder (**right**).

The internal nozzle configuration features a double-curved surface designed to achieve a gradual transition from the 25 mm extruder diameter to the 8 mm nozzle tip over a height of 40 mm, ensuring a smooth and continuous profile. It was aimed to minimize the internal volume of the nozzle as increased material volumes between the extruder's auger and the nozzle's exit would result in extrusion process delays. All interior obstacles inside the nozzle, such as the three bridges mounting the inner nozzle in the center, were placed symmetrically along a vertical center axis to ensure equal flow dynamics in all printing directions. Moreover, all interior edges were chamfered to avoid the formation of material deposits on the inside, which could dry during the printing process and clog the nozzle.

The ratio of extruded clay to filament diameter is important, as this enables the filament to be successfully pulled along during the printing process. If the material is flowing insufficiently quickly, the filament does not have enough tension to unroll evenly. In phase two, the best results were achieved when using a nozzle with an inner nozzle diameter of 2.5 mm with a wall thickness of 0.75 mm, an outer nozzle with an 8 mm diameter, and a vertical distance D of 5 mm between the closings of the nozzles.

To eliminate the requirement for a filament feeding mechanism, it was necessary to design the nozzle in a manner that would enable the flow of clay material to naturally draw the reinforcing filament forward. After the initial tests, it became apparent that the distance (D) between the end of the inner nozzle and the end of the outer nozzle determined whether the filament would be pulled firmly enough and whether the encasing of the filament with clay would create air-filled hollow spaces or not. Efforts were invested to avoid the latter outcome, as hollow areas tended to collapse once further weight was added in the course of the printing, resulting in a loss of the desired geometry. The goal was to ensure that the clay could close its own material flow after being divided by the inner nozzle and its mounts.

The placement of the filament within the printed extrusion was also observed to be defined by the described distance (D). Having both nozzles ending at the same height causes the filament to be placed above the extrusion and results in a loose composite, which was not desired, as it would not have a positive impact on the bridging ability of single extrusions (Figure 3, left). It was aimed to position the filament at the central axis of the extrusion, which was successfully realized by having a distance (D) of 5 mm (Figure 3, right).

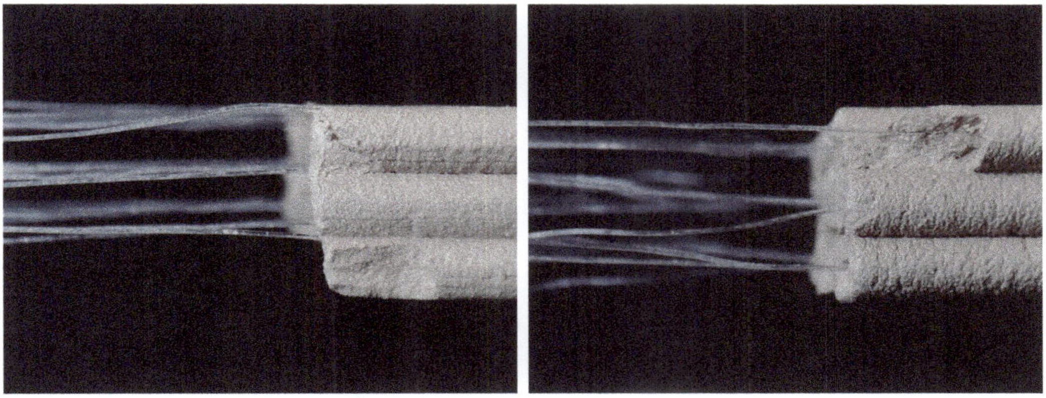

Figure 3. Pictures showing effects of the inner nozzle geometry on printed pieces in terms of filament placement above (**left**) and inside the extrusion (**right**). All samples were fabricated using the developed hardware setup.

A Prusa i3 MK3S+ 3D printer (Prusa Research a.s., Prague, Czech Republic) with a 0.4 mm nozzle was used to produce all tested nozzles (Figure 4, left). Acrylonitrile butadiene styrene copolymers (abbreviated as ABS) with the product designation PrimaValue from 3D Prima were chosen due to their high durability, allowing for thinner nozzle parts and, thereby, larger sections of extrusion, and their ability to be smoothed. This material has a density of 1.03 g/cm^3 and a melting point of 245 °C.

Figure 4. Production of ABS printing the nozzles (**left**) and smoothing them with acetone (**right**).

To ensure the formation of a smooth surface on the inside, the printed nozzles were smoothed through exposure to acetone vapor [25] since any remains of surface texture originating from the layer-based additive manufacturing would have increased the friction of the clay mixture and the filament (Figure 4, right). Fibrous filaments, such as mineral fibers, tend to clog the nozzle when exposed to a rough guiding channel. This exposure was achieved by pouring 100 mL of acetone into an airtight vessel that also contained the nozzle so that the solvent covered the base of the vessel without having contact with the printed nozzle. After 15 to 30 min with no ventilation, the surface was smooth while the overall geometry had still been preserved.

2.2. *Software Setup*

Rhinoceros 3D (McNeel Europe S.L. C/de Roger de Flor, 32, 08018 Barcelona, Spain) modeling software (7 SR 20) was used to provide the data regarding the movement and extrusion of the printer. Printing paths were drawn as lines either manually or parametrically by using the visual programming language Grasshopper (Figure 5, left). A digital model of the hardware setup was also used to simulate printing procedures to avoid collisions with the newly attached equipment. Based on the authors' previous experience and knowledge about the exact material behavior, the appearance and shrinkage of the printed object could be simulated and adjusted in advance.

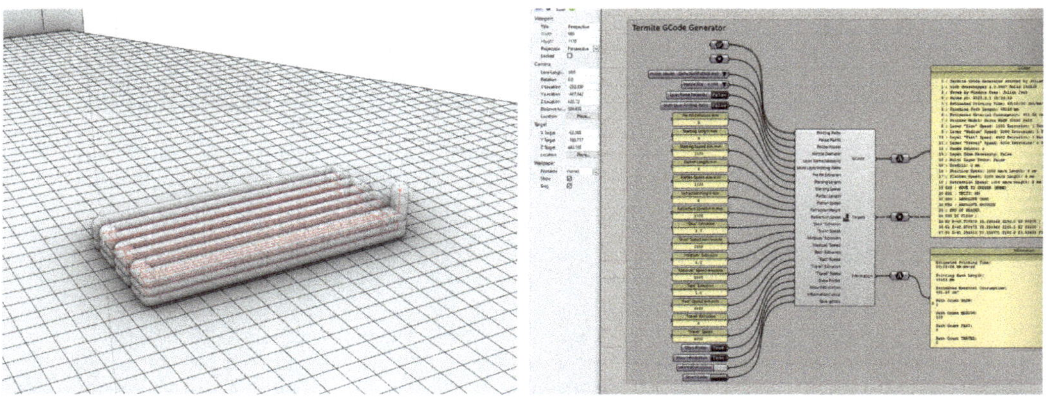

Figure 5. Images showing the printing paths in Rhino 3D (**left**) and the G-code generator in Grasshopper (**right**).

Developing a new software was necessary for the direct transmission of Rhinoceros 3D geometry into a G-code [26] in order to 3D print the composite material. Grasshopper was used to create a software plugin named Termite [27] that could generate G-codes containing information about the toolpath, pause positions, and the extrusion rate directly in the environment used for designing the individual objects (Figure 5, right). Thereby, designing and providing the machine data was directly achieved with a single software, allowing a highly efficient workflow. The printing paths were created through curves that were discretized and divided into parts, whereby the control points were transferred into a G-code with specific printing speeds and material flow values. A single-lined printing path was introduced for the test prints and the succeeding test specimens. At the end of each printing path, a pause function was implemented to manually cut the string and give the machine a signal to continue before returning it to its home position by inserting an unconditional stop function into the G-code. As the string had to be cut manually, all objects were designed from a continuous printing path.

2.3. *Experiment 1: Bridging Ability*

To test the bridging ability (i.e., the ability to print in mid-air), clay was printed in between two elevated printing beds separated by a configurable distance ranging from 0 to 300 mm (Figure 6, left and right). The composition of the material employed in this study remained consistent with previous experiments, featuring a clay powder-to-water weight ratio of 10:3. For reinforcing, a black corded polyamide filament was used to achieve a visual differentiation between the clay and the filament material.

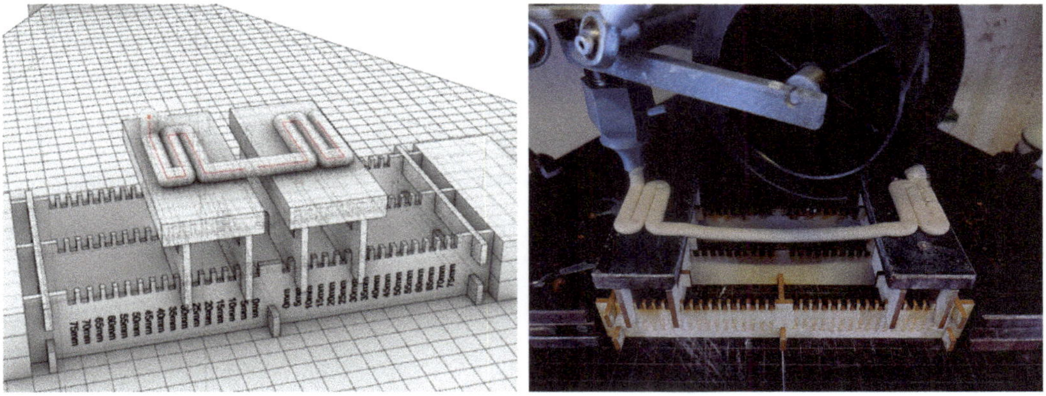

Figure 6. The experimental setup (**left**) increased the bridging ability during the testing procedure (**right**).

To start the printing process, parameters such as the material pressure, material amount being extruded, and printing speed were set to allow for the best possible printing in mid-air. The air pressure on the material in the tank was set at a constant value of 5 bar, which enabled enough material to be fed into the extruder to achieve an appropriate amount of extrusion. The material extrusion was set to a factor supplying enough material for mid-air extrusion to produce a circular strand following the printing path with minimal sagging and to maintain an extrusion diameter of 8 mm (based on a nozzle diameter of 8 mm). Less material being extruded would have caused the extrusion to break in mid-air, while increasing the material extrusion above the ideal value would have caused the material to sag further, resulting in a mismatch between the printing path and the extruded material. The printing speed was set to 4500 mm per minute to achieve optimal results, considering the viscosity of the given material mixture. Any speed lower than this would have increased the chance of breaking the extrusion while it is being printed in mid-air, while any speed higher than this would have prevented the clay from establishing a sufficient connection to the printing bed below. After the optimal settings were determined, all tests were carried out without changing the values.

2.4. Experiment 2: Tensile Strength

After assessing the improvement in performance by reinforcing the wet clay during printing, the effect of the reinforcement on the mechanical properties of the material after the clay had dried was characterized. Four types of samples were tested: (1) unreinforced clay, (2) reinforced clay containing extruded polyamide filaments with a plain texture, a tensile strength of 28 N, and a diameter of 0.23 mm, (3) reinforced clay containing corded polyamide filaments with a fibrous surface, a tensile strength of 250 N, and a diameter of 0.8 mm, and (4) reinforced clay containing mineral silicate filaments with a fibrous surface, a tensile strength of 30 N, and a diameter of 0.46 mm. The latter was chosen due to its resistance up to 1400 °C.

All testing samples had identical dimensions of 100 × 50 × 12 mm once they had been dried (approximately 48 h after printing) and sanded, including three layers of extrusion, each containing seven parallel filaments along the longer dimension of the sample. The settings for printing, such as speed, material pressure, and flow factor, had consistent values throughout the process of printing all samples. Uniaxial tensile strength tests were conducted using a Shimadzu AG-X plus testing machine ((Shimadzu, Kyōto, Japan)). The samples were mounted between two rough grippers (Figure 7, left) so that the test length, l_0, was 80 mm, and the cross-sectional area A_0 was $50 \times 12 = 600$ mm^2. Loads were applied in the displacement-controlled mode, with strain rates of 25% per minute. Roughly

20 samples were tested from each type, which allowed for the statistical treatment of the results. The failure of a sample was defined as the presence of cracks in the clay, resulting in a relaxation of the composite (Figure 7, right).

Figure 7. Pictures showing the testing procedure (**left**) and a sample after failure (**right**).

2.5. Exemplary 1:1 Architectural Prototype

Finally, to demonstrate the architectural application, an exemplary 1:1 brick module was fabricated, which would not have otherwise been possible using conventional 3D printing methods with clay. The dimensions of each module were 25.5 cm × 9 cm × 8 cm, containing 10 bridges of 11 cm and 4 bridges that were 6 cm in length, which hung freely during the printing process (Figure 8, left). Polyamide filament was used as a reinforcement. Based on the values that enabled an optimal printing result for bridging elements to be achieved, as described in Section 2.3, the speed of printing the segments in mid-air was set at 4500 mm per minute, while the speed of printing the standard layer-on-layer segments of the paths was set at 3000 mm per minute. This ensured the establishment of a safe connection between the wall and bridge elements, as well as increased the overall precision of the process. Once printed, the modules were slowly dried in a chamber to evenly dry the wall and bridge segments and to avoid cracking. This prototype represents an example of a 3D-printed brick-like module that contains an inner bracing without closed chambers as, for example, a 2D-extruded brick would necessarily have. The novel developed method clearly offers new, advantageous ways of 3D printing clay (Figure 8, right).

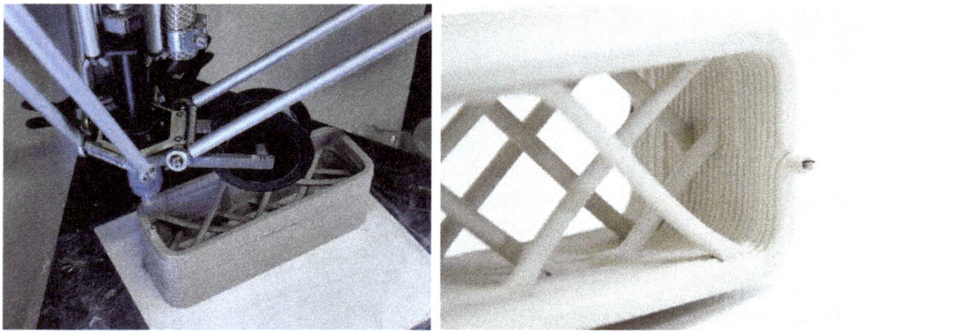

Figure 8. Pictures of exemplary 3D printing of a brick stiffened by reinforced bridges (**left**) and the final prototype showcasing the potential of filament-reinforced 3D printing (**right**).

3. Results and Discussion

3.1. Results of Experiment 1: Bridging Ability

The following results depict findings from tests regarding the enhanced bridging capability during extrusion. As the distance between the two printing beds was increased by 5 mm for each iteration of the experiment, the maximum distance that could be bridged by this material mixture without any additional reinforcement was observed to be 50 mm. A test distance was labeled as successful once it was possible to bridge two times out of three without a significant number of cracks forming on the surface in order to ensure reliable production results. The maximum distance that could be successfully bridged by the filament-reinforced clay was 280 mm. The results show an increased bridging ability of 460% under consistent testing conditions.

Cracking was observed to occur on the first printing bed's edge at all times, as the transition from being pressed against a solid surface to being ejected in mid-air resulted in a sudden change in the pressure within the extruded material (Figure 9). This pressure change led to a large concentration of stress, which tended to break off the extrusion. If sudden cracking of the clay extrusion occurred, the filament kept the printed object in place and only allowed for minimal displacement. This finding suggests that otherwise brittle, clay-based building elements could be given the quality of non-catastrophic behavior by reinforcing them with filaments, protecting them in the event of a collapse. Furthermore, having continuous reinforcement within the extrusion was shown to decrease the chance of crack formations during drying, while the overall stability of the printed structure was increased.

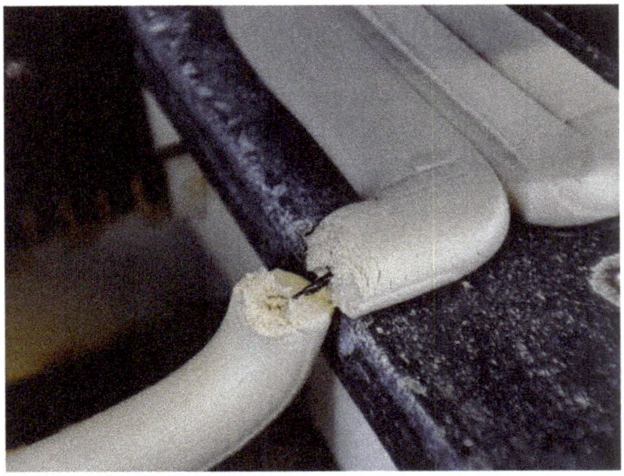

Figure 9. Close-up picture displaying the failure behavior during the printing process.

3.2. Results of Experiment 2: Tensile Strength

Following the drying of test specimens, an evaluation was conducted to assess the impact of reinforcement on the mechanical properties of the material. The uniaxial tensile stresses are obtained by dividing the recorded tensile force, F, by the cross-sectional area A_0, $\sigma = F/A_0$. Uniaxial tensile strains, ε, arise from the applied displacements, Δl, as $\varepsilon = \Delta l/l_0$. In this way, stress-strain diagrams (σ-ε diagrams) were obtained and analyzed (Figure 10). Interestingly, after observing an initial, virtually linear increase in the stress as the strain increased, a plateau was detected at stress levels of around 13 kPa for both the unreinforced and the reinforced samples. After the plateau had been reached, the stresses increased again until failure occurred. The mechanical reason for the observed plateau is not clear. Compression or shear tests performed on raw clay reveal a similar plateau [28,29], but the increase seen afterward in this work has not been detected in other studies. Structural effects may play a role here, such as the clamping of the samples, as the fracture mostly occurred close to the clamps or re-alignment of the filaments at the clamped ends. Moreover, a slip between the machine and the sample may have occurred.

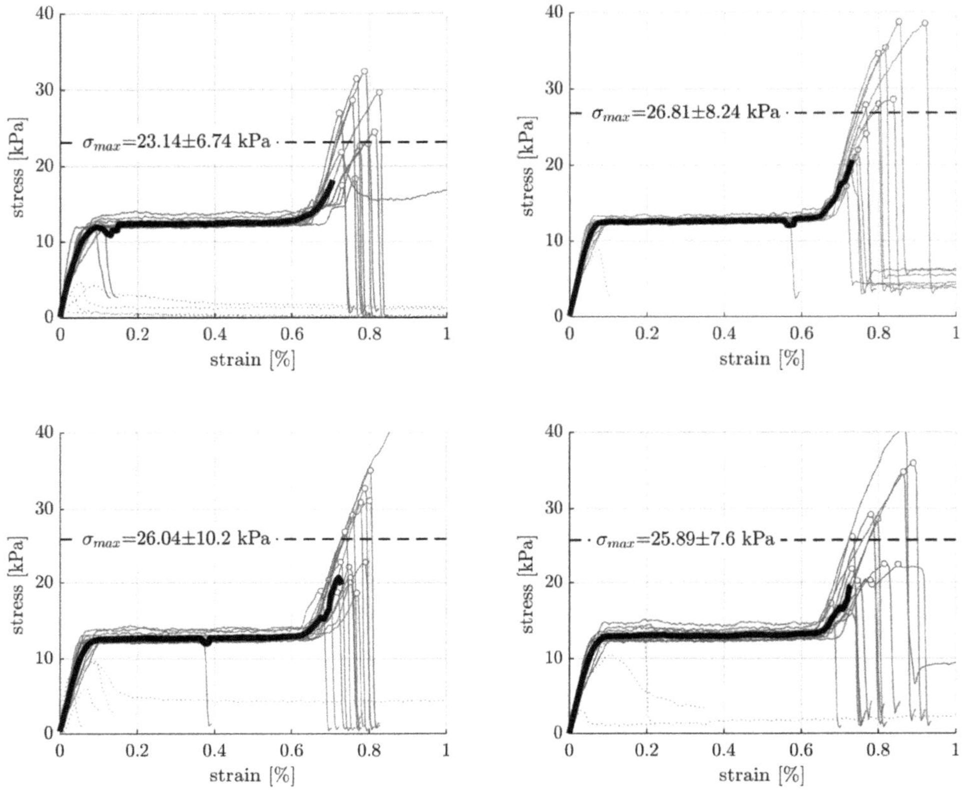

Figure 10. Stress-strain diagrams for the unreinforced (**top left**) as well as reinforced samples with plain polyamide filaments (**top right**), corded polyamide filaments (**bottom left**), and mineral silicate filaments (**bottom right**), showing the individual curves (gray lines) and a moving average (black bold line). Reinforcing increases the tensile strength by roughly 15% and additionally reduces the number of outlier samples (dotted gray lines) with rather low strengths due to pre-existing cracks.

After discussing the qualitative shape of the σ-ε diagrams, a quantitative comparison between the unreinforced and reinforced samples was performed. The average tensile strength (=maximum stress), σ_{max}, of the unreinforced samples was 23.14 kPa, with a standard deviation of 6.74 kPa. The reinforced samples appear to be slightly stronger, whereby the type of reinforcement had only a very minor effect on the tensile strength. Described in more detail, the strength of the samples with straight polyamide filaments is the highest (26.81 $\pm$ 8.24 kPa), followed by the samples with corded polyamide filaments (26.04 $\pm$ 10.2 kPa), followed by those with the mineral silicate filaments (25.89 $\pm$ 7.60 kPa). These results show that the average tensile strength of the reinforced sample could be improved by roughly 15%.

Notably, outliers were removed from the calculations of averages, standard derivations, and moving averages. A stress-strain curve is considered an outlier if the sample fails before reaching the plateau. In mathematical terms, all tests for which the maximum stress measured is below the plateau stress of 13 kPa are considered outliers. Regarding these samples, failure occurred at much lower maximum stresses and, more noticeably, at much lower ultimate strain levels, and sometimes even before any load had been applied.

Pre-existing microcracks resulting from the severe shrinkage might be one reason for the observed early failure. Interestingly, as many as five outliers were observed for the unreinforced samples, while only one was observed for the samples reinforced with straight polyamide filaments. These results show that the reinforcement also helps to prevent these cracks and supports the 3D-printed structure during the drying process.

Finally, an interesting observation can be reported when comparing the behavior of the samples with corded polyamide filaments with the behavior of those with plain polyamide filaments. After failure (which occurred at roughly the same stress levels in both), the crack opening for samples with the corded filaments (Figure 11, left) is more than one order of magnitude larger than the crack opening observed for samples with the plain filament (Figure 11, right). This results from the debonding of the corded filament from the clay, which creates an essentially hollow, cylinder-like structure.

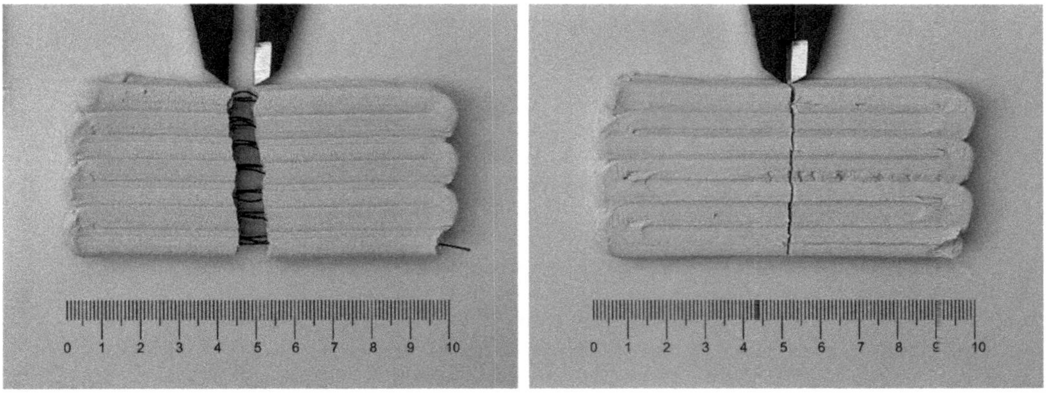

Figure 11. Comparison between crack openings after failure for corded polyamide-reinforced composite with a rough surface (**left**) and plain polyamide-reinforced composite with a flat surface (**right**).

4. Conclusions

In order to integrate reinforcement into the 3D printing of clay, we successfully developed hardware and software components that can be used to produce 3D-printed clay objects, including a while-printing linear filament reinforcement. The study findings support the conclusion that filament-reinforced 3D printing of clay is a viable solution that can be applied to increase the tensile strength of unfired clay in both a wet and a dry state. The filament material can be chosen based on the object's final purpose, ranging from increased support while drying to allow for the production of more lightweight constructions to increased tensile strength in the dried clay objects. The best results regarding multiple criteria were achieved when using extruded polyamide filament-reinforced clay with a plain texture and a diameter of 0.23 mm.

The new method increased the ability of linear extrusions to span bridges while printing by 460%. Furthermore, when comparing the unfired clay test specimens without and with while-printing filament reinforcement, an increase of roughly 15% in the tensile strength values along the axis of extrusion was observed.

Currently, the main limitation of the newly developed method is the necessity to have one continuous printing path. Further, once the lifespan of the composite has ended, the material needs to be separated into its individual components, making up- or recycling purposes more difficult. However, linear while-printing reinforcement presents an alternative to the conventional methods used to increase tensile strength, such as adding fibers to the clay mixture. Industrial production seems feasible for more versatile lightweight designs, including bracing structures, in a high-performance sector of the building industry.

5. Outlook

If this research is contextualized regarding its potential for future implementation, the developed building principle clearly has the potential to be used to construct large-scale, freeform structures. The enhancement in bridging ability enables the 3D printing of more structurally efficient objects with added bracings, as well as facilitating the creation of more porous designs. Examples of this capability can be observed in Section 2.5 and in the woven clay extrusions developed at Harvard University [30]. In this study, additional research directions and questions were identified.

Further research has to be conducted to determine how polyamide filaments, such as extruded polyamide with a plain texture, are able to handle clay shrinkage without exhibiting large displacements after breaking. The overall aim of such research would be to adapt the gained knowledge to other more sustainable filament materials, such as natural fibers [31–33], in order to address known recycling issues of composite materials [34,35]. As filaments with a plain texture have proven to be more suitable, further material testing is planned, including tests with optical fibers such as glass fibers, polycarbonate, or polymethyl methacrylate.

The next step will be to introduce an automated filament-cutting mechanism, which will be controlled by embedding additional signals in the G-code. Once the process is fully automated, the necessity to compose the design in one continuous printing path will be eliminated. Furthermore, it is conceivable to incorporate a sewing-like mechanism to enhance the tensile strength between layers [36].

To increase the tensile strength of fired high-performance elements, a high temperature-resistant filament, such as materials used for the aerospace industry with a tensile strength exceeding the one of fired clay, needs to be found with reasonable accessibility.

Another notable finding that emerged from this research is the use of the method developed for filament-reinforced printing to introduce other materials, such as lightweight porous foam, flammable organic matter, or gel-like support material [37], through the inner nozzle in order to fill the center volume of the extrusion (Figure 12, left) resulting in a co-extrusion process [38]. It is conceivable that such a process enables controlled material distribution within the extrusion [39]. Preliminary tests involving different clay types, flammable filaments to produce continuous cavities (Figure 12, right), and organic substrates for mycelial growth [40] have been carried out and successfully tested for their applicability.

Figure 12. Possible alternatives to using the filament-guiding nozzles to insert other materials, such as flammable organic matter (**left**) or to print hollows (**right**).

Author Contributions: Conceptualization, J.J.; Methodology, J.J.; Software, J.J.; Resources, L.G.; data curation, M.K.; writing—review and editing, H.V. and J.S.; supervision, M.S.; funding acquisition, M.S. All authors have read and agreed to the published version of the manuscript.

Funding: This work was funded by the Austrian Science Fund (FWF) project F77 (SFB "Advanced Computational Design") and Open Access Funding by the Austrian Science Fund (FWF).

Institutional Review Board Statement: Not applicable.

Informed Consent Statement: Not applicable.

Data Availability Statement: Data available on request.

Acknowledgments: We are most grateful to the Institute of Technology and Testing of Construction Materials, Graz University of Technology, for conducting the tensile strength tests and rheological tests.

Conflicts of Interest: The authors declare no conflict of interest.

References

1. Bechthold, M.; Kane, A.; King, N. *Ceramic Material Systems: In Architecture and Interior Design*; Birkhäuser: Basel, Switzerland, 2015; ISBN 978-3-03821-024-5.
2. Seibold, Z.; Hinz, K.; López, J.; Alonso, N.; Mhatre, S.; Bechthold, M. Ceramic Morphologies: Precision and Control in Paste-Based Additive Manufacturing. In Proceedings of the 38th Annual Conference of the Association for Computer Aided Design in Architecture, Mexico City, Mexico, 18–20 October 2018; pp. 350–357.
3. Witte, D. *Clay Printing: The Fourth Generation Brickwork*; Springer Vieweg: Wiesbaden, Germany, 2022; ISBN 978-3-658-37160-9.
4. Bechthold, M. Ceramic Prototypes—Design, Computation, and Digital Fabrication. *Inf. Constr.* **2016**, *68*, 4–6. [CrossRef]
5. Wolf, A.; Rosendahl, P.; Knaack, U. Additive Manufacturing of Clay and Ceramic Building Components. *Autom. Constr.* **2022**, *133*, 6–13. [CrossRef]
6. Alonso-Cañon, S.; Blanco-Fernandez, E.; Castro-Fresno, D.; Yoris-Nobile, A.; Castañón Jano, L. Reinforcements in 3D Printing Concrete Structures. *Arch. Civ. Mech. Eng.* **2022**, *23*, 25. [CrossRef]
7. Sangiorgio, V.; Parisi, F.; Fieni, F.; Parisi, N. The New Boundaries of 3D-Printed Clay Bricks Design: Printability of Complex Internal Geometries. *Sustainability* **2022**, *14*, 598. [CrossRef]
8. Kontovourkis, O.; Tryfonos, G. Robotic 3D Clay Printing of Prefabricated Non-Conventional Wall Components Based on a Parametric-Integrated Design. *Autom. Constr.* **2020**, *110*, 17–18. [CrossRef]
9. Alonso, J.; Sotorrío Ortega, G.; Carabaño, J.; Olsson, N.; Tenorio, J. 3D Claying: 3D Printing and Recycling Clay. *Crystals* **2023**, *13*, 375. [CrossRef]
10. Carr, M.; Wang, Y.; Ghayoomi, M.; Newell, P. Effects of 3D Printing on Clay Permeability and Strength. *Transp. Porous Media* **2023**, *148*, 503–504. [CrossRef]
11. Naughton, P.; Grennan, R. The Strength of Fibre Reinforced Clays. In *Proceedings of the XVI ECSMGEGeotechnical Engineering for Infrastructure and Development*; CE Publishing: Edinburgh, UK, 2015; pp. 3255–3260.
12. Mirzababaei, M.; Mohamed, M.; Arulrajah, A.; Anggraini, V. Practical Approach to Predict the Shear Strength of Fibre-Reinforced Clay. *Geosynth. Int.* **2017**, *25*, 5–6. [CrossRef]
13. Ismail, K.; YAP, T.C.; Ahmed, R. 3D-Printed Fiber-Reinforced Polymer Composites by Fused Deposition Modelling (FDM): Fiber Length and Fiber Implementation Techniques. *Polymers* **2022**, *14*, 4659. [CrossRef]
14. Mirjan, A.; Mata-Falcón, J.; Rieger, C.; Herkrath, J.; Kaufmann, W.; Gramazio, F.; Kohler, M. *Mesh Mould Prefabrication*; Springer International Publishing: Cham, Switzerland, 2022; pp. 31–36. ISBN 978-3-031-06115-8.
15. Reinforced Clay Printing. Available online: https://www.iaacblog.com/programs/reinforced-clay-printing-3d-printing-pla-printing/ (accessed on 11 September 2023).
16. Fiber Reinforcement in 3D Printing with Clay. Available online: https://www.iaacblog.com/programs/fiber-reinforcement-3d-printing-clay/ (accessed on 11 September 2023).
17. Asprone, D.; Menna, C.; Bos, F.; Salet, T.; Mata-Falcón, J.; Kaufmann, W. Rethinking Reinforcement for Digital Fabrication with Concrete. *Cem. Concr. Res.* **2018**, *112*, 2–8. [CrossRef]
18. Bos, F.; Ahmed, Z.; Jutinov, E.; Salet, T. Experimental Exploration of Metal Cable as Reinforcement in 3D Printed Concrete. *Materials* **2017**, *10*, 1314. [CrossRef] [PubMed]
19. Mierzwiński, D.; Łach, M.; Gądek, S.; Lin, W.-T.; Tran, H.; Korniejenko, K. A Brief Overview of the Use of Additive Manufacturing of Con-Create Materials in Construction. *Acta Innov.* **2023**, 28–29. [CrossRef]
20. Rajendran, S.; Palani, G.; Kanakaraj, A.; Shanmugam, V.; Gądek, S.; Korniejenko, K.; Marimuthu, U. Metal and Polymer Based Composites Manufactured Using Additive Manufacturing—A Brief Review. *Polymers* **2023**, *15*, 2564. [CrossRef] [PubMed]
21. Yang, H.-Q.; Klug, C.; Schmitz, T. Fiber-Reinforced Clay: An Exploratory Study on Automated Thread Insertion for Enhanced Structural Integrity in LDM. *Ceramics* **2023**, *6*, 1365–1383. [CrossRef]
22. Ekinci, A.; Abki, A.; Mirzababaei, M. Parameters Controlling Strength, Stiffness and Durability of a Fibre-Reinforced Clay. *Int. J. Geosynth. Ground Eng.* **2022**, *8*, 6. [CrossRef]

23. Neef, T.; Mechtcherine, V. Simultaneous Integration of Continuous Mineral-Bonded Carbon Reinforcement Into Additive Manufacturing With Concrete. *Open Conf. Proc.* **2022**, *1*, 74–75. [CrossRef]
24. Gosch, L.; Jauk, J.; Vašatko, H.; Šamec, E.; Raffaelli, M.; Rutzinger, S.; Stavric, M. Fabricating Lightweight Ceramics by Spraying Clay on Knitted Structures. *Int. J. Archit. Comput.* **2022**, *20*, 693–706. [CrossRef]
25. Khan, M.; Mishra, S. Minimizing Surface Roughness of ABS-FDM Build Parts: An Experimental Approach. *Mater. Today Proc.* **2020**, *26*, 1557–1566. [CrossRef]
26. Brown, A.; Beer, D. Development of a Stereolithography (STL) Slicing and G-Code Generation Algorithm for an Entry Level 3-D Printer. In Proceedings of the IEEE AFRICON Conference, Pointe-Aux-Piments, Mauritius, 9–12 September 2013; pp. 1–5.
27. Termite Food4Rhino. Available online: https://www.food4rhino.com/en/app/termite (accessed on 11 September 2023).
28. Wang, J.; Gee-Clough, D. Deformation and Failure in Wet Clay Soil: Part 1, Stress-Strain Relationships. *J. Agric. Eng. Res.* **1993**, *54*, 37–55. [CrossRef]
29. Sachan, A. Effect of Intermediate Microfabric on Shear Strength and Strain Localization Response of Kaolin Clay Under Compression and Extension Loading. *Geotech. Geol. Eng.* **2012**, *31*, 213–228. [CrossRef]
30. Friedman, J.; Kim, H.; Mesa, O. Experiments in Additive Clay Depositions. *Rob/Arch 2014 Robot. Fabr. Archit. Art Des.* **2014**, *2014*, 261–272. [CrossRef]
31. Galán-Marín, C.; Rivera-Gómez, C.; Petric, J. Clay-Based Composite Stabilized with Natural Polymer and Fibre. *Constr. Build. Mater.* **2010**, *24*, 1462–1468. [CrossRef]
32. Rajeshkumar, G.; Seshadri, S.A.; Ramakrishnan, S.; Sanjay, M.R.; Siengchin, S.; Nagaraja, K.C. A Comprehensive Review on Natural Fiber/Nano-clay Reinforced Hybrid Polymeric Composites: Materials and Technologies. *Polym. Compos.* **2021**, *42*, 3687–3701. [CrossRef]
33. Chand, N.; Mohammed, F. *Natural Fibers and Their Composites*; Woodhead Publishing: Sawston, UK, 2021; pp. 1–59. ISBN 978-0-12-818983-2.
34. Shuaib, N.A.; Mohamed Sultan, A.A.; Ismail, S.; Samat, A.; Omar, N.; Azmi, A.; Mativenga, P. *Recycling of Composite Materials*; Springer: Berlin, Germany, 2021; pp. 532–535. ISBN 978-3-030-71437-6.
35. Milošević, J.; Brajković, J.; Josifovski, A.; Žujović, M.; Zivkovic, M. Reimagining Materiality: 3D Printing With Recycled Clay for Circular Architectural Elements. In *Proceedings of the On Architecture—Challenges in Design*; Sustainable Urban Society Association—STRAND: Belgrade, Serbia, 2023; p. 1.
36. Jacquet, Y.; Perrot, A. Sewing Concrete Device—Combining In-Line Rheology Control and Reinforcement System for 3D Concrete Printing. *Materials* **2023**, *16*, 5110. [CrossRef] [PubMed]
37. Keating, S. From Bacteria to Buildings: Additive Manufacturing Outside the Box. PhD Thesis. PhD Thesis, Massachusetts Institute of Technology, Cambridge, MA, USA, 2016.
38. Han, C. *Coextrusion*; Oxford Academic: New York, NY, USA, 2006; pp. 379–423. ISBN 978-0-19-518783-0.
39. Ahmed, Z.; Bos, F.; Brunschot, M.C.A.J.; Salet, T. On-Demand Additive Manufacturing of Functionally Graded Concrete. *Virtual Phys. Prototyp.* **2020**, *15*, 194–210. [CrossRef]
40. Jauk, J.; Gosch, L.; Vašatko, H.; Christian, I.; Klaus, A.; Stavric, M. MyCera. Application of Mycelial Growth within Digitally Manufactured Clay Structures. *Int. J. Archit. Comput.* **2022**, *20*, 31–40. [CrossRef]

materials

MDPI

Article

Increasing the Compressive Strength of Helicoidal Laminates after Low-Velocity Impact upon Mixing with 0° Orientation Plies and Its Analysis

Zhefeng Yu *, Xin Du, Rui Liu, Qiwu Xie, Xiaojing Zhang and Qiao Zhu

Aerospace Structure Research Center, School of Aeronautics and Astronautics, Shanghai Jiao Tong University, Shanghai 200240, China
* Correspondence: yuzf@sjtu.edu.cn; Tel.: +86-021-34206387

Abstract: The helicoidal laminate is a kind of nature inspired fiber reinforced polymer, and the ply orientation affects their mechanical properties for engineering structural applications. A variety of helicoidal laminates with uniform and non-linear pitch angles mixed with additional 0° plies are fabricated to investigate the impact resistance through low-velocity impact and after-impact compression tests. Additionally, helicoidal laminates with constant pitch angles, quasi-isotropic laminate, and cross-ply laminates are also fabricated for a comparative study. The impact characteristics and the compressive strength are analyzed in view of the impact model, shear stress distribution, and fracture toughness. The results suggest that 10° or 20° are the better basic pitch angles before mixing 0° orientation plies. The 0° orientation plies may affect the contact stiffness, bending stiffness, damage extent, and compressive modulus. The compressive strength reaches the highest in tests on two samples with different percentages of 0° orientation plies and ply setups. Bending stiffness also dominates the impact response. The analysis on the laminate parameters provides ideas to improve the residual strength of helicoidal laminate.

Keywords: helicoidal composite; low velocity impact; compressive strength; laminate design

1. Introduction

Composite materials such as fiber reinforced polymer (FRP) laminates are widely used in engineering structures with lightweight and high safety requirements, such as aerospace structures. Numerous composite structures are susceptible to impact from runway debris, hailstones, dropping tools, etc. The delamination caused by the low-velocity impact is generally invisible or barely visible; however, it reduces the flexural rigidity, thus leading to premature buckling failure under the in-plane compressive load [1,2]. Therefore, improving the damage resistance against impacts in FRP laminates is highly significant. Lamination sequences inspired by naturally occurring impact-resistant micro-structures, such as Bouligand structures [3] or helicoidal composite [4], have been investigated in the past years. A helicoidal composite with uniform inter-ply angles quasi-isotropic (QI) but some laminates are required to have directionally advantageous performance in stiffness or strength, such as the wing skin subjected to single directional bending. The helicoidal composites mixed with additional 0° plies were investigated in this study on its impact performance and residual strength under compression after impact (CAI).

Bouligand-featured structures can be found in a variety of animals, including mantis shrimp [5], fish scales [6], and the exoskeleton of arthropods, such as beetles [7]. The dactyl appendages of the mantis shrimps can endure forces of 0.4–1.5 kN, and the last segment of the appendage consists of several repetitions of the Bouligand unit stacked up along the thickness [5]. Each Bouligand unit of the dactyl appendage (Figure 1) contains a helicoidal layup with very small pitch angles (from 6.2° to 1.6°) [8,9] for a rotation of 180° inside each unit. Research in the field of biomechanics has shown that cracks growing in the matrix of the helicoidal composite specimens follow the fiber orientations and lead to the formation of twisted cracks [5]. A twisted crack growing in the helicoidal architecture amplifies the crack surface per unit volume, thereby enhancing energy dissipation and stress relaxation in the composite without leading to catastrophic failure. Some interlaminar cracks still exist; however, they are regarded as the sub-critical delamination rather than the large delamination occurring in ordinary composite.

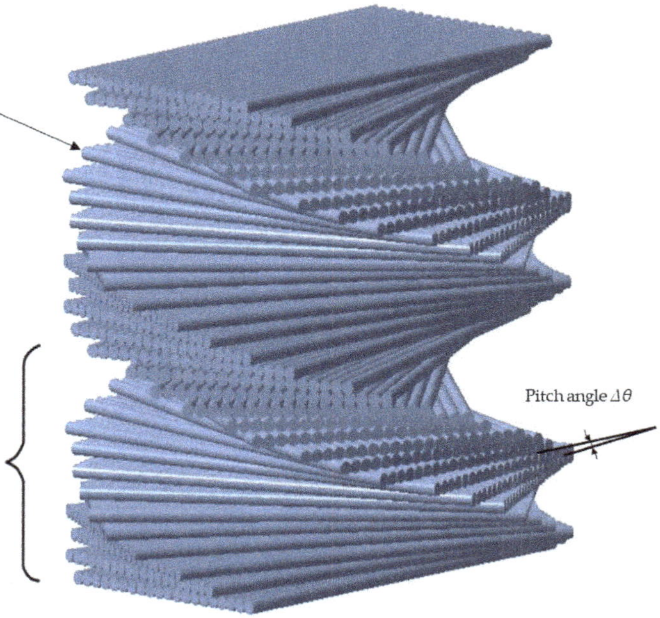

Figure 1. 3D schematic of the Bouligand structure in the periodic region of the dactyl club.

Significant progress has been made in understanding the damage growth in Bouligand structures and studies have been conducted toward engineering materials. Suksangpanya et al. [10,11] developed a theoretical model for the local stress intensity factors at the crack front of twisting cracks formed within the Bouligand structure. Their study revealed that the changes in the local fracture mode at the crack front lead to a reduction of the local strain energy release rate, thereby increasing the energy required to propagate the crack, which is quantified by the local toughening factor. The theoretical values of the energy release rate and local stress intensity factors were validated using 3D simulations of the asymptotic crack front field. Mencattelli et al. [12] conducted a finite element (FE) simulation on thin-ply carbon fiber reinforced polymer (CFRP) laminates, thereby showing that the loci of maximum intralaminar shear stress infers the distribution of shearing matrix cracks in laminates with different pitch angles. Additionally, they investigated the effects of the pitch angle and ply thickness on the resistance of Bouligand CFRP laminates against the twisting cracks formed at the back face of the laminate subjected to low-velocity impact (LVI) due to the tensile stresses originating from bending.

In addition to the matrix cracks, penetration resistance is another key characteristic of composite panels. Using drop-weight impact tests, Grunenfelder et al. [8] demonstrated that the helicoidal composite panels exhibit higher penetration resistance with a wider-spread energy dissipation compared with the unidirectional and [0/±45/90] composite panels. The sinusoidally architected helicoidal laminate also plays a role in energy dissipation and increasing the load-bearing capability [13]. Another study conducted by Mencattelli et al. [14] reported that the laminates with a small pitch angle (2.5°, 5°, and

10°) have an overwhelming advantage over those with a pitch angle of 45° in terms of energy dissipation, peak load, and penetration load under quasi-static-indentation test. Liu et al. [15] reported that the resistance to the transverse load of helicoidal composite is related to the ratio of inter-ply angle to ply thickness and the strain energy release rate; they verified this using helicoidal laminates fabricated from unidirectional carbon-epoxy, Kevlar-epoxy, and glass fiber-epoxy prepregs [16]. The effect of pitch angle on the penetration energy of 3D printed helicoidal composite was studied by Liu et al. [17]; they found that the pitch angle of 5° is optimum.

According to the three-point-bending test, the resistance to the growth of cracks at the back face of the helicoidal laminate is less when the twist angle is small [14], so the inter-ply angle on the back face should be larger. Liu et al. [18] studied the effect of the non-uniform inter-ply angles on the penetration resistance and reported that the helicoidal configuration with a larger pitch angle (10°) at the bottom half and smaller pitch angle (5°) at the top half of the laminate consistently outperforms the other helicoidal laminates. Jiang et al. [19] conducted a numerical study on the helicoidal composite laminates with non-linear rotation angle; their study indicated that the opposite movement of adjacent layers may be hindered if they have different orientations.

In addition to the penetration resistance, residual strength under CAI is an important property of the laminates used in damage tolerance structures. Grunenfelder et al. [8] conducted CAI tests on the composite laminate; herein, the specimens were held in place by upper and lower clamp plates, with 76 mm diameter circular holes at the center, and an impact energy of 100 J was adopted. The compressive residual strength of specimen with a 7.8° pitch angle was 13 % lower than that of the QI specimen, whereas it was 16 % and 18 % higher than that of QI specimens with 16.3° and 25.7° pitch angles, respectively. Pinto et al. [20] reported that the symmetric helicoidal laminate has a better CAI strength compared with QI laminate; however, the pure helicoidal stacking configuration (in which the orientation of the plies goes from 0° to 360°) does not. In the study conducted by Ginzburg et al. [21], the residual compressive strength of CFRP helicoidal composite after high energy (80 J) impact was found to be better than that of the corresponding quasi-isotropic laminate; however, the residual compressive strengths of CFRP helicoidal laminates at lower impact energies (40 J) were lower than those of QI and cross-ply (CP) laminates. Mencattelli et al. [12] analyzed the compressive strength of samples with different pitch angles; however, no statistically significant trends were found.

Furthermore, studies on CAI of helicoidal laminates have also been conducted, and some studies demonstrate improvement in damage tolerance against foreign object impact by optimizing the mismatch angle between plies [22,23], bending stiffness [24], and thickness of ply [25–27] and relationship between pitch angle and thickness [28]. The CAI strength increases with the proportion of 0° plies (i.e., plies aligned with the loading direction) to some extent [29] for ordinary laminates. Sasikumar et al. [30] mixed thick plies of 0° with thin plies to fabricate a hybrid laminate, thereby yielding a 40% increase in CAI strength over the thin-ply baseline laminate. However, the proportion of 0° plies is constant if the helicoidal structure has a uniform pitch angle; hence, more 0° plies should be mixed into the laminate to increase its stiffness and strength in 0° orientation.

The compressive strength of laminates is generally tested according to ASTM D7136/D7136M-15 [31] and ASTM D7137/D7137M-17 [32], which were followed in this study. Impacts causing damage in samples are defined as low-velocity impacts when the contact time between the impactor and plate is sufficiently large to allow all wave reflections from the boundaries [33,34]; therefore, the LVI test in this study can be treated as a quasi-static problem. The performance of helicoidal composite under LVI could be describe with lumped-parameters model or energy-balance model [35,36]. In the model, the mass is mainly that of the impactor, and the stiffness is the contact stiffness between the impactor and laminate, the bending stiffness, and the membrane stiffness of the laminate. The contact stiffness is relative to the modulus and local strength under the concentrated load [37–39]. The bending stiffness is relative to the bending stiffness matrix coefficients of the laminates [40–43]. These material parameters relative to the ply setup need to be considered to study the properties of the composite laminate to resist LVI or vibrational load [44,45].

In this study, helicoidal composite materials were mixed with additional 0° plies in order to increase the CAI strength, and the pitch angles of some plies were designed to make the laminates symmetric for the fixed ply number. Subsequently, helicoidal laminates with non-linear twist angle (NLTA) were yielded. Additionally, helicoidal laminates with constant pitch angle, QI laminates, and CP laminates were fabricated for a comparative investigation. The damage modes (after LVI and CAI tests) and residual strength were analyzed based on impact theory, shear stress simulation, and strain energy release rate.

2. Materials and Experimental Setup

2.1. Specimens Design

Helicoidal, QI, and CP laminates were designed and fabricated using unidirectional T700/SKRK51 carbon-epoxy prepreg, the mechanical properties of which are shown in Table 1. The average ply thickness is 68.5 μm. The total number of layers used in this study was fixed at 73 for each plate leading to an overall plate thickness of 5 mm. The minimum pitch angle $\Delta\theta$ was set to 2.5°. The configurations and abbreviation used to denote them are provided in Table 2, where H denotes the helicoidal laminate, QI denotes the quasi-isotropic laminate, and CP denotes the cross-ply laminate. We manufactured 2 samples for each type of laminate. All of the layups were symmetric to avoid the thermally induced twist in the laminate in the case of the non-symmetric layup. Layups H1, H2, and H3 consisted of a uniform pitch angle; therefore, the plies had a linear twist angle (LTA). The other helicoidal layups were mixed with more 0° plies in addition to the origin 0° plies of the LTA laminate to maintain the continuity. The ply orientation of H4, H5, and H6 are demonstrated in Figure 2.

Table 1. Mechanical properties of T700/SKRK51 carbon-epoxy prepreg.

Mechanical Properties	Value
0°Tensile modulus E_1 (GPa)	125.3
90°Tensile modulus E_2 (GPa)	8.4
0°Tensile strength (MPa)	2500
0°Compressive strength (MPa)	780
90°Tensile strength (MPa)	60
0°Interlaminar shear strength (MPa)	88.5

Herein, H4 had a basic pitch angle of 5°. Five consecutive 0° plies were on two surfaces, and one was in the middle plane (180°). To achieve a symmetric sequence, a pitch angle of 10° was applied to plies 20–23 and 52–55. Layup H5 had a basic pitch angle of 10° and 12 plies of 0° (180°). A pitch angle of 20° was applied to plies 11, 12, 29, 45, 46, 63, and 64 to achieve a symmetric sequence. Layup H6 had pitch angles of 2.5°, 5°, and 10°. Eleven plies were of 0° (180°), and the pitch angle was 2.5° for the plies close to the 0° (180°) plies, which increased the stiffness in that direction while maintaining a non-zero pitch angle.

The extensional stiffness coefficients and bending stiffness coefficients were calculated, see Table 2. Specimen H1, H2, and H3 are QI laminates according to the extensional stiffness coefficients. More 0° plies increase A_{11} for specimen H4, H5, and H6, while CP-73 has the maximum A_{11} due to possessing the most 0° plies.

The plies laid in the required configurations were cured using an autoclave at 125 °C and 5 bars for 90 min. The cured laminations were cut into 150 mm × 100 mm squares to meet the testing standard ASTM D7136.

Table 2. Sample specifications.

Samples	Stacking Sequence	Extensional Stiffness Coefficients (MPa)			Coefficients of Bending Stiffness Matrix (N·m)						The Critical Load of Buckling (N/mm)
		A_{11}	A_{12}	A_{66}	D_{11}	D_{12}	D_{22}	D_{66}	D_{16}	D_{26}	$N_{x,cr}$
H1	$[0/5/10/\ldots/170/175/\overline{180}]_s$	272	264	92	682	160	454	182	172	107	2115
H2	$[0/10/20/\ldots/160/170/180/10/20/\ldots/170/\overline{180}]_s$	272	264	92	611	164	517	186	85	54	2348
H3	$[0/20/40/\ldots/160/180/20/40/\ldots/160/180/20/\ldots/160/180/20/\ldots/160/\overline{180}]_s$	272	264	92	593	165	533	187	40	27	2385
H4	$[0/0/0/0/0/5/10/15/20/\ldots/60/65/70/80/90/100/110/115/120/\ldots/165/170/175/\overline{180}]_s$	304	233	91	907	126	297	148	125	68	1771
H5	$[0/0/0/10/20/30/\ldots/60/70/90/110/120/\ldots/150/160/170/180/0/0/10/20/30/\ldots/60/70/90/110/120/\ldots/150/160/170/\overline{180}]_s$	336	204	90	790	147	372	169	66	36	2101
H6	$[0/0/0/0/0/2.5/5/7.5/10/12.5/15/20/25/30/35/40/50/60/\ldots/120/130/140/145/150/155/160/165/167.5/170/172.5/175/177.5/\overline{180}]_s$	415	159	73	1030	96	234	118	104	35	1569
QI-73	$[(0/45/90/-45)_9/0/(-45/90/45/0)_9]$	272	264	92	589	165	537	187	12	12	1901
CP-73	$[(0/90)_{36}/0]$	340	332	24	725	28	675	50	0	0	2394

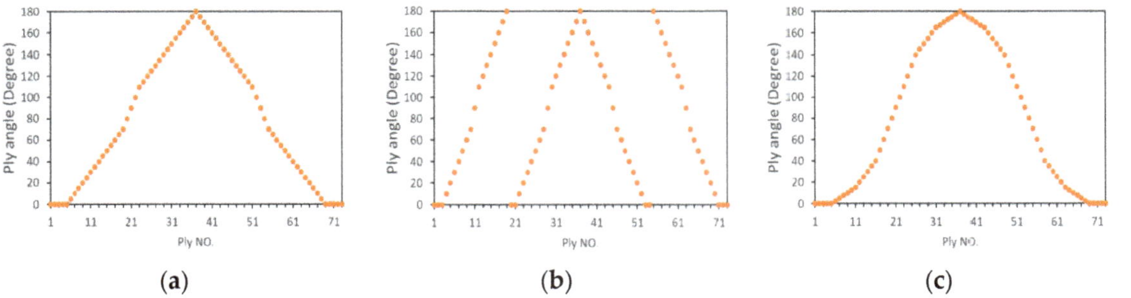

Figure 2. Orientation of the plies for samples (**a**) H4, (**b**) H5, and (**c**) H6.

2.2. Test Equipment

Low-velocity-impact (LVI) tests were conducted using Instron CEAST 9350 drop-weight tower (Figure 3a) following ASTM D7136. The impactor mass was 5.277 kg and the hemispherical tip was 12.7 mm in diameter. To achieve a reasonable amount of damage, each sample was impacted with an impact energy of 20 J, i.e., 4 J/mm along the thickness. The time history of impact load, displacement of the impactor, and energy absorbed by the specimen were recorded by the control system of the tester. After the LVI test, the dent depth was measured using a vernier micrometer. The computed tomography (CT) images were obtained from Zeiss Xradia 520 Versa X-ray microscope (Zeiss, Oberkochen/Jena, Germany) (Figure 3c).

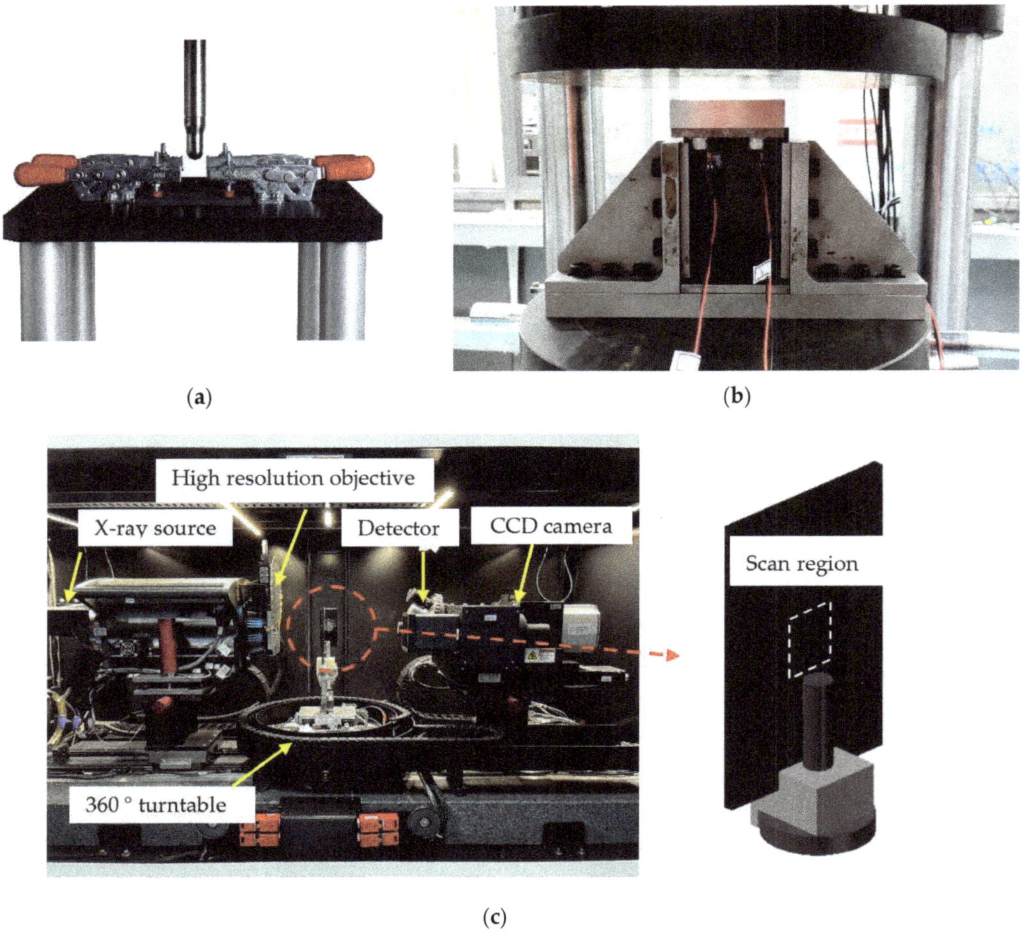

Figure 3. Experimental setup: (**a**) LVI test, (**b**) CAI test, and (**c**) Zeiss Xradia 520 Versa X-ray microscope for CT imaging and installation of specimen.

A set of compression after impact (CAI) tests was performed on the samples following the testing standard ASTM D7137 (Figure 3b). Samples were loaded by a 300 kN load cell tester with a constant displacement rate of 1.25 mm/min. The strains during the test were measured by four strain gauges instrumented on the front and back sides of the samples to verify that the initial alignment of the sample was within 10% and that no global buckling occurred. The ultimate compressive residual strength and effective compressive modulus were recorded in the test direction.

3. Results and Discussion

3.1. Experimental Results

The outputs of the operating system were the time history of the impact load, curve of impact load versus the displacement of the impactor, and time history of energy absorbed by the specimen (see Figure 4), which are used to analyze the characteristics of different specimens in Section 4. The images of the impact site and distal face are shown in Figure 5, some of which reflect the internal damage. The ultrasonic C-scanning images were obtained using GE Phasor XS, as shown in Figure 6, where the blue area denotes the delamination region to be discussed in Sections 3.3 and 3.4. The side cross-sectional views of CT images are shown in Figure 7.

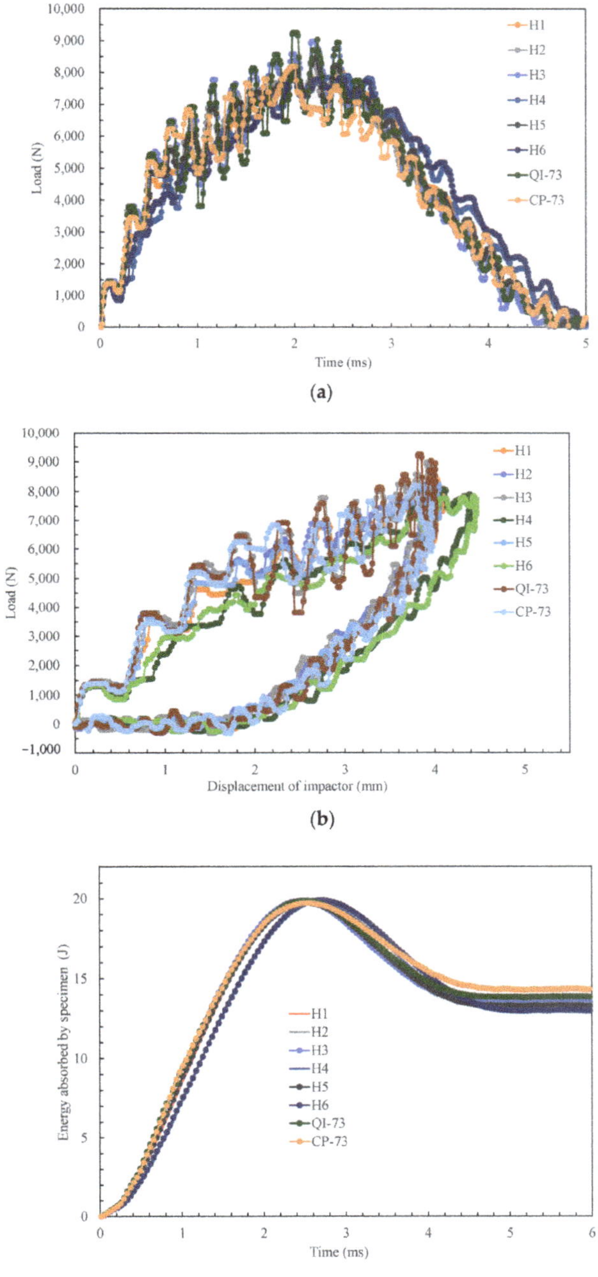

Figure 4. Experimental data of low-velocity test: (**a**) time histories of the impact load, (**b**) displacement of impactor vs. impact load, and (**c**) time histories of the energy absorbed by specimen.

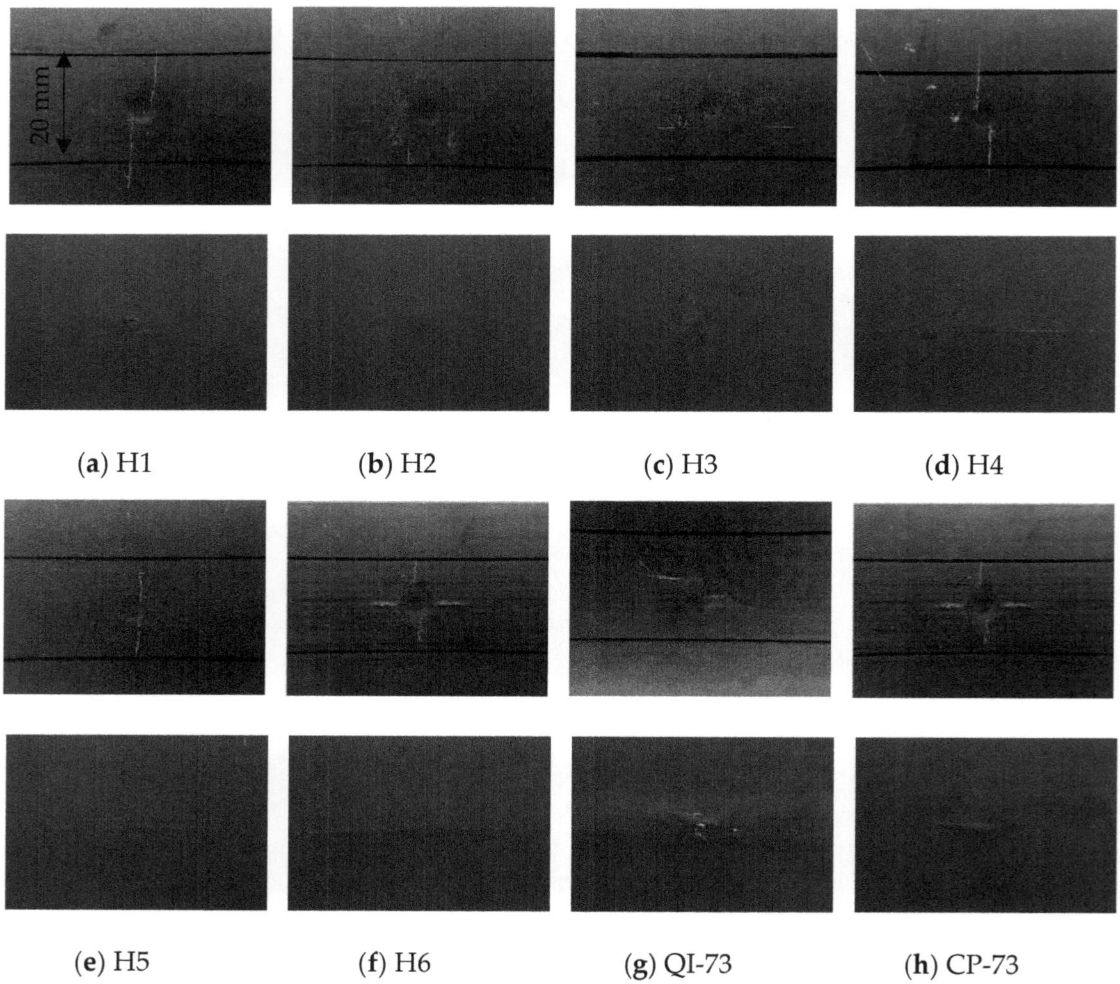

Figure 5. Images of the impact site (**upper**) and distal surface (**lower**) of specimens.

The peak load, maximum displacement of the impactor, final energy absorbed due to sample damage, and the indentation measured after the impact test are shown in Figure 8a–d, respectively. Furthermore, the bending stiffness of each specimen was calculated to reveal the characteristics of the laminate under LVI. The CAI strength was plotted against the bending stiffness and compressive modulus, as shown in Figure 9a,b, respectively.

According to the experimental results, the impact load, damage area, strain energy release rate, and CAI characteristics are analyzed in the following subsections, respectively.

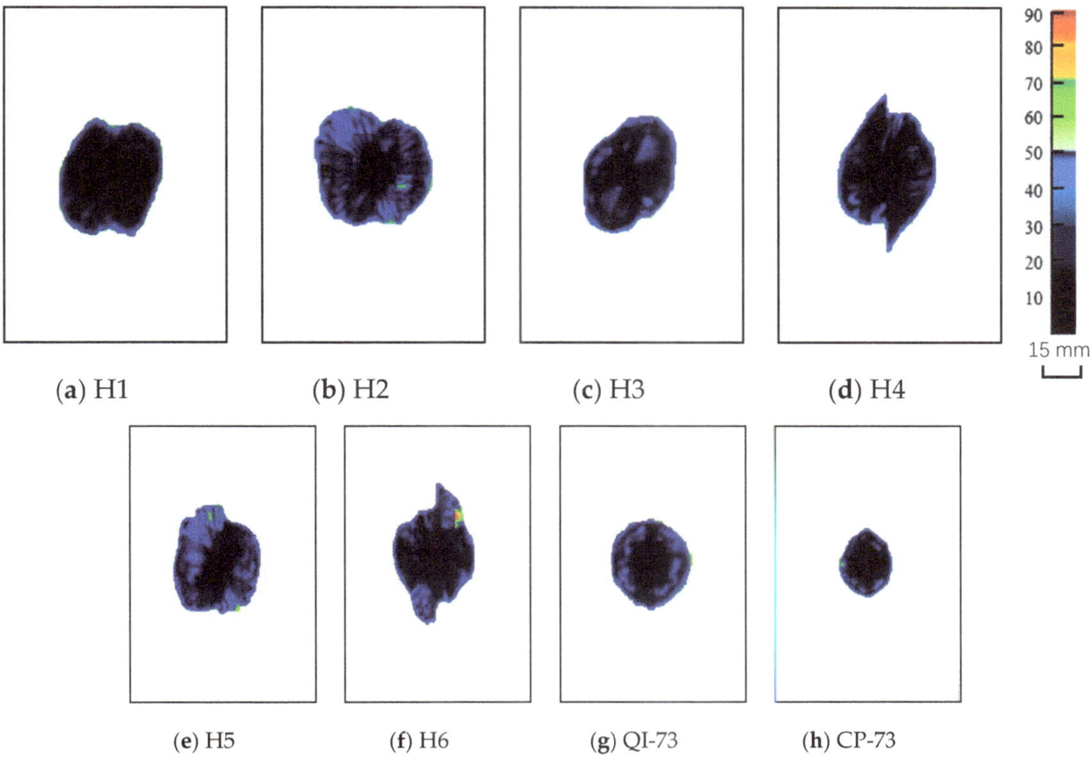

Figure 6. Ultrasonic C-scan images of the composite laminates.

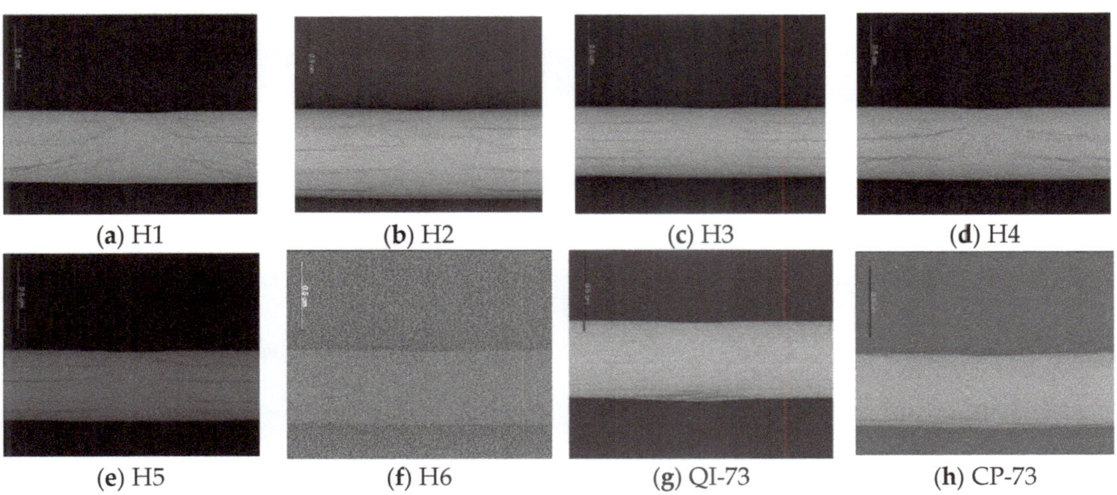

Figure 7. Side cross-sectional view of the central area of laminates.

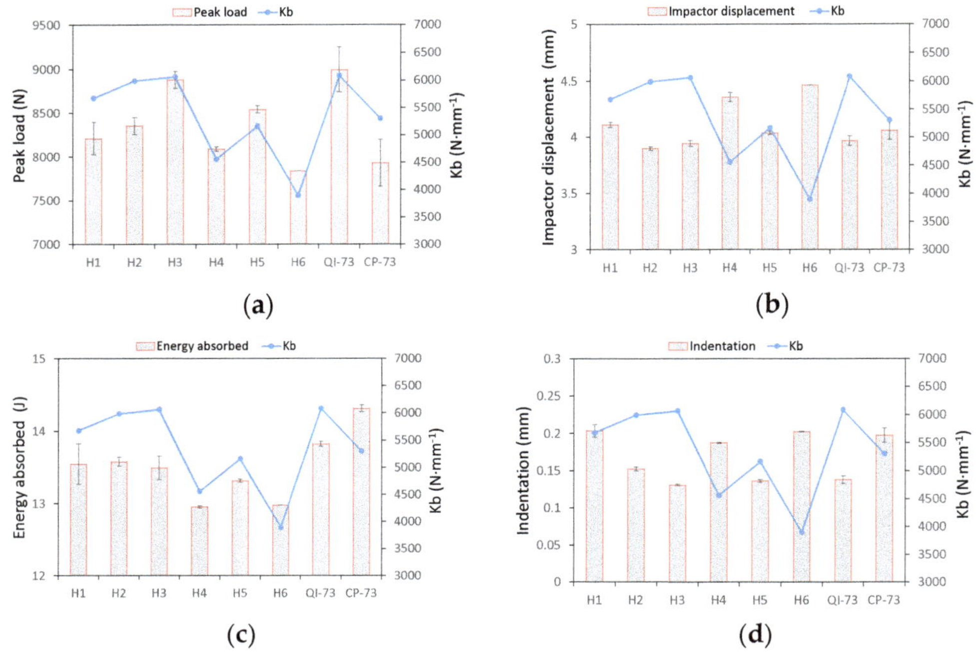

Figure 8. Relationships between (**a**) peak load, (**b**) impactor displacement, (**c**) energy absorbed, and (**d**) indentation of impact tests and K_b of samples.

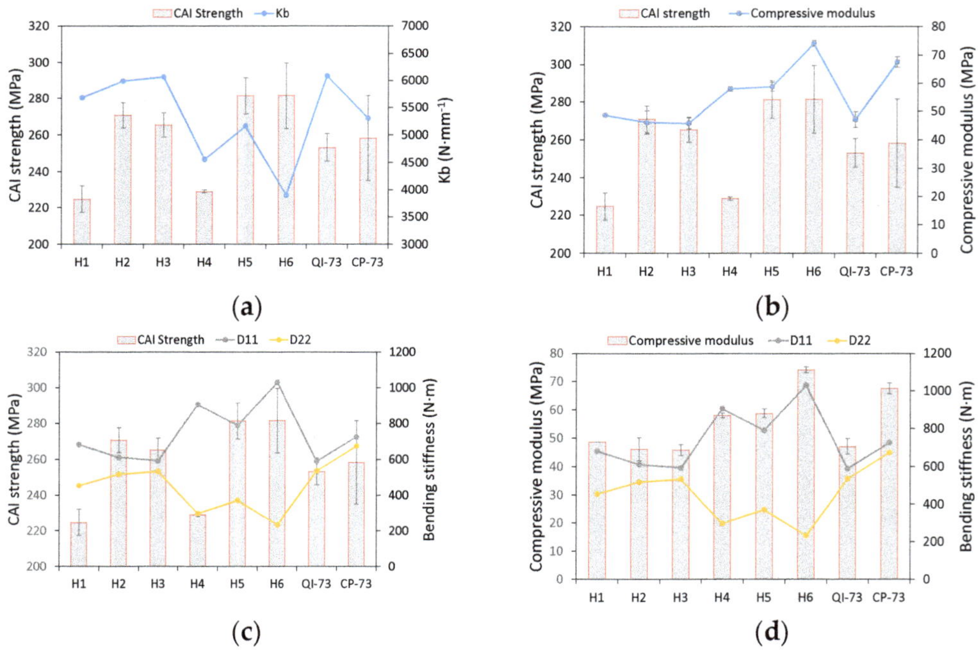

Figure 9. Relationship between CAI strength and (**a**) K_b, (**b**) compressive modulus, (**c**) bending stiffness, and (**d**) relationship between compressive modulus and bending stiffness.

3.2. Impact Load

According to the lumped-parameter model [36], the impact load is affected by stiffness. The stiffness is relative to the bending stiffness, membrane stiffness, and contact stiffness, and the bending stiffness is dominated by the stack sequence. The spring stiffness in the lumped-parameter model related to the laminate bending stiffness (i.e., the main part of transverse stiffness when the plate is normally applied a concentrated load) is denoted by K_b (in N/m) and may be expressed from Navier's solution [43]:

$$K_b = \frac{\overline{A}\pi^4}{4b^2} \frac{1}{\sum_{m=1}^{\infty} \sum_{n=1}^{\infty} \frac{1}{\left[D_{11}\left(\frac{m}{\overline{A}}\right)^4 + \frac{2(D_{12}+2D_{66})m^2n^2}{\overline{A}^2} + D_{22}n^4\right]}} \quad (1)$$

where $\overline{A} = a/b$, a and b are the length and width of the rectangle boundary of the laminate sample, and are equal to 125 mm and 75 mm, respectively, D_{ij} represents the elements of the bending stiffness matrix; and m and n are odd. Equation (1) implies that the coefficients D_{16} and D_{26} are zero; however, they are not equal to zero for some helicoidal samples used in this study. Though Equation (1) is not accurate for all samples, it was employed to evaluate the effect of the ply setup on the impact response. For each laminate, K_b was calculated and plotted (Figures 8 and 9).

The transverse stiffness affects the response when the laminate is subjected to a concentrated load. Figure 8a shows that the trend of peak load follows that of K_b, and it is similar to that of the energy absorption due to damage (Figure 8c) as a higher load leads to serious damage. A higher peak load coincides with a lower impactor displacement when the impact energy is the same; therefore, the trend of impactor displacement is opposite to that of K_b (Figure 8b). It can be seen from curves shown in Figure 4b that H4 and H6 have a lower peak load and a larger displacement, which also have the lower K_b.

The indentation is affected primarily by transverse modulus and matrix cracks. It decreases with an increase in the pitch angle of the LTA laminates: H1, H2, and H3, as shown in Figure 8d, because the small angle plies are easy to be pushed away by the hemispheric impactor tip [19]. Consequently, H4 and H5 also have a high indentation, because of five 0° plies arranged on the surfaces. The specimen CP-73 has a higher indentation, which may be caused by local damage. A higher indentation decreases the contact load, thus leading to the opposite trend of the two parameters for each specimen, as shown in Figure 8a,d. Figure 8d indicates that the trend of indentation is also opposite to that of K_b. Even though a high stiffness causes a higher peak load and a deeper indentation, the indentation is more dependent on the ply setup near the surface; hence, a small pitch angle leads to deeper indentation.

The trend of CAI strength is irrelative to that of bending stiffness (Figure 9a) and compressive modulus (Figure 9b), which is a result of the complicated damage initiation and its propagation.

The transverse stiffness which allows laminate to resist the impact load, and to which the peak load is mainly relative, was studied. Transverse stiffness is defined as the ratio of the impact load and the displacement of the impactor measured in the test; it is also a result of bending stiffness, membrane stiffness, and contact stiffness. During impact, delamination and other damages occur, thus leading to a decrease in transverse stiffness. Therefore, two values of transverse stiffness corresponding to before and after delaminating, namely K_0 and K_d, respectively, are introduced to demonstrate the behaviors of the laminate. The load curve versus the displacement of the impactor on H1, having two segments of different slopes in the loading procedure, is shown in Figure 10. The slopes were determined by linear fitting. The stiffness of each laminate and the residual stiffness in percentage is shown in Figure 11a. A comparison between transverse stiffness and bending stiffness is shown in Figure 11b.

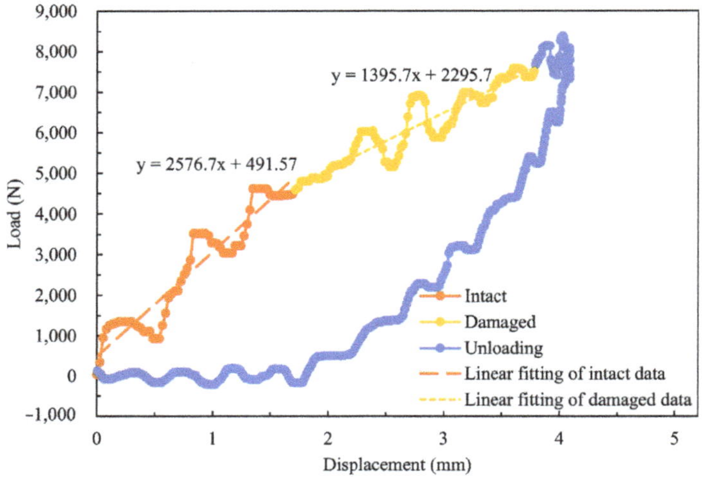

Figure 10. Load–displacement curve for the determination of transverse stiffness for sample H1.

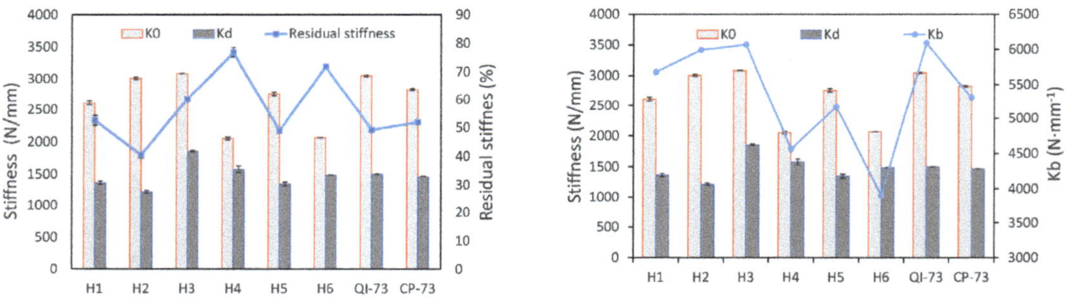

(a) (b)

Figure 11. Tendency of transverse stiffness with (**a**) residual stiffness in percentage and (**b**) K_b.

The transverse stiffness K_0 increases with an increase in twist angle of LTA specimens: H1, H2, and H3 (Figure 11a); however, the residual stiffness, K_d, does not. Specimens H4 and H6 have a higher percentage of residual stiffness, because the initial stiffness is low and more impact energy is absorbed by elastic deflection rather than by damage growth (Figure 8c). The LTA laminates, namely H1, H2, H3, QI-73, and CP-73, possess higher values of K_0. Laminate H5 also has a high K_0 because the plies in 0° direction were set on a depth of 2/7 and 5/7 beside the two surfaces, and the contact stiffness is high when a small indentation is observed (Figure 8d). These factors make H5 possess a higher transverse stiffness than the other two NLTA specimens. The higher transverse stiffness of H5 allows it to absorb more energy compared with H4 and H6 upon damage growth (Figure 8c). Additionally, the percentage of residual stiffness in H5 is lower, which indicates serious damage (Figure 11a); K_0 goes with K_b roughly, however K_d does not (Figure 11b).

3.3. Damage Area

The helicoidal damage is evident from the ultrasonic C-scan images of H1 (Figure 6a; the figure reveals two sectors of damage, which are the projections of two cycles of twisting Bouligand-like cracks shown by the CT image (Figure 7a)). Similarly, four and eight cycles of twisting Bouligand-like cracks are presented in H2 and H3, respectively; hence, the boundary of their C-scan images is smooth.

The C-scan images of NLTA specimen, H4, and H6 indicate that the damage is extended along the 0° direction owing to multiple 0° layers on the damage-prone upper and lower surfaces. The 3D reconstruction of the CT image of H4, shown in Figure 12a, shows multiple cracks in the 0° direction on the back. Additionally, a large crack on the distal face is observed in the images shown in Figure 5d. These cracks expand along the 0° layer, and then develop in a spiral shape. The cracks in the 0° direction form a relatively large projection area in the C-scan image, so the number of continues 0° plies on the impact surface should be limited.

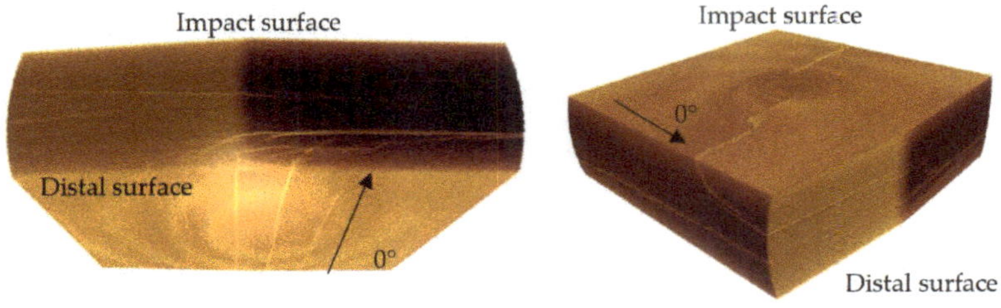

(a) (b)

Figure 12. Cracks on (**a**) the distal face of H4 and (**b**) on the impact face of H1.

The specimen with the smallest pitch angle, H1, has no such long crack on the back of the impact point. A crack in the 90° orientation appears near the impact point on the surface, see Figure 5a. The CT scan image, shown in Figure 12b, indicates that there is only one crack on the impact surface that develops into a spiral crack after expanding to a certain depth.

The damage areas from C-scanning of QI-73 and CP-73 are small; however, Figure 8c indicates that as these two test pieces absorbed more energy due to damage, the damage is widely distributed. The CT scan section image near the impact point shown in Figure 7 shows that the delamination of QI-73 and CP-73 is very dense. The delamination of spiral laminates is related to the number of cycles and H1 has only two cycles. However, the crack could easily propagate to the adjacent matrix owing to the small pitch angle. Specimen H2 has four cycles of delamination, whereas H3 has eight cycles. The non-linear layups, H4 and H6, have delamination similar to that of H1 and also have scattered cracks. The delamination position of specimen H5 is similar to that of H2.

3.4. Analysis on Damage Shape with Strain Energy Release Rate and Shear Stress

The role of pitch angle in the growth of twisting Bouligand-like cracks was analyzed using numerical methods. Suksangpanya et al. [10,11] developed a model for local stress intensity factors along the crack front of a continuously twisting crack in a Bouligand medium, originating from an initially flat crack from the distal face due to the tensile stresses developed from bending (Figure 5 in [10]). Mencattelli et al. [12] computed the strain energy release rate for the growth of twisting Bouligand-like cracks using this model and demonstrated the relationship between the damage feature and pitch angle and the ply thickness. The distribution of maximum τ_{23} stress calculated using the FE model shows the damage distribution in the Bouligand CFRP thin-ply laminates. In this section, the FE results are also presented.

According to Figure 5 in [10], the twisting of the crack front can be represented by the angle (ϕ) between Z and z', and for a fiber-reinforced laminate with an individual ply thickness of t_{ply}, it is given by:

$$\phi = X\left(\frac{\Delta\theta}{t_{\text{ply}}}\right) \tag{2}$$

In addition, kinking of the crack is represented by the effective kink angle α^*, which is defined as the angle between y^* and Y, where y^* is obtained by rotating y about X by an angle—ϕ, and can be expressed as:

$$\alpha^* = \cos^{-1}\left\{\frac{\cos\phi\left[1+\tan^2\phi\right]}{\sqrt{\left[(Z\Delta\theta/t_{\text{ply}})\sec^2\phi\right]^2+1+\tan^2\phi}}\right\} \tag{3}$$

In the context of four assumptions taken in their study, Suksangpanya et al. [10] provided the energy release rate (G) at each point on the twisting crack front, and normalized to the strain energy release rate (G_0) in the case of the crack remaining flat, as follows:

$$\frac{G}{G_0} = \left(\frac{k'_I}{K_I}\right)^2 + \left(\frac{k'_{II}}{K_I}\right)^2 + \frac{1}{1-v}\left(\frac{k'_{III}}{K_I}\right)^2 \tag{4}$$

where v is the Poisson ratio of the isotropic medium, and the local stress intensity factors along the crack front k_{I}, k_{II}, and k_{III} are expressed as:

$$\begin{aligned}\frac{k'_I}{K_I} &= \cos\left(\frac{\alpha^*}{2}\right)\left[\cos^2\left(\frac{\alpha^*}{2}\right)\cos^2\phi + 2v\sin^2\phi\right]\\ \frac{k'_{II}}{K_I} &= \sin\left(\frac{\alpha^*}{2}\right)\cos^2\left(\frac{\alpha^*}{2}\right)\cos\phi\\ \frac{k'_{III}}{K_I} &= \cos\left(\frac{\alpha^*}{2}\right)\sin\phi\cos\phi\left[\cos^2\left(\frac{\alpha^*}{2}\right) - 2v\right]\end{aligned} \tag{5}$$

Assuming Z = 0.5 mm, the normalized energy release rate from the 1st to 37th ply of every sample was calculated (Figure 13). Evidently, the energy release rate decreased with an increase in pitch angle H1, H2, and H3. A higher pitch angle tends to resist the propagation of intra-lamina cracks, thus leading to more catastrophic mechanisms of failure such as delamination and fiber breaks. Samples H4, H5, and H6 possessed several consecutive 0° plies, thus leading to a higher energy release rate at the relative layers of consecutive 0° plies. Except at consecutive 0° plies, the energy release rates of H4 and H5 coincided with that of H1 and H2, respectively. H6 had a pitch angle of 2.5° for some plies, thus leading to a higher energy release rate than others at most positions, followed by H4, with a pitch angle of 5°. The energy absorption due to damage in H4 and H6 is lower than that of other specimens (Figure 8c) owing to their higher energy dissipation due to scatter cracks and lower transverse stiffness.

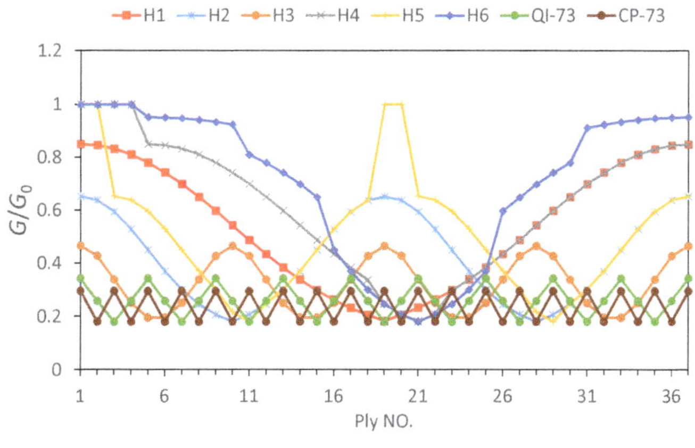

Figure 13. Energy release rate (G/G_0) vs. X evaluated at Z = 0.5 mm for each ply of the specimen with a different ply sequence.

The distribution of the maximum intra-ply shear stress ($|\tau_{23}|$) within the test samples was used to demonstrate the delamination development following the method outlined by Mencattelli et al. [12]. Six static linear elastic models were developed for each helicoidal composite specimen using the FE software Abaqus 2020. Geometries of both the laminate and indenter used were the same as the ones used in actual testing conditions. The material properties assigned to each lamina are reported in Table 1. Solid elements (C3D8R) were used to model the laminate (ply-by-ply using lamination theory) and a rigid surface for the indenter (Figure 14), with a minimum size of 1.4 mm. The mesh was refined in the central region of the laminate, with a minimum size of 0.2 mm after mesh convergence analysis. Analysis showed that large mesh sizes, such as 0.4 mm, result in location discrepancies in the maximum shear stress at each layer, especially if the deviation angle is small. Integration points for adjacent layers may overlap, resulting in approximations and reduced accuracy, with a maximum stress point to a distance error of up to 19%. A frictionless hard contact was created between the upper surface of the composite laminate and the outer surface of the indenter.

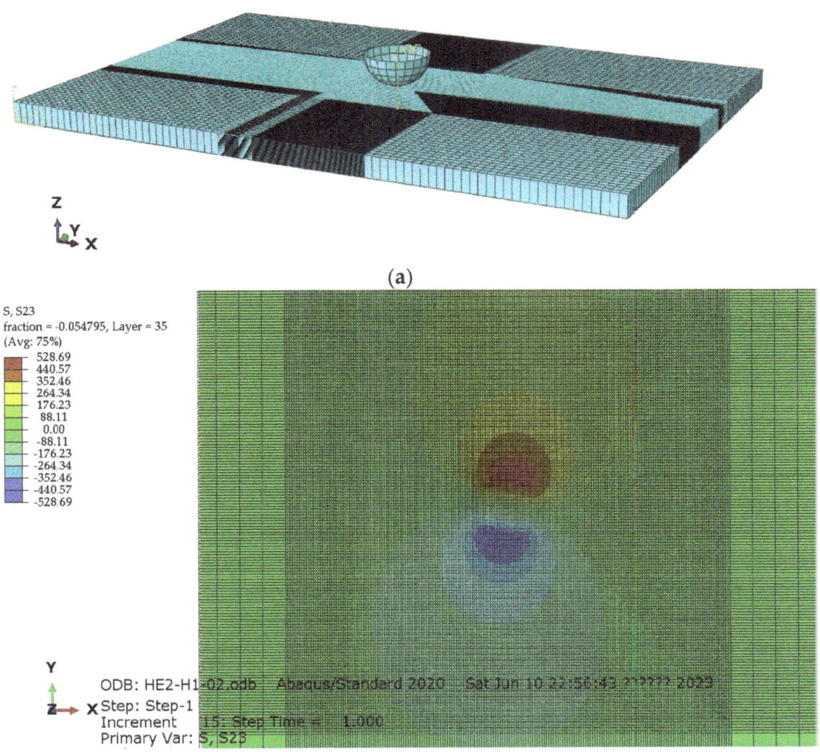

Figure 14. FE model of the specimen for the simulation of quasi-static indentation test: (**a**) global view to the model and (**b**) shear stress distribution of a ply in H1.

Post-processing was performed using a Python script to extract the maximum shear stress and corresponding the integration point locations from each ply of the laminate. The locations of the maximum $|\tau_{23}|$ are plotted in 3D mode and also projected to the plane of the laminate (Figure 15), where 0° ply is aligned with the x-axis. For each ply, Figure 15 shows the likely location of the shearing cracks, i.e., the points with the maximum $|\tau_{23}|$. In each ply, $|\tau_{23}|$ exhibits two identical maximum values. The loci of the points with maximum $|\tau_{23}|$ form a double helix in H1, H2, and H3 (Figure 15a–c). The double helices are symmetric about the laminate symmetry plane. Some drastic changes in the locations of the maximum $|\tau_{23}|$ in H4 and H6 are due to the consecutive 0° plies. When observed perpendicularly to the plane of the laminate, the out profile of the loci of the maximum $|\tau_{23}|$ reflects the shape of C-scan image shown in Figure 6. The sector delaminated region is obvious for H1 and H2, and the protruding part on C-scan image of H4 and H6 due to the multiple 0° ply is clearly shown by $|\tau_{23}|$.

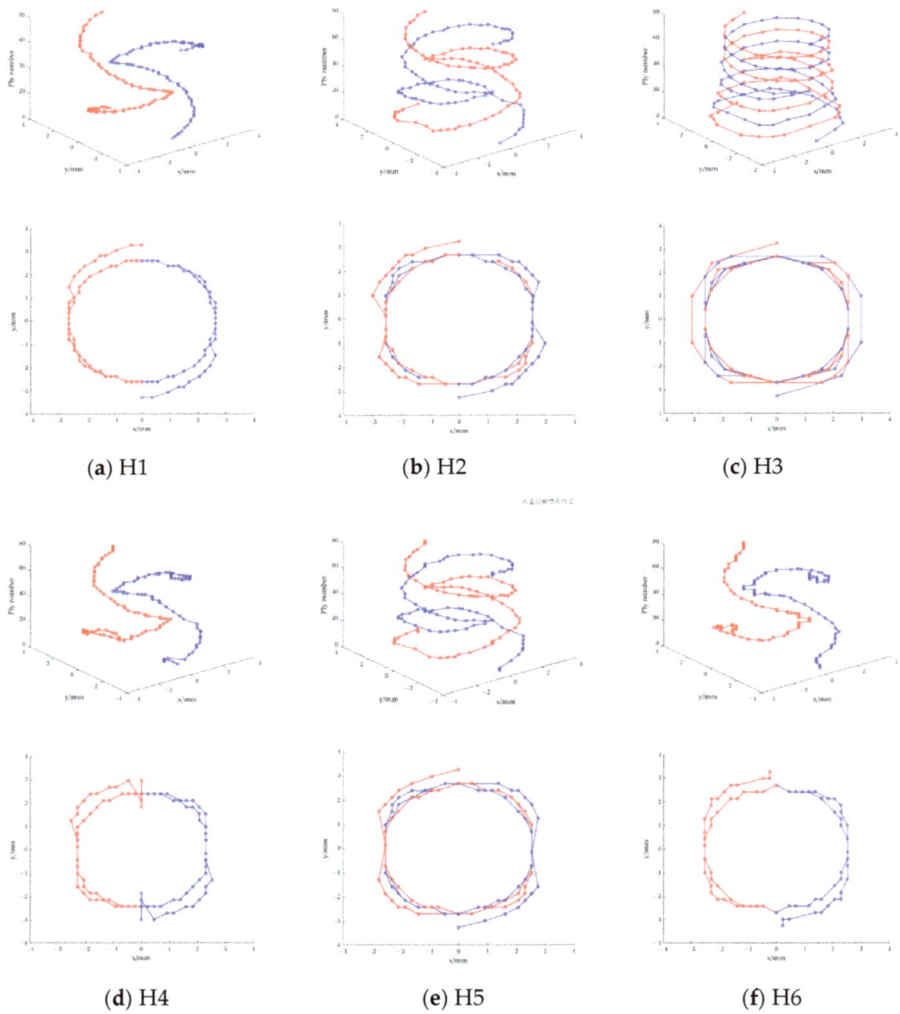

Figure 15. Loci of maximum $|\tau_{23}|$ (**upper**) and its projection on the plane of the laminates (**lower**).

3.5. CAI Characteristics

Figure 9a indicates that the relationship between CAI strength and K_d is uncertain. The laminates with high compressive modulus possess high CAI strength except for H1 and H4. The pitch angle of H1 is 5°, whereas H4 has two pitch angles of 5° and 10°. The constraints between the plies with small pitch angles are weak. The indentation of H1 is also greater than that of H2 and H3 due to the weak interlaminar constraints, which means the more serious local damage or fiber distortion; hence, it is more susceptible to damage during compression. This could be regarded as a characteristic of small pitch-angled composites.

CAI strength is less affected by the extent of impact damage here, as shown in Figure 8c. The energy absorbed due to damage is nearly equal for H1, H2, and H3; however, their CAI strengths (and that of H4 and H6) are different.

To further investigate the characteristics of specimens, the critical buckling load, $N_{x,cr}$, applied on the 150 mm × 100 mm specimen with four edges, simply supported along the longer edge, was computed using the FE model, as listed in the last row of Table 2. Note that the value of the critical load is meaningful only numerically because the stresses corresponding to the buckling loads have exceeded the strength of the material in the stability analysis of plates of such sizes. H2 and H3 have a higher buckling load than H1 and therefore have the higher CAI strength under the same damage extent.

Figure 9b shows that the compressive modulus of H4 is higher than that of H2 and H3 because H4 has more 0° plies; however, its CAI strength is lower. Samples H1, H2, H3, and QI-73 are quasi-isotropic laminates; thus, the compressive modulus is similar. However, the CAI strengths of H2 and H3 are higher than that of QI-73, thereby indicating the better performance of helicoidal composites with 10° and 20° pitch angles.

The mean CAI strengths of H5 and H6 are higher than those of QI-73 and CP-73. Its value for H5 is slightly higher than that of H6; however, the dispersion of CAI strength in H6 is higher. H5 has pitch angles of 10° and 20°, whereas H6 has 2.5°, 5°, and 10°. H6 has the highest compressive modulus and lower transverse stiffness (Figure 11a) to resist the impactor. This results in a low peak load (Figure 8a) and more energy absorption by elastic deformation than by damage (Figure 8c), eventually leading to a high CAI strength.

The stiffness, k_b, which is relative to transverse stiffness, is calculated from the bending stiffness coefficients D_{11} and D_{22}. Therefore, CAI strength and compressive modulus are plotted against D_{11} and D_{22} (Figure 9c,d). A definite relationship between CAI and bending stiffness coefficients is not obtained from Figure 9c. In Figure 9d, the compressive modulus roughly follows D_{11}, as they are strongly affected by the number of plies in the 0° orientation.

The transverse stiffness and peak load of H5 are higher than that of H6. The CAI strength and energy absorbed due to damage are also higher in H5 compared with H6; however, the compressive modulus of H5 is lower than that of H6. This is attributed to the proper pitch angles of 10° and 20°, and the enhancement of CAI strength by 0° orientation plies in H5 despite being less in number compared to H6. The critical buckling loads are 2100.8 N/mm and 1569.2 N/mm for sample H5 and H6, respectively; the results revealed that H5 is superior to H6 in terms of structural stability.

4. Conclusions

Helicoidal composite laminates made from carbon-epoxy prepreg with plies reinforcing the loading capability in 0° direction were studied using low-velocity impact tests and compressive tests. In this study, we can find that:

- The laminates having pitch angles of 10° and 20° were found to possess higher CAI strengths. However, having a large number of 0° plies yielded a larger damage area because of the easy propagation of cracks in them. The continuous 0° plies on the impact surface also yielded larger indentation, thereby leading to a decrease in the transverse stiffness and the impact load. Lower impact load resulted in less damage, which could be estimated from the energy absorption.
- The specimen with the plies of 10° and 20° pitch angles and continuous 0° plies (H5) yielded the highest CAI strength despite more damage than the one with more plies in 0° and a higher compressive modulus (H6), because a small pitch angle (5°) should be avoided due to the weak constraint between adjacent plies. Therefore, a set of proper pitch angles (10° and 20°) and proper arrangement of 0° plies are the key parameters for CAI strength.
- Analysis on the strain energy release rate and maximum $|\tau_{23}|$ could be beneficial in the prediction of damage characteristics.

- The first ply in all specimens is of 0°; therefore, a method should be proposed to evaluate the effect of the first ply direction on its performance. Non-symmetric and non-uniform layup with low coupling stiffness could be optimized to obtain better performance.
- The mechanism of ply pitch angle affecting impact load, indentation, damage, CAI strength, and critical buckling load needs further investigation to obtain the optimal ply setups for a variety of services.

Author Contributions: Conceptualization, writing—original draft, writing—review and editing, and funding acquisition, Z.Y.; investigation and experiment, X.D., Q.X. and X.Z.; simulation and validation, R.L. and Q.Z. All authors have read and agreed to the published version of the manuscript.

Funding: This research was funded by National Natural Science Foundation of China, grant number 12072199.

Institutional Review Board Statement: Not applicable.

Informed Consent Statement: Not applicable.

Data Availability Statement: Not applicable.

Conflicts of Interest: The authors declare no conflict of interest in re-searching, authorship, and/or publication of this article.

References

1. Abrate, S. Impact on Laminated Composite Materials. *Appl. Mechan. Rev.* **1991**, *44*, 155–190. [CrossRef]
2. Nilsson, K.-F.; Asp, L.E.; Alpman, J.E.; Nystedt, L. Delamination Buckling and Growth for Delaminations at Different Depths in a Slender Composite Panel. *Int. J. Solids Struct.* **2001**, *38*, 3039–3071. [CrossRef]
3. Bouligand, Y. Twisted Fibrous Arrangements in Biological Materials and Cholesteric Mesophases. *Tissue Cell* **1972**, *4*, 189–217. [CrossRef] [PubMed]
4. Apichattrabrut, T.; Ravi-Chandar, K. Helicoidal Composites. *Mechan. Adv. Mater. Struct.* **2006**, *13*, 61–76. [CrossRef]
5. Weaver, J.C.; Milliron, G.W.; Miserez, A.; Evans-Lutterodt, K.; Herrera, S.; Gallana, I.; Mershon, W.J.; Swanson, B.; Zavattieri, P.; DiMasi, E.; et al. The Stomatopod Dactyl Club: A Formidable Damage-Tolerant Biological Hammer. *Science* **2012**, *336*, 1275–1280. [CrossRef] [PubMed]
6. Zimmermann, E.A.; Gludovatz, B.; Schaible, E.; Dave, N.K.N.; Yang, W.; Meyers, M.A.; Ritchie, R.O. Mechanical Adaptability of the Bouligand-Type Structure in Natural Dermal Armour. *Nat. Commun.* **2013**, *4*, 2634. [CrossRef] [PubMed]
7. Cheng, L.; Wang, L.; Karlsson, A.M. Mechanics-Based Analysis of Selected Features of the Exoskeletal Microstructure of *Popillia japonica*. *J. Mater. Res.* **2009**, *24*, 3253–3267. [CrossRef]
8. Grunenfelder, L.K.; Suksangpanya, N.; Salinas, C.; Milliron, G.; Yaraghi, N.; Herrera, S.; Evans-Lutterodt, K.; Nutt, S.R.; Zavattieri, P.; Kisailus, D. Bio-Inspired Impact-Resistant Composites. *Acta Biomater.* **2014**, *10*, 3997–4008. [CrossRef]
9. Yaraghi, N.A.; Guarín-Zapata, N.; Grunenfelder, L.K.; Hintsala, E.; Bhowmick, S.; Hiller, J.M.; Betts, M.; Principe, E.L.; Jung, J.; Sheppard, L.; et al. A Sinusoidally Architected Helicoidal Biocomposite. *Adv. Mater.* **2016**, *28*, 6835–6844. [CrossRef]
10. Suksangpanya, N.; Yaraghi, N.A.; Kisailus, D.; Zavattieri, P. Twisting Cracks in Bouligand Structures. *J. Mechan. Behav. Biomed. Mater.* **2017**, *76*, 38–57. [CrossRef]
11. Suksangpanya, N.; Yaraghi, N.A.; Pipes, R.B.; Kisailus, D.; Zavattieri, P. Crack Twisting and Toughening Strategies in Bouligand Architectures. *Int. J. Solids Struct.* **2018**, *150*, 83–106. [CrossRef]
12. Mencattelli, L.; Pinho, S.T. Realising Bio-Inspired Impact Damage-Tolerant Thin-Ply CFRP Bouligand Structures via Promoting Diffused Sub-Critical Helicoidal Damage. *Compos. Sci. Technol.* **2019**, *182*, 107684. [CrossRef]
13. Mencattelli, L.; Pinho, S.T. Herringbone-Bouligand CFRP Structures: A New Tailorable Damage-Tolerant Solution for Damage Containment and Reduced Delaminations. *Compos. Sci. Technol.* **2020**, *190*, 108047. [CrossRef]
14. Mencattelli, L.; Pinho, S.T. Ultra-Thin-Ply CFRP Bouligand Bio-Inspired Structures with Enhanced Load-Bearing Capacity, Delayed Catastrophic Failure and High Energy Dissipation Capability. *Compos. Part A Appl. Sci. Manufact.* **2020**, *129*, 105655. [CrossRef]
15. Liu, J.L.; Lee, H.P.; Tan, V.B.C. Effects of Inter-Ply Angles on the Failure Mechanisms in Bioinspired Helicoidal Laminates. *Compos. Sci. Technol.* **2018**, *165*, 282–289. [CrossRef]
16. Liu, J.L.; Lee, H.P.; Lai, K.S.; Tan, V.B.C. Bio-Inspired Laminates of Different Material Systems. *J. Appl. Mechan.* **2020**, *87*, 31007. [CrossRef]
17. Liu, J.L.; Lim, E.W.L.; Sun, Z.P.; Wang, J.; Tay, T.E.; Tan, V.B.C. Improving Strength and Impact Resistance of 3D Printed Components with Helicoidal Printing Direction. *Int. J. Impact Eng.* **2022**, *169*, 104320. [CrossRef]

18. Liu, J.L.; Lee, H.P.; Kong, S.H.R.; Tan, V.B.C. Improving Laminates through Non-Uniform Inter-Ply Angles. *Composites Part A Appl. Sci. Manufact.* **2019**, *127*, 105625. [CrossRef]
19. Jiang, H.; Ren, Y.; Liu, Z.; Zhang, S.; Lin, Z. Low-Velocity Impact Resistance Behaviors of Bio-Inspired Helicoidal Composite Laminates with Non-Linear Rotation Angle Based Layups. *Compos. Struct.* **2019**, *214*, 463–475. [CrossRef]
20. Pinto, F.; Iervolino, O.; Scarselli, G.; Ginzburg, D.; Meo, M. Bioinspired Twisted Composites Based on Bouligand Structures. In Proceedings of the SPIE Smart Structures and Materials + Nondestructive Evaluation and Health Monitoring, Las Vegas, NE, USA, 15 April 2016; Martín-Palma, R.J., Lakhtakia, A., Knez, M., Eds.; SPIE: Las Vegas, NE, USA, 2016; p. 97970E.
21. Ginzburg, D.; Pinto, F.; Iervolino, O.; Meo, M. Damage Tolerance of Bio-Inspired Helicoidal Composites under Low Velocity Impact. *Compos. Struct.* **2017**, *161*, 187–203. [CrossRef]
22. Sebaey, T.A.; González, E.V.; Lopes, C.S.; Blanco, N.; Maimí, P.; Costa, J. Damage Resistance and Damage Tolerance of Dispersed CFRP Laminates: Effect of the Mismatch Angle between Plies. *Compos. Struct.* **2013**, *101*, 255–264. [CrossRef]
23. Sebaey, T.A.; González, E.V.; Lopes, C.S.; Blanco, N.; Costa, J. Damage Resistance and Damage Tolerance of Dispersed CFRP Laminates: Effect of Ply Clustering. *Compos. Struct.* **2013**, *106*, 96–103. [CrossRef]
24. Wagih, A.; Maimí, P.; Blanco, N.; García-Rodríguez, S.M.; Guillamet, G.; Issac, R.P.; Turon, A.; Costa, J. Improving Damage Resistance and Load Capacity of Thin-Ply Laminates Using Ply Clustering and Small Mismatch Angles. *Compos. Part A Appl. Sci. Manufact.* **2019**, *117*, 76–91. [CrossRef]
25. Sebaey, T.A.; González, E.V.; Lopes, C.S.; Blanco, N.; Costa, J. Damage Resistance and Damage Tolerance of Dispersed CFRP Laminates: The Bending Stiffness Effect. *Compos. Struct.* **2013**, *106*, 30–32. [CrossRef]
26. Sebaey, T.A.; Mahdi, E. Using Thin-Plies to Improve the Damage Resistance and Tolerance of Aeronautical CFRP Composites. *Compos. Part A Appl. Sci. Manufact.* **2016**, *86*, 31–38. [CrossRef]
27. Saito, H.; Morita, M.; Kawabe, K.; Kanesaki, M.; Takeuchi, H.; Tanaka, M.; Kimpara, I. Effect of Ply-Thickness on Impact Damage Morphology in CFRP Laminates. *J. Reinforced Plast. Compos.* **2011**, *30*, 1097–1106. [CrossRef]
28. Yuan, Y.; Zhang, X.; Li, X.; Zhang, Q.; Yin, Q.; Liu, W.; Zhang, Z. Manipulating Impact Damage Modes in Composite Laminates by Helical Pitch Angle and Ply Thickness. *Eng. Fract. Mechan.* **2022**, *265*, 108383. [CrossRef]
29. Nettles, A.T.; Sabo, S. Compression after Impact Strength of Thin Laminates with Various Percentage of 0° Plies. *J. Compos. Mater.* **2014**, *48*, 345–354. [CrossRef]
30. Sasikumar, A.; Trias, D.; Costa, J.; Blanco, N.; Orr, J.; Linde, P. Effect of Ply Thickness and Ply Level Hybridization on the Compression after Impact Strength of Thin Laminates. *Compos. Part A Appl. Sci. Manufact.* **2019**, *121*, 232–243 [CrossRef]
31. *ASTM D7136*; D30 Committee Test Method for Measuring the Damage Resistance of a Fiber-Reinforced Polymer Matrix Composite to a Drop-Weight Impact Event. ASTM International: West Conshohocken, PE, USA, 2020.
32. *ASTM D7137*; D30 Committee Test Method for Compressive Residual Strength Properties of Damaged Polymer Matrix Composite Plates. ASTM International: West Conshohocken, PE, USA, 2017.
33. Olsson, R. Mass Criterion for Wave Controlled Impact Response of Composite Plates. *Compos. Part A Appl. Sci. Manufact.* **2000**, *31*, 879–887. [CrossRef]
34. Zhu, Q.; Yu, Z. A Perturbation-Based Model for the Prediction of Responses Involving Delamination during Small Mass Impacts on Orthotropic Composite Plates. *Compos. Sci. Technol.* **2021**, *208*, 108754. [CrossRef]
35. Abrate, S. *Impact on Composite Structures*; Cambridge University Press: Cambridge, MA, USA, 1998; ISBN 0-521-47389-6.
36. Abrate, S. Modeling of Impacts on Composite Structures. *Compos. Struct.* **2001**, *51*, 129–138. [CrossRef]
37. Johnson, K.L.; Johnson, K.L. *Contact Mechanics*; Cambridge University Press: Cambridge, MA, USA, 1987; ISBN 0-521-34796-3.
38. Tan, T.M.; Sun, C.T. Use of Statical Indentation Laws in the Impact Analysis of Laminated Composite Plates. *J. Appl. Mech.* **1985**, *52*, 6–12. [CrossRef]
39. Yu, Z.; Gao, S. Increase of Contact Radius Due to Deflection in Low Velocity Impact of Composite Laminates and Prediction of Delamination Threshold Load. *Compos. Struct.* **2016**, *147*, 286–293. [CrossRef]
40. Olsson, R. Analytical Prediction of Damage Due to Large Mass Impact on Thin Ply Composites. *Compos. Part A Appl. Sci. Manufact.* **2015**, *72*, 184–191. [CrossRef]
41. Belardi, V.G.; Fanelli, P.; Vivio, F. Bending Analysis with Galerkin Method of Rectilinear Orthotropic Composite Circular Plates Subject to Transversal Load. *Compos. Part B Eng.* **2018**, *140*, 250–259. [CrossRef]
42. Belardi, V.G.; Fanelli, P.; Vivio, F. Application of the Ritz method for the bending and stress analysis of thin rectilinear orthotropic composite sector plates. *Thin Walled Struct.* **2023**, *183*, 110374. [CrossRef]
43. Barbero, E.J. *Introduction to Composite Materials Design*; CRC Press: Boca Raton, FL, USA, 2010; ISBN 0-429-10947-4.
44. Li, D.H.; Liu, Y.; Zhang, X. Low-velocity impact responses of the stiffened composite laminated plates based on the progressive failure model and the layerwise/solid-elements method. *Compos. Struct.* **2014**, *110*, 249–275. [CrossRef]
45. Wang, J.; Xiao, J. Analytical solutions of bending analysis and vibration of rectangular nano laminates with surface effects. *Appl. Math. Modell.* **2022**, *110*, 663–673. [CrossRef]

materials

MDPI

Article

Optimization Design and Nonlinear Bending of Bio-Inspired Helicoidal Composite Laminated Plates

Taoye Lu [1,2,3], Hui-Shen Shen [2], Hai Wang [2], Xiuhua Chen [1,2,*] and Miaolin Feng [3]

1 Zhe Jiang Key Laboratory of General Aviation Operation Technology, General Aviation Institute of Zhejiang Jiande, Jiande 311600, China; lutaoye@sjtu.edu.cn
2 School of Aeronautics and Astronautics, Shanghai Jiao Tong University, Shanghai 200240, China; hsshen@sjtu.edu.cn (H.-S.S.); wanghai601@sjtu.edu.cn (H.W.)
3 School of Ocean and Civil Engineering, Shanghai Jiao Tong University, Shanghai 200240, China; mlfeng@sjtu.edu.cn
* Correspondence: chenxiuhua@sjtu.edu.cn

Abstract: Inspired by the bionic Bouligand structure, helicoidal carbon fiber-reinforced polymer composite (CFRPC) laminates have been proven to own outstanding out-of-plane mechanical properties. This work aims to further explore the excellent bending characteristics of helicoidal CFRPC laminated plates and find out the optimal helicoidal layup patterns. The optimization design of laminated plates stacked with single-form and combination-form helicoidal layup sequences are carried out by using the finite element method (FEM) and adaptive simulated annealing (ASA) optimization algorithm on the Isight platform. Then, the nonlinear bending responses of optimal helicoidal CFRPC laminated plates are investigated via the FEM for the first time. The helicoidal CFRPC laminated plates under three different types of boundary conditions subjected to transverse uniformly distributed load are considered. The numerical results reveal that the combination-form helicoidal layup sequences can decrease the dimensionless bending deflection of laminated plates by more than 5% compared with the quasi-isotropic plate and enhance the out-of-plane bending characteristics of CFRPC laminated plates effectively. The boundary conditions can significantly influence the nonlinear bending responses of helicoidal CFRPC laminated plates.

Keywords: Bouligand structure; simulation-based optimization; large deflection; finite element model

1. Introduction

The Bouligand structure [1], which can be commonly found in various crustacean exoskeletons and insect cuticles [2–4], is the natural helicoidal structure and contributes to high transverse load resistance. Carbon fiber-reinforced polymer composite (CFRPC) laminated plates, stacked by designed unidirectional plies with different layup angles, are widely used in aerospace and marine engineering. Motivated by the natural Bouligand structure, the CFRPC laminated plates are designed to have bio-inspired helicoidal layups, enhancing out-of-plane mechanical performance in recent years.

Many researchers have examined the outstanding out-of-plane mechanical properties of helicoidal composite laminated plates. Apichattrabrut and Ravi-Chandar [5] concluded that the smaller angle change between adjacent plies will lead to lower interlaminar shear stress, which results in better damage tolerance of helicoidal CFRPC laminates than unidirectional and cross-ply laminates under tension, bending and impact load. Cheng et al. [6–8] first explained the mechanical responses of the exoskeleton of selected crustaceans with helicoidal structures. They showed the improvement in the bending and shear properties of helicoidal composite laminates via experiments and theoretical studies. Liu et al. [9,10] pointed out that, for the thicker helicoidal plates, a smaller rotation angle may promote the transverse propagation of cracks and not always result in a higher peak load under transverse load testing [9]. Jiang et al. [11] firstly introduced three nonlinear helicoidal layup arrangements—recursive, exponential and semicircular—into composite laminated plates and proved the enhancement of the impact resistance of recursive and exponential helicoidal plates with increasing rotation angles. Mencattelli and Pinho [12,13] demonstrated the feasibility of using Bouligand structures to improve the out-of-plane mechanical properties of thin-ply and ultra-thin-ply CFRPCs via experimental and numerical investigations. Mohamed et al. [14] compared the bending properties of CFRPC laminated plates stacked with conventional layups, one kind of linear layup and Fibonacci helicoidal layups. In their studies, the influence of the lamination scheme of the plates was discussed numerically.

The effect of helicoidal layups on the enhancement of the out-of-plane properties of laminated plates has been widely proved. However, the comparison works in the published literatures were only conducted on predefined helicoidal patterns. Thus, the helicoidal CFRPC laminated plates with various rotation angles need to be investigated more comprehensively. In order to obtain helicoidal laminated plates with better bending behaviors, optimization work is introduced in this paper. The common optimization objective for the bending problem is minimizing the deflection. Topology optimization was applied by Pedersen [15] to choose the optimal laminated plates subjected to prestresses. Based on a genetic algorithm used together with the finite element method (FEM), Almeida and Awruch [16] obtained optimal composite laminated structures with minimum weight and deflection under a transverse load. Monte et al. [17] used the direct search method to optimize the layup sequences of composite plates. However, to the knowledge of the authors, the layup optimization on bending behaviors of helicoidal CFRPC laminated plates is still lacking. In this paper, the optimization process is performed using the global optimization algorithm, adaptive simulated annealing (ASA) optimization algorithm [18], and the optimization objective of minimizing the bending deflection of helicoidal CFRPC laminated plates is considered.

Additionally, in actual aerospace application, many static problems of composite laminated plates cannot be adequately solved via the classical thin-plate theory only. The geometrically nonlinear responses and transverse shear effect in the large-deflection region of laminated plates have been considered in many studies [19–25]. Based on the first-order shear deformation theory (FSDT), Reddy [26] systematically investigated the nonlinear bending of composite laminated plates. Malekzadeh and Setoodeh [27] successfully carried out the prediction of large deformation on moderately thick laminated plates. For the more accurate shear stress distribution, the higher-order shear deformation theory (HSDT) is developed. Based on Reddy's third-order shear deformation theory (TSDT) [28], a large amount of research on nonlinear bending problems of laminated plates has been reported [29–33]. At the same time, various non-polynomial HSDTs [34–36] have been established to analyze the nonlinear responses of laminated plates in order to establish high computational efficiency. The FEM is frequently used to solve the differential equations and deal with laminated plates under different boundary conditions efficiently [37–39]. Nowadays, the FEM has been widely used in the analysis of various mechanical problems for composite structures [40–43]. However, the nonlinear bending responses of helicoidal CFRPC laminated plates remain unclear, and this also motivates the present study.

In the present work, the out-of-plane bending responses of bio-inspired helicoidal CFRPC laminated plates are examined. Unlike the previous research, the layup optimizations of single-form and combination-form helicoidal patters are conducted, and the nonlinear bending responses of optimal helicoidal CFRPC laminated plates are investigated for the first time. The optimization objective of minimizing the bending deflection for higher out-of-plane bending stiffness under three different boundary conditions is carried out using the ASA algorithm on Isight v5.9-5 software. The calculations of the dimensionless bending deflection in linear analysis for optimization and the central bending deflection and moment for nonlinear bending analysis are carried out by using the FEM with ABAQUS. In this paper, 4 kinds of single-form and 12 types of combination-form linear (L) and nonlinear—i.e., Recursive (R), Fibonacci (F) and Exponential (E)—helicoidal layup configurations are designed. All the helicoidal laminated plates are compared with the conventional quasi-isotropic one, and the optimal helicoidal CFRPC laminated plates are selected for the following nonlinear bending analysis.

2. Modelling of Helicoidal CFRPC Laminated Plates

2.1. Layup Configuration

In this paper, CFRPC laminated plates stacked with two types of single-form and combination-form helicoidal layup configurations are considered. As shown in Figure 1, the helicoidal CFRPC laminated plates contain 16 plies; θ_i refers to the angle between the unidirectional carbon fiber and X-axis, and the rotation angle $\theta_i - \theta_{i-1}$ refers to the angle difference between adjacent plies. As mentioned before, one linear (L) rotation angle and three nonlinear rotation angles patterns, i.e., recursive (R), Fibonacci (F) and exponential (E), are considered in the present study. The L-helicoidal pattern is the most commonly used in previous studies. Apart from that, some nonlinear helicoidal patterns have been reported recently for the reason that in natural biological helicoidal structures, the rotation angles are always non-equal. Thus, linear and nonlinear helicoidal layup patterns, i.e., the R and E helicoidal patterns proposed by Jiang et al. [11] and the F helicoidal pattern designed by Wang et al. [44], which have been proven to provide outstanding out-of-plane mechanical properties for laminates, are considered in this paper. Thereafter, four kinds of single-form helicoidal layup configurations that rotate according to one of the L, R, F or E helicoidal patterns, and twelve kinds of combination-form helicoidal layup configurations that rotate according to two of the L, R, F or E helicoidal patterns, are designed. The expressions of these helicoidal layup configurations can be expressed as Table 1:

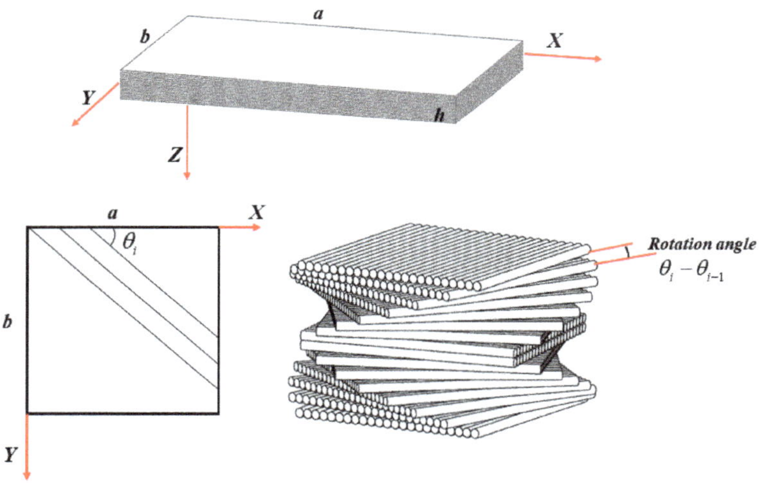

Figure 1. Coordinate system of a helicoidal CFRPC laminated plate.

Table 1. The single-form and combination-form helicoidal layup configurations.

	Single-form helicoidal layup configurations
L	$[0/\alpha/2\alpha/3\alpha/4\alpha/5\alpha/6\alpha/7\alpha]_s \Rightarrow \theta_1 = 0, \theta_i - \theta_{i-1} = \alpha$
R	$[0/\beta/3\beta/6\beta/10\beta/15\beta/21\beta/28\beta]_s \Rightarrow \theta_1 = 0, \theta_i - \theta_{i-1} = \beta(i-1)$
F	$[0/\gamma/\gamma/2\gamma/3\gamma/5\gamma/8\gamma/13\gamma]_s \Rightarrow \theta_1 = 0, \theta_2 = \gamma, \theta_i = \theta_{i-2} + \theta_{i-1}$
E	$[\xi/\xi^2/\xi^3/\xi^4]_{2s} \Rightarrow \theta_i = \xi^i$
	Combination-form helicoidal layup configurations
L + R	$[0/\alpha/2\alpha/3\alpha/0/\beta/3\beta/6\beta]_s$
L + F	$[0/\alpha/2\alpha/3\alpha/0/\gamma/\gamma/2\gamma]_s$
L + E	$[0/\alpha/2\alpha/3\alpha/\xi/\xi^2/\xi^3/\xi^4]_s$

Table 1. *Cont.*

R + L	$[0/\beta/3\beta/6\beta/0/\alpha/2\alpha/3\alpha]_s$
R + F	$[0/\beta/3\beta/6\beta/0/\gamma/\gamma/2\gamma]_s$
R + E	$[0/\beta/3\beta/6\beta/\xi/\xi^2/\xi^3/\xi^4]_s$
F + L	$[0/\gamma/\gamma/2\gamma/0/\alpha/2\alpha/3\alpha]_s$
F + R	$[0/\gamma/\gamma/2\gamma/0/\beta/3\beta/6\beta]_s$
F + E	$[0/\gamma/\gamma/2\gamma/\xi/\xi^2/\xi^3/\xi^4]_s$
E + L	$[\xi/\xi^2/\xi^3/\xi^4/0/\alpha/2\alpha/3\alpha]_s$
E + R	$[\xi/\xi^2/\xi^3/\xi^4/0/\beta/3\beta/6\beta]_s$
E + F	$[\xi/\xi^2/\xi^3/\xi^4/0/\gamma/\gamma/2\gamma]_s$

For the L-helicoidal layup configuration, the rotation angle $\theta_i - \theta_{i-1}$ is constant and equal to α. For the R-helicoidal layup configuration, $\theta_i - \theta_{i-1}$ are in an arithmetic sequence with the tolerance of β. For the F-helicoidal layup configuration, the angle θ_i follows a Fibonacci sequence starting with 0 and γ. For the E-helicoidal layup configuration, the angle θ_i changes exponentially with the base of ξ.

2.2. Design Optimization

In this paper, the bending responses of square bio-inspired helicoidal CFRPC laminated plates subjected to transverse uniformly distributed load are discussed. As reported by Mohamed et al. [14], the boundary conditions (BCs) have significant effects on the bending responses of helicoidal CFRPC laminated plates. In the current study, three types of BCs, namely CCCC, SSSS and CSCS, are considered, as shown in Figure 2. The letter "C" represents the clamped BC and "S" means the simply supported BC.

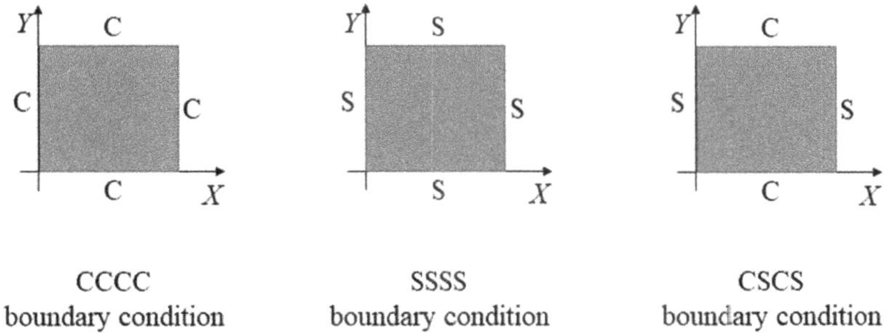

Figure 2. Plates with CCCC, SSSS and CSCS boundary conditions.

In the process of optimizing the bending responses of CFRPC laminated plates, the optimal helicoidal layup configurations are searched for to achieve a higher out-of-plane stiffness for the CFRPC laminated plates. The layup optimization problem of the helicoidal laminated plates is non-differentiability, thus, the global optimization algorithm ASA [18] is applied here. Based on a probabilistic function, each new optimal result during the optimization process is probabilistically accepted, so the inferior results can also be accepted in certain probability. The above strategy prevents the ASA optimization process from falling into local optimization. The optimization problem can be formulated as:

$$\begin{cases} Minimize\ \overline{W} \\ s.t. \begin{cases} \theta_i < 2\pi \\ layup \in [L+R,\ L+F,\ L+E,\ R+L,\ R+F,\ R+E, \\ \qquad F+L,\ F+R,\ F+E,\ E+L,\ E+R,\ E+F] \end{cases} \end{cases} \tag{1}$$

in which, the angle θ_i of each ply is constrained to less than 360°. For the single-form helicoidal patterns, the number of positive integers of α, β, γ and ξ are limited, so the ASA optimization is only conducted for the combination-form helicoidal patterns, where two of the positive integers of α, β, γ and ξ are specified as the design variables. $\overline{W}$ is the dimensionless bending deflection [28] and can be expressed as:

$$\overline{W} = 10^2 W_{mid} \frac{E_{22}h^3}{a^4 q_0} \quad (2)$$

where W_{mid} is the displacement along the thickness direction of the center point on the middle plane. E_{22} is Young's modulus normal to the fiber direction. a and h are the length and thickness of the plate, respectively. q_0 is the intensity of the applied uniformly distributed load:

$$q(x, y) = q_0 \quad (3)$$

The optimization program is operated on the Isight platform. The bending deflection of the laminated plates subjected to transverse uniformly distributed load are obtained in linear analysis using the parametric FEM in ABAQUS. Then, the nonlinear bending analysis is conducted on the optimal helicoidal laminated plates. Figure 3 shows the process of helicoidal layup optimization and nonlinear bending analysis. Isight firstly inputs the new design variables to the FEM. Subsequently, ABAQUS returns the calculated bending deflection to the ASA method. The optimal solution according to the optimization objective of minimizing $\overline{W}$ will be updated, and the new design variables are again input to the FEM for the next optimization cycle. Finally, based on ASA together with the FEM, the nonlinear bending behaviors of the selected optimal helicoidal laminated plates and the QI $[0/45/90/-45]_{2s}$ ones are compared using ABAQUS.

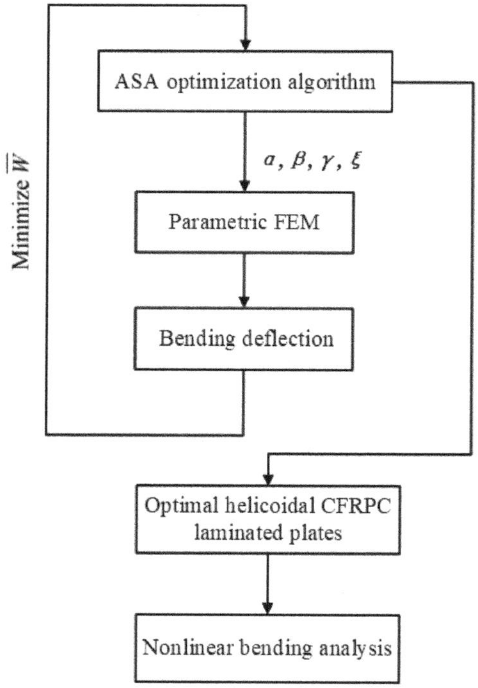

Figure 3. The process of helicoidal layup optimization and nonlinear bending analysis.

2.3. Finite Element Modelling

As shown in Figure 1, consider a square bio-inspired helicoidal CFRPC laminated plate with the geometric size of length a = 224 mm, width b = 224 mm and thickness h = 2.24 mm.

The plate is symmetrical and contains 16 plies with an identical thickness of 0.14 mm. A cartesian coordinate system is defined, and its origin point is set at the corner on the middle plane, where X, Y and Z axes are defined along the length, width and thickness directions of the plate.

The bending analyses of CFRPC laminated plates are carried out by using the FEM with ABAQUS. The laminated plate is established using the composite layup editor tool in ABAQUS and is meshed with eight-node brick element C3D8. The material properties of each CFRPC ply are E_{11} = 120 GPa, E_{22} = E_{33} = 11 GPa, G_{12} = G_{13} = 6.478 GPa, G_{23} = 2.518 GPa, ν_{12} = ν_{13} = 0.33 GPa and ν_{23} = 0.35 GPa [39]. The degrees of freedom of the nodes for these three BCs in the FEM are presented in Equation (4).

CCCC:

$$U_X = U_Y = U_Z = 0\begin{cases} \text{at } Y = 0, b \\ \text{at } X = 0, a \end{cases} \tag{4a}$$

SSSS:

$$U_X = U_Y = U_Z = 0\begin{cases} \text{at } Y = 0, b \text{ and } Z = 0 \\ \text{at } X = 0, a \text{ and } Z = 0 \end{cases} \tag{4b}$$

CSCS:

$$U_X = U_Y = U_Z = 0\begin{cases} \text{at } Y = 0, b \\ \text{at } X = 0, a \text{ and } Z = 0 \end{cases} \tag{4c}$$

where U_X, U_Y and U_Z represent the translation displacements along the X, Y and Z directions, respectively.

The mesh convergence study should be performed before the FE analysis. The dimensionless bending deflection $\overline{W}$ of QI CFRPC laminated plates with different mesh sizes under CCCC, SSSS and CSCS boundary conditions from the FEM are listed in Table 2. Thus, according to the results, in the following bending analysis, all the CFRPC laminated plates are meshed with centrally densified grids and the minimum mesh size is 0.56 × 0.56 × 0.56 mm, as shown in Figure 4.

Table 2. Mesh convergence study for the dimensionless bending deflection $\overline{W}$ of the square QI CFRPC laminated plates under CCCC, SSSS and CSCS boundary conditions.

BC	Minimum Mesh Size (mm)	Nodes	$\overline{W}$	Difference
CCCC	2.24 × 2.24 × 0.56	51,005	0.3044	-
	1.12 × 1.12 × 0.56	96,605	0.3109	(0.3109−0.3044)/0.3044 ≈ 2.14%
	0.56 × 0.56 × 0.56	171,125	0.3130	(0.3130−0.3109)/0.3109 ≈ 0.68%
SSSS	2.24 × 2.24 × 0.56	51,005	1.0603	-
	1.12 × 1.12 × 0.56	96,605	1.0891	(1.0891−1.0603)/1.0603 ≈ 2.72%
	0.56 × 0.56 × 0.56	171,125	1.0985	(1.0985−1.0891)/1.0891 ≈ 0.86%
CSCS	2.24 × 2.24 × 0.56	51,005	0.5027	-
	1.12 × 1.12 × 0.56	96,605	0.5149	(0.5149−0.5027)/0.5027 ≈ 2.43%
	0.56 × 0.56 × 0.56	171,125	0.5189	(0.5189−0.5149)/0.5149 ≈ 0.78%

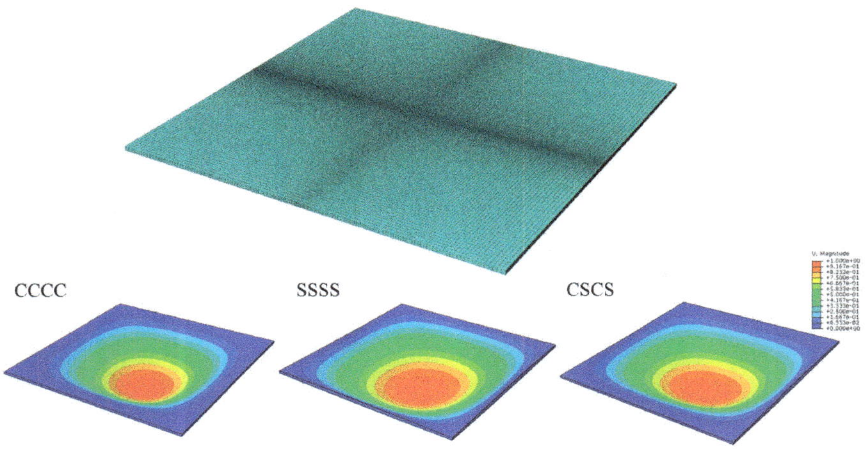

Figure 4. Schematic diagram of FE analysis of the laminated plate and deflection distribution of laminated plates under CCCC, SSSS and CSCS boundary conditions.

3. Layup Optimization and Nonlinear Bending Analysis

In this section, the layup optimization for minimum bending deflection and the nonlinear bending analysis of helicoidal CFRPC laminated plates with optimal single-form and combination-form helicoidal layups are conducted based on the ASA optimization algorithm (on Isight) together with the FEM (in ABAQUS). Comparison works are carried out with the counterpart QI plate under CCCC, SSSS and CSCS boundary conditions subjected to a transverse uniformly distributed load. The prediction of the bending deflection in the optimization process is carried out in linear analysis. Additionally, in the subsequently nonlinear bending simulation, the geometrical nonlinearity is considered in ABAQUS/Standard with the Nlgeom option turned on.

3.1. Validation Studies

Example 1: In this case, the linear central deflection of conventional laminated plates UD ($[0]_{32}$) and QI ($[0/45/90/-45]_{4s}$) together with helicoidal laminated plates LH ($[0/24/48/72/96/120/144/168/192/216/240/264/288/312/336/360]_s$) and FH ($[0/10/10/20/30/50/80/130/210/340/190/170/360/170/170/340]_s$) under CCCC and SSSS boundary conditions subjected to transverse uniformly distributed load are calculated and compared in Table 2. The plates have a/h = 100 and a/b = 1, and the material properties used here are: E_{11}/E_{22} = 25, E_{22} = E_{33}, G_{12} = G_{13} = $0.5E_{22}$, G_{23} = $0.2E_{22}$, ν_{12} = ν_{13} = ν_{23} = 0.25. In the present FE simulation, the boundary conditions are set as mentioned in Section 2.3. As shown in Table 3, good agreements can be observed between the dimensionless bending deflection obtained from the present FEM and those obtained by Mohamed [14].

Table 3. Comparison of the dimensionless bending deflection $\overline{W}$ of the square composite laminated plates under CCCC and SSSS boundary conditions subjected to a uniformly distributed load.

	CCCC		SSSS	
	Mohamed [14]	**Present (FEM)**	**Mohamed [14]**	**Present (FEM)**
UD	0.1340	0.1387	0.6528	0.6911
QI	0.1535	0.1566	0.5018	0.5301
LH	0.1548	0.1570	0.5111	0.5347
FH	0.1541	0.1579	0.5691	0.5996

Example 2: In this example, a 16-ply square $[0/45/-45/90]_4$ laminated plate with the geometric size of $a/b = 1$ and $a/h = 100$ is considered. The CCCC, SSSS and CSCS boundary conditions are set as mentioned in Section 2.3. The material properties [45] are: $E_{11}/E_{22} = 25$, $E_{22} = E_{33}$, $G_{12} = G_{13} = 0.5E_{22}$, $G_{23} = 0.2E_{22}$, $\nu_{12} = \nu_{13} = \nu_{23} = 0.25$. The nonlinear bending load–central deflection curves of the composite lamianted plates are presented in Figure 5, from which a good agreement can be observed again.

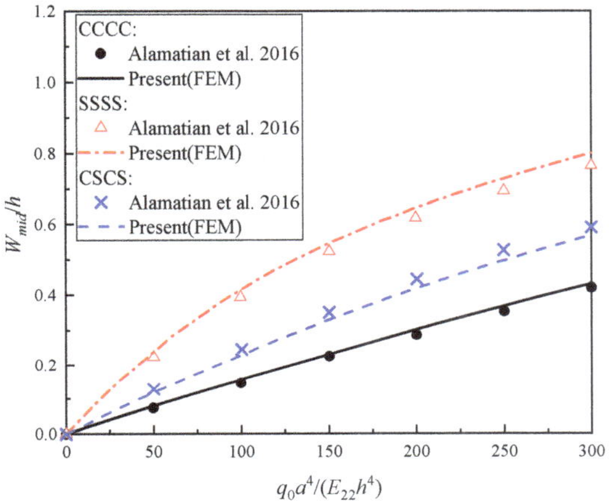

Figure 5. Comparison of the nonlinear bending load–central deflection curves of composite laminated plates under CCCC, SSSS and CSCS boundary conditions subjected to a transverse uniformly distributed load with Alamatian et al. [45].

3.2. Optimization of Helicoidal Layups for Minimum Bending Deflection

The layup optimization studies for better out-of-plane bending responses of helicoidal CFRPC laminated plates are carried out. The dimensionless bending deflection $\overline{W}$ of symmetrical square helicoidal CFRPC laminated plates stacked with either only one (single-form) or a combination of two (combination-form) of L, R, F and E helicoidal arrangements are compared with the conventional QI, i.e., $[0/45/90/-45]_{2s}$ subjected to atransverse uniformly distributed load.

Tables 4–6 show the dimensionless bending deflection $\overline{W}$ of optimal bio-inspired helicoidal laminated plates with a layup sequence of only one of the L, R, F or E helicoidal arrangements (called single-form) under CCCC, SSSS and CSCS boundary conditions. For these single-form helicoidal plates with limited design variables of α, β, γ and ξ, the bending deflection of all four types of helicoidal laminates subjected to a transverse uniformly distributed load are calculated and listed in Tables 4–6.

Table 4. The dimensionless bending deflection $\overline{W}$ of the optimal bio-inspired helicoidal CFRPC laminated plates with a single-form helicoidal layup pattern under the CCCC boundary condition.

Designation	Configuration	$\overline{W}$	Relative Change
QI	$[0/45/90/-45]_{2s}$	0.3130	-
L	$[0/1/2/3/4/5/6/7]_s$	0.2934	−6.261% *
R	$[0/1/3/6/10/15/21/28]_s$	0.2950	−5.743%
F	$[0/1/1/2/3/5/8/13]_s$	0.2933	−6.278%
E	$[2/4/8/16]_{2s}$	0.2974	−4.973%

* Relative change comparing to QI plate = 100% $(\overline{W} - 0.3130)/0.3130$.

Table 5. The dimensionless bending deflection $\overline{W}$ of the optimal bio-inspired helicoidal CFRPC laminated plates with a single-form helicoidal layup pattern under the SSSS boundary condition.

Designation	Configuration	$\overline{W}$	Relative Change
QI	$[0/45/90/-45]_{2s}$	1.0985	-
L	$[0/51/102/153/204/255/306/357]_s$	1.0945	−0.365%
R	$[0/12/36/72/120/180/252/336]_s$	1.2051	9.706%
F	$[0/27/27/54/81/135/216/351]_s$	1.1868	8.041%
E	$[3/9/27/81]_{2s}$	1.3031	18.626%

Table 6. The dimensionless bending deflection $\overline{W}$ of the optimal bio-inspired helicoidal CFRPC laminated plates with a single-form helicoidal layup pattern under the CSCS boundary condition.

Designation	Configuration	$\overline{W}$	Relative Change
QI	$[0/45/90/-45]_{2s}$	0.5189	-
L	$[0/43/86/129/172/215/258/301]_s$	0.5183	−0.113%
R	$[0/12/36/72/120/180/252/336]_s$	0.6498	25.228%
F	$[0/27/27/54/81/135/216/351]_s$	0.6769	30.444%
E	$[4/16/64/256]_{2s}$	0.6157	18.645%

The dimensionless bending deflections $\overline{W}$ for optimal single-form helicoidal laminated plates under the CCCC boundary condition in Table 4 show that the F helicoidal laminated plate with ($[0/1/1/2/3/5/8/13]_s$) has the lowest bending deflection, $\overline{W}$ = 0.2933, which is 6.278% lower than that of the QI plate. The bending deflection of the optimal L, R and E helicoidal plates decreased by 6.261%, 5.743% and 4.973%, respectively, compared to the QI plate. It can be observed from the results in Table 4 that the single-form helicoidal laminated plates with lower rotation angles tend to have higher out-of-plane bending stiffness under the CCCC boundary condition subjected to a transverse uniformly distributed load.

For the minimum dimensionless bending deflection $\overline{W}$ of CFRPC laminated plates under the SSSS boundary condition, Table 5 gives the optimal single-form helicoidal layups. Among the four single-form helicoidal laminated plates, the L helicoidal laminated plate with ($[0/51/102/153/204/255/306/357]_s$) decreases the bending deflection by 0.365%. However, for the other three single-form helicoidal layups, no improvement in the out-of-plane bending stiffness can be observed, with the bending deflection of these plates being higher than that of the QI plate. It is suggested that the enhancement of the out-of-plane bending stiffness of the helicoidal CFRPC laminated plates stacked with single-form helicoidal arrangements under the SSSS boundary condition subjected to a transverse uniformly distributed load are limited.

Table 6 lists the optimal helicoidal laminated plates stacked with single-form helicoidal layups for the minimum dimensionless bending deflection $\overline{W}$ under the CSCS boundary condition. Similar to the observation for plates under the SSSS boundary condition, only the L-helicoidal laminated plate with ($[0/43/86/129/172/215/258/301]_s$) can decrease the bending deflection by 0.113%, and the other three single-form helicoidal plates cannot decrease the bending deflection. It is suggested that the single-form helicoidal arrangements do not contribute significantly to improving the out-of-plane bending stiffness of CFRPC laminated plates under the CSCS boundary condition subjected to a transverse uniformly distributed load.

The optimal combination-form helicoidal CFRPC laminated plates for minimum dimensionless bending deflection $\overline{W}$ under the CCCC, SSSS and CSCS boundary conditions subjected to a uniformly distributed load are listed in Tables 7–9. For these CFRPC laminated plates stacked with two of the L, R, F or E helicoidal layup patterns, the ASA optimization algorithm operated on the Isight platform together with ABAQUS is applied.

Table 7. The dimensionless bending deflection $\overline{W}$ of the optimal bio-inspired helicoidal CFRPC laminated plates with a combination-form helicoidal layup pattern under the CCCC boundary condition.

Designation	Configuration	$\overline{W}$	Relative Change
QI	$[0/45/90/-45]_{2s}$	0.3130	-
L + R	$[0/1/2/3/0/1/3/6]_s$	0.2932	−6.313%
L + F	$[0/1/2/3/0/178/178/356]_s$	0.2932	−6.326%
L + E	$[0/1/2/3/2/4/8/16]_s$	0.2934	−6.243%
R + L	$[0/1/3/6/0/1/2/3]_s$	0.2935	−6.225%
R + F	$[0/1/3/6/0/178/178/356]_s$	0.2935	−6.236%
R + E	$[0/1/3/6/2/4/8/16]_s$	0.2938	−6.139%
F + L	$[0/179/179/358/0/1/2/3]_s$	0.2931	−6.353%
F + R	$[0/179/179/358/0/1/3/6]_s$	0.2931	−6.349%
F + E	$[0/179/179/358/2/4/8/16]_s$	0.2933	−6.305%
E + L	$[2/4/8/16/0/1/2/3]_s$	0.2970	−5.116%
E + R	$[2/4/8/16/0/1/3/6]_s$	0.2970	−5.105%
E + F	$[2/4/8/16/0/174/174/348]_s$	0.2968	−5.175%

Table 8. The dimensionless bending deflection $\overline{W}$ of the optimal bio-inspired helicoidal CFRPC laminated plates with a combination-form helicoidal layup pattern under the SSSS boundary condition.

Designation	Configuration	$\overline{W}$	Relative Change
QI	$[0/45/90/-45]_{2s}$	1.0985	-
L + R	$[0/111/222/333/0/47/141/282]_s$	1.0899	−0.785%
L + F	$[0/46/92/138/0/135/135/270]_s$	1.0871	−1.037%
L + E	$[0/112/224/336/4/16/64/256]_s$	1.1047	0.570%
R + L	$[0/50/150/300/0/36/72/108]_s$	1.0429	−5.059%
R + F	$[0/50/150/300/0/45/45/90]_s$	1.0376	−5.542%
R + E	$[0/50/150/300/4/16/64/256]_s$	1.0515	−4.276%
F + L	$[0/121/121/242/0/32/64/96]_s$	1.0898	−0.791%
F + R	$[0/121/121/242/0/44/132/264]_s$	1.0881	−0.944%
F + E	$[0/121/121/242/4/16/64/256]_s$	1.0998	0.122%
E + L	$[4/16/64/256/0/119/238/357]_s$	1.2649	15.148%
E + R	$[3/9/27/81/0/1/3/6]_s$	1.2690	15.527%
E + F	$[4/16/64/256/0/136/136/272]_s$	1.2394	12.826%

The optimization results in Table 7 show that all the combination-form helicoidal layup patterns can improve the out-of-plane bending stiffness under the CCCC boundary condition compared to the QI plate. Among them, the optimal (F + L) helicoidal plate of $[0/179/179/358/0/1/2/3]_s$ has the lowest dimensionless bending deflection $\overline{W}$ of 0.2931, which is 6.353% lower than that of the QI plate. All the combination-form helicoidal layups can improve the out-of-plane bending stiffness of the CFRPC plate by more than 5% compared to the QI plate. Additionally, under the CCCC boundary condition, the smaller the angle between the adjacent plies, the lower the bending deflection of the laminated plate subjected to a transverse uniformly distributed load. The lower rotation angles of the helicoidal laminated plates enhance the bending stiffness D_{11} obviously, which leads to the lower bending deflection.

Table 9. The dimensionless bending deflection $\overline{W}$ of the optimal bio-inspired helicoidal CFRPC laminated plates with combination-form helicoidal layup pattern under the CSCS boundary condition.

Designation	Configuration	$\overline{W}$	Relative Change
QI	$[0/45/90/-45]_{2s}$	0.5189	-
L + R	$[0/42/84/126/0/43/129/258]_s$	0.5183	−0.128%
L + F	$[0/41/82/123/0/103/103/206]_s$	0.5085	−2.006%
L + E	$[0/42/84/126/4/16/64/256]_s$	0.5260	1.365%
R + L	$[0/43/129/258/0/118/236/354]_s$	0.5251	1.193%
R + F	$[0/43/129/258/0/101/101/202]_s$	0.5186	−0.064%
R + E	$[0/44/132/264/4/16/64/256]_s$	0.5359	3.275%
F + L	$[0/113/113/226/0/47/94/141]_s$	0.4911	−5.360%
F + R	$[0/113/113/226/0/43/129/258]_s$	0.4940	−4.812%
F + E	$[0/114/114/228/4/16/64/256]_s$	0.4996	−3.730%
E + L	$[4/16/64/256/0/119/238/357]_s$	0.5941	14.485%
E + R	$[4/16/64/256/0/44/132/264]_s$	0.6024	16.076%
E + F	$[4/16/64/256/0/112/112/224]_s$	0.5849	12.720%

For combination-form helicoidal CFRPC laminated plates under the SSSS boundary condition, the optimal layup patterns (L + R), (L + F), (R + L), (R + F), (R + E), (F + L) and (F + R) can decrease the dimensionless bending deflection $\overline{W}$ of laminated plates, as shown in Table 8. The lowest $\overline{W}$ of 1.0376, 5.542% lower than the QI plate, is achieved by the (R + F) plate with ($[0/50/150/300/0/45/45/90]_s$). Compared to the optimal single-form helicoidal laminated plates under the SSSS boundary condition in Table 5, the improvement in the out-of-plane bending stiffness of plates with combination-form helicoidal layups of (R + L), (R + F) and (R + E) is much more obvious. Compared to the QI ($[0/45/90/-45]_{2s}$) plate with zero A_{16} and A_{26}, the optimal helicoidal laminated plates have the nonzero A_{16} and A_{26} that offers more shear stiffness than the QI plate.

The results of the layup optimization of combination-form helicoidal patterns for achieving minimum bending deflection for CFRPC laminated plates under the CSCS boundary condition subjected to a transverse uniformly distributed load are presented in Table 9. It can be seen from the optimal helicoidal configurations that (L + R), (L + F), (R + F), (F + L), (F + R) and (F + E) can decrease the $\overline{W}$ of laminated plates compared to the QI plate. The (F + L) plate with ($[0/113/113/226/0/47/94/141]_s$) has the lowest $\overline{W}$, 0.4911, which is 5.360% lower than the QI plate. Additionally, compared to the single-form helicoidal laminated plates under the CSCS boundary condition in Table 6, the combination-form (F + L), (F + R) and (F + E) helicoidal layups can enhance the out-of-plane bending stiffness of plates much more effectively.

3.3. Nonlinear Bending Behaviors of Optimal Helicoidal CFRPC Laminated Plates

In the previous subsection, the dimensionless bending deflection $\overline{W}$ of the QI and helicoidal laminated plates for linear analysis are calculated. The optimal single-form and combination-form helicoidal CFRPC laminated plates are selected to predict the nonlinear bending responses which consider the geometrical nonlinearity. For the laminated plates under the CCCC boundary condition, the nonlinear bending responses of the optimal single-form F-helicoidal plate of $[0/1/1/2/3/5/8/13]_s$ and combination-form (F + L) helicoidal plate of $[0/179/179/358/0/1/2/3]_s$ are compared with that of the QI plate. For the laminated plates under the SSSS boundary condition, the optimal single-form L-helicoidal plate of $[0/51/102/153/204/255/306/357]_s$ and combination-form (R + F) plate of $[0/50/150/300/0/45/45/90]_s$ are selected. Additionally, for the laminated plates under the CSCS boundary condition, the optimal single-form L-helicoidal plate of

$[0/43/86/129/172/215/258/301]_s$ and combination-form (F + L) plate of $[0/113/113/226/0/47/94/141]_s$ are selected. All the plates have the same geometric size, with a = 224 mm, b = 224 mm and h = 2.24 mm, and are subjected to a transverse uniformly distributed load. In the following figures, W_{mid} and M_X represent the deflection of the center point on the middle plane and the central bending moment of the plate, respectively.

Figure 6 illustrates the nonlinear bending response of the optimal helicoidal plates and QI plate under the CCCC boundary condition. It can be observed that the F and (F + L) helicoidal plates have very close curves which are lower than QI plate for both the load–central deflection and the load–bending moment curves. During the nonlinear bending process, the optimal F and (F + L) helicoidal plates maintain the higher out-of-plane bending stiffness under the CCCC boundary condition subjected to a uniformly distributed load.

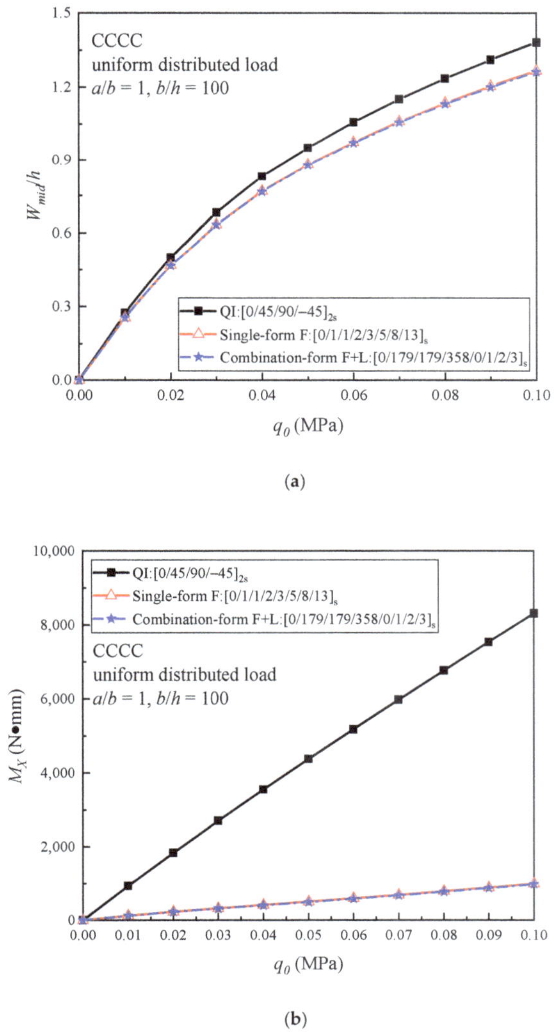

Figure 6. Comparison of (**a**) the load–central deflection curves; and (**b**) the load–bending moment curves of the QI and optimal helicoidal CFRPC laminated plates under the CCCC boundary condition subjected to a uniformly distributed load.

In Figure 7, the relationships between the applied load and the central deflection together with the applied load and the bending moment are presented. The results show that the load–central deflection curves of these three plates are close under the SSSS boundary condition, while the enhancement of the out-of-plane bending properties is not obvious for the optimal helicoidal laminated plates compared with the QI plate during the nonlinear bending process. On the other hand, the load–bending moment curves of the optimal plates are lower than those of the QI plate.

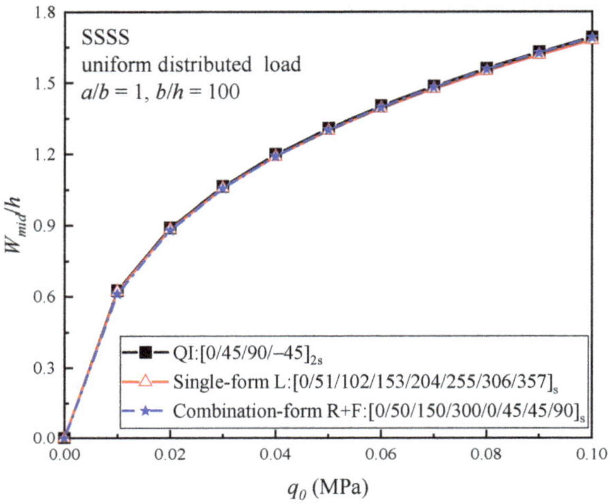

(**a**)

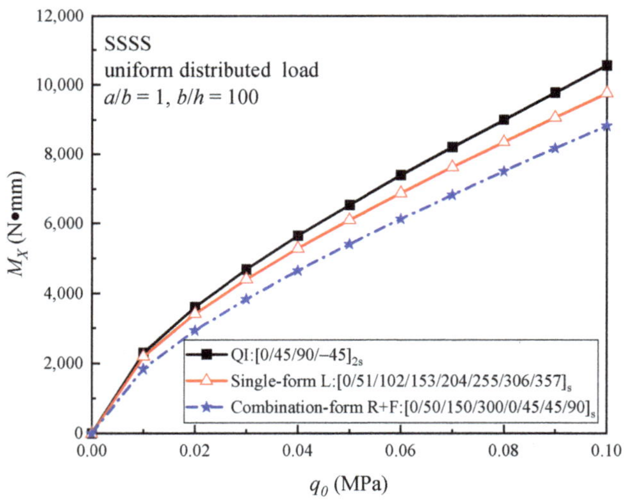

(**b**)

Figure 7. Comparison of (**a**) the load–central deflection curves; and (**b**) the load–bending moment curves of the QI and optimal helicoidal CFRPC laminated plates under the SSSS boundary condition subjected to a uniformly distributed load.

The nonlinear bending responses of the optimal helicoidal laminated plates under the CSCS boundary condition are given in Figure 8. It can be seen from the results that, among the three plates, the (F + L) plate has the lowest load–central deflection curve, but the highest load–bending moment curve. The load–central deflection curves of the optimal L helicoidal plate and QI plate are very close, while the load–bending moment curve of the L plate is slightly higher than that of the QI plate.

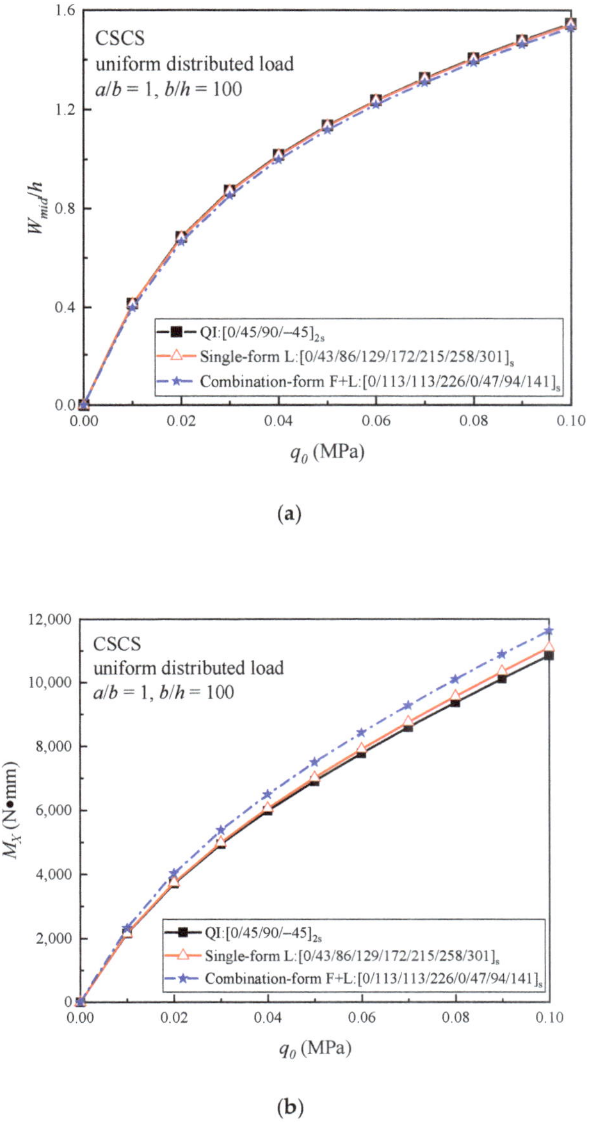

Figure 8. Comparison of (**a**) the load–central deflection curves; and (**b**) the load–bending moment curves of the QI and optimal helicoidal CFRPC laminated plates under the CSCS boundary condition subjected to a uniformly distributed load.

4. Conclusions

The optimization design and nonlinear bending analysis of bio-inspired helicoidal CFRPC laminated plates under CCCC, SSSS and CSCS boundary conditions subjected to a transverse uniformly distributed load have been presented. The purpose of layup optimization is to find the optimal CFRPC laminated plates stacked with single-form and combination-form bio-inspired helicoidal layups that own the best out-of-plane mechanical properties and which demonstrate the lowest bending deflection when subjected to a transverse uniformly distributed load. The global optimization algorithm ASA based on Isight and associated with the FEM operated in ABAQUS is applied. For the first time, the nonlinear bending responses of the optimal helicoidal CFRPC laminated plates are studied. The notable results are summarized below.

(a) Under the CCCC boundary condition, both the single-form and combination-form helicoidal layups with the lower rotation angles can decrease the bending deflection.
(b) Among the optimal CCCC helicoidal laminated plates, the F helicoidal plate ($[0/1/1/2/3/5/8/13]_s$) has the lowest dimensionless bending deflection score, 0.2933, which is 6.278% lower than the QI plate, and the (F + L) helicoidal plate ($[0/179/179/358/0/1/2/3]_s$) has the lowest bending deflection of 0.2931, which is 6.353% lower than the QI plate.
(c) Under SSSS and CSCS boundary conditions, the enhancement of the out-of-plane bending properties of the CFRPC laminated plates using single-form helicoidal layup arrangements are limited, while by stacking with the combination-form helicoidal patterns, the linear dimensionless bending deflection of the CFRPC laminated plates can be obviously decreased.
(d) For the optimal SSSS helicoidal laminated plates, the lowest bending deflection of 1.0376, which is 5.542% lower than the QI plate, is achieved by the (R + F) helicoidal plate ($[0/50/150/300/0/45/45/90]_s$). Additionally, for the optimal CSCS helicoidal laminated plates, the (F + L) helicoidal plate ($[0/113/113/226/0/47/94/141]_s$) has the lowest bending deflection of 0.4911, which is 5.360% lower than the QI plate.
(e) The boundary conditions have significant effects on the nonlinear bending responses of helicoidal laminated plates.

Author Contributions: Conceptualization, H.-S.S. and H.W.; methodology, T.L.; software, T.L.; validation, T.L., H.-S.S. and X.C.; formal analysis, T.L.; investigation, H.-S.S.; resources, X.C.; data curation, X.C.; writing—original draft preparation, T.L. and H.-S.S.; writing—review and editing, H.W., X.C. and M.F.; visualization, T.L. and H.-S.S.; supervision, H.W., X.C. and M.F.; project administration, X.C. and M.F.; funding acquisition, X.C. and M.F. All authors have read and agreed to the published version of the manuscript.

Funding: This research was supported by Zhe Jiang Key Laboratory of General Aviation Operation Technology (General Aviation Institute of Zhejiang Jiande), NO: JDGA2020-13.

Institutional Review Board Statement: Not applicable.

Informed Consent Statement: Not applicable.

Data Availability Statement: The data presented in this study are available on request from the corresponding author.

Conflicts of Interest: The authors declare no conflict of interest.

References

1. Bouligand, Y. Twisted fibrous arrangements in biological materials and cholesteric mesophases. *Tissue Cell* **1972**, *4*, 192–217. [CrossRef] [PubMed]
2. Fabritius, H.O.; Sachs, C.; Triguero, P.R.; Raabe, D. Influence of structural principles on the mechanics of a biological fiber-based composite material with hierarchical organization: The Exoskeleton of the lobster homarus americanus. *Adv. Mater.* **2009**, *21*, 391–400. [CrossRef]
3. Zimmermann, E.A.; Gludovatz, B.; Schaible, E.; Dave, N.K.N.; Yang, W.; Meyers, M.A. Mechanical adaptability of the Bouligand-type structure in natural dermal armour. *Nat. Commun.* **2013**, *4*, 2634. [CrossRef] [PubMed]

4. Suksangpanya, N.; Yaraghi, N.A.; Kisailus, D.; Zavattieri, P. Twisting cracks in Bouligand structures. *J. Mech. Behav. Biomed. Mater.* **2017**, *76*, 38–57. [CrossRef] [PubMed]
5. Apichattrabrut, T.; Ravi-Chandar, K. Helicoidal Composites. *Mech. Adv. Mater. Struct.* **2006**, *13*, 61–76. [CrossRef]
6. Cheng, L.; Wang, L.; Karlsson, A.M. Image analyses of two crustacean exoskeletons and implications of the exoskeletal microstructure on the mechanical behavior. *J. Mater. Res.* **2008**, *23*, 2854–2872. [CrossRef]
7. Cheng, L.; Wang, L.; Karlsson, A.M. Mechanics-based analysis of selected features of the exoskeletal microstructure of Popillia japonica. *J. Mater. Res.* **2009**, *24*, 3253–3267. [CrossRef]
8. Cheng, L.; Thomas, A.; Glancey, J.L.; Karlsson, A.M. Mechanical behavior of bio-inspired laminated composites. *Compos. Pt. A-Appl. Sci. Manuf.* **2011**, *42*, 211–220. [CrossRef]
9. Liu, J.L.; Lee, H.P.; Tan, V.B.C. Effects of inter-ply angles on the failure mechanisms in bioinspired helicoidal laminates. *Compos. Sci. Technol.* **2018**, *165*, 282–289. [CrossRef]
10. Liu, J.L.; Lee, H.P.; Kong, S.H.R.; Tan, V.B.C. Improving laminates through non-uniform inter-ply angles. *Compos. Pt. A-Appl. Sci. Manuf.* **2019**, *127*, 105625. [CrossRef]
11. Jiang, H.; Ren, Y.; Liu, Z.; Zhang, S.; Lin, Z. Low-velocity impact resistance behaviors of bio-inspired helicoidal composite laminates with non-linear rotation angle based layups. *Compos. Struct.* **2019**, *214*, 463–475. [CrossRef]
12. Mencattelli, L.; Pinho, S.T. Realising bio-inspired impact damage-tolerant thin-ply CFRP Bouligand structures via promoting diffused sub-critical helicoidal damage. *Compos. Sci. Technol.* **2019**, *182*, 107684. [CrossRef]
13. Mencattelli, L.; Pinho, S.T. Ultra-thin-ply CFRP Bouligand bio-inspired structures with enhanced loadbearing capacity, delayed catastrophic failure and high energy dissipation capability. *Compos. Pt. A-Appl. Sci. Manuf.* **2020**, *129*, 105655. [CrossRef]
14. Mohamed, S.A.; Mohamed, N.; Eltaher, M.A. Bending, buckling and linear vibration of bio-inspired composite plates. *Ocean Eng.* **2022**, *259*, 111851. [CrossRef]
15. Pedersen, N.L. Topology optimization of laminated plates with prestress. *Comput. Struct.* **2003**, *80*, 559–570. [CrossRef]
16. Almeida, F.S.; Awruch, A.M. Design optimization of composite laminated structures using genetic algorithms and finite element analysis. *Compos. Struct.* **2009**, *88*, 443–454. [CrossRef]
17. Monte, S.M.; Infante, V.; Madeira, J.F.; Moleiro, F. Optimization of fibers orientation in a composite specimen. *Mech. Adv. Mater. Struct.* **2017**, *24*, 410–416. [CrossRef]
18. Ingber, L. Adaptive simulated annealing (ASA): Lessons learned. *Control Cybern.* **1996**, *25*, 33–54.
19. Shen, H.-S. Nonlinear bending of shear deformable laminated plates under transverse and in-plane loads and resting on elastic foundations. *Compos. Struct.* **2000**, *50*, 131–142. [CrossRef]
20. Shen, H.-S. Non-linear bending of shear deformable laminated plates under lateral pressure and thermal loading and resting on elastic foundations. *J. Strain Anal. Eng. Des.* **2000**, *35*, 93–108. [CrossRef]
21. Shen, H.-S. Nonlinear bending analysis of unsymmetric cross-ply laminated plates with piezoelectric actuators in thermal environments. *Compos. Struct.* **2004**, *63*, 167–177. [CrossRef]
22. Shen, H.-S. Hygrothermal effects on the nonlinear bending of shear deformable laminated plates. *J. Eng. Mech.-ASCE* **2002**, *128*, 493–496. [CrossRef]
23. Kumar, V.; Dewangan, H.C.; Sharma, N.; Panda, S.K.; Mahmoud, S.R. Nonlinear deflection characteristics of damaged composite structure theoretical prediction and experimental verification. *Structures* **2023**, *52*, 410–421. [CrossRef]
24. Belardi, V.G.; Fanelli, P.; Vivio, F. Application of the Ritz method for the bending and stress analysis of thin rectilinear orthotropic composite sector plates. *Thin-Walled Struct.* **2023**, *183*, 110374. [CrossRef]
25. Lu, T.; Shen, H.-S.; Wang, H.; Chen, X.; Feng, M. Nonlinear bending and thermal postbuckling of thermoplastic composite laminated plates under temperature variation. *Thin-Walled Struct.* **2023**, *183*, 110322. [CrossRef]
26. Reddy, J.N. *Mechanics of Laminated Composite Plates and Shells: Theory and Analysis*, 2nd ed.; CRC Press: Boca Raton, FL, USA, 2004.
27. Malekzadeh, P.; Setoodeh, A. Large deformation analysis of moderately thick laminated plates on nonlinear elastic foundations by DQM. *Compos. Struct.* **2007**, *80*, 569–579. [CrossRef]
28. Reddy, J.N. A simple higher order shear deformation theory for laminated composite plates. *J. Appl. Mech.* **1984**, *51*, 745–753. [CrossRef]
29. Putcha, N.S.; Reddy, J.N. A refined mixed shear flexible finite element for the nonlinear analysis of laminated plates. *Compos. Struct.* **1986**, *22*, 529–538. [CrossRef]
30. Nayak, A.K.; Shenoi, R.A.; Moy, S.S.J. Transient response of composite sandwich plates. *Compos. Struct.* **2004**, *64*, 249–267. [CrossRef]
31. Golmakani, M.E.; Kadkhodayan, M. Nonlinear bending analysis of annular FGM plates using higher-order shear deformation plate theories. *Compos. Struct.* **2011**, *93*, 973–982. [CrossRef]
32. Upadhyay, A.K.; Shukla, K.K. Large deformation flexural behavior of laminated composite skew plates: An analytical approach. *Compos. Struct.* **2012**, *94*, 3722–3735. [CrossRef]
33. Keleshteri, M.M.; Asadi, H.; Aghdam, M.M. Nonlinear bending analysis of FG-CNTRC annular plates with variable thickness on elastic foundation. *Thin-Walled Struct.* **2019**, *135*, 453–462. [CrossRef]
34. Karama, M.; Afaq, K.S.; Mistou, S. A new theory for laminated composite plates. *Proc. Inst. Mech. Eng. L* **2009**, *223*, 53–62. [CrossRef]

35. Grover, N.; Maiti, D.K.; Singh, B.N. A new inverse hyperbolic shear deformation theory for static and buckling analysis of laminated composite and sandwich plates. *Compos. Struct.* **2013**, *95*, 667–675. [CrossRef]
36. Nguyen, T.N.; Thai, C.H.; Nguyen-Xuan, H. On the general framework of high order shear deformation theories for laminated composite plate structures: A novel unified approach. *Int. J. Mech. Sci.* **2016**, *110*, 242–255. [CrossRef]
37. Dash, P.; Singh, B.N. Geometrically nonlinear bending analysis of laminated composite plate. *Commun. Nonlinear Sci. Numer. Simul.* **2010**, *15*, 3170–3181. [CrossRef]
38. Amabili, M.; Reddy, J.N. The nonlinear, third-order thickness and shear deformation theory for statics and dynamics of laminated composite shells. *Compos. Struct.* **2020**, *244*, 112265. [CrossRef]
39. Lu, T.; Chen, X.; Wang, H.; Zhang, L.; Zhou, Y. A novel experiment-based approach on nonlinear bending analysis of thermoplastic composite laminates. *Compos. Pt. B-Eng.* **2021**, *209*, 108611. [CrossRef]
40. Chen, Y.; Hou, S.; Fu, K.; Han, X.; Ye, L. Low-velocity impact response of composite sandwich structures: Modelling and experiment. *Compos. Struct.* **2017**, *168*, 322–334. [CrossRef]
41. Liu, H.; Falzon, B.G.; Dear, J.P. An experimental and numerical study on the crush behaviour of hybrid unidirectional/woven carbon-fibre reinforced composite laminates. *Int. J. Mech. Sci.* **2019**, *164*, 105160. [CrossRef]
42. Liu, H.; Zhou, J.; Kong, X.; Li, S. Fracture behaviour of fibre-reinforced composite materials subjected to shear loading: An experimental and numerical study. *Int. J. Light. Mater. Manuf.* **2023**, *6*, 108–116. [CrossRef]
43. Chen, Y.; Klingler, A.; Fu, K.; Han, X.; Ye, L. 3D printing and modelling of continuous carbon fibre reinforced composite grids with enhanced shear modulus. *Eng. Struct.* **2023**, *286*, 116165. [CrossRef]
44. Wang, C.; Su, D.; Xie, Z.; Wang, H.; Hazell, P.J.; Zhang, Z.; Zhou, M. Dynamic behaviour of Bio-inspired heterocyclic aramid Fibre-reinforced laminates subjected to Low-velocity Drop-weight impact. *Compos. Pt. A-Appl. Sci. Manuf.* **2022**, *153*, 106733. [CrossRef]
45. Alamatian, J.; Rezaeepazhand, J. Nonlinear bending analysis of variable cross-section laminated plates using the dynamic relaxation method. *J. Mech. Sci. Technol.* **2016**, *30*, 783–788. [CrossRef]

materials

MDPI

Article

Study of Hybrid Composite Joints with Thin-Ply-Reinforced Adherends

Farin Ramezani [1], Ricardo J. C. Carbas [1,2,*], Eduardo A. S. Marques [2] and Lucas F. M. da Silva [2]

[1] Instituto de Ciência e Inovação Em Engenharia Mecânica e Engenharia Industrial (INEGI), Rua Dr. Roberto Frias, 4200-465 Porto, Portugal

[2] Departamento de Engenharia Mecânica, Faculdade de Engenharia (FEUP), Universidade Do Porto, Rua Dr. Roberto Frias, 4200-465 Porto, Portugal

* Correspondence: rcarbas@fe.up.pt

Abstract: It has been demonstrated that a possible solution to reducing delamination in a unidirectional composite laminate lies in the replacement of conventional carbon-fibre-reinforced polymer layers with optimized thin-ply layers, thus creating hybrid laminates. This leads to an increase in the transverse tensile strength of the hybrid composite laminate. This study investigates the performance of a hybrid composite laminate reinforced by thin plies used as adherends in bonded single lap joints. Two different composites with the commercial references Texipreg HS 160 T700 and NTPT-TP415 were used as the conventional composite and thin-ply material, respectively. Three configurations were considered in this study: two reference single lap joints with a conventional composite or thin ply used as the adherends and a hybrid single lap. The joints were quasi-statically loaded and recorded with a high-speed camera, allowing for the determination of damage initiation sites. Numerical models of the joints were also created, allowing for a better understanding of the underlying failure mechanisms and the identification of the damage initiation sites. The results show a significant increase in tensile strength for the hybrid joints compared to the conventional ones as a result of changes in the damage initiation sites and the level of delamination present in the joint.

Keywords: composite joints; thin ply; single lap joints

1. Introduction

Composites usually consist of two main components known as the matrix, which provides the cohesion of the material, and the reinforcement, such as fibres, which provides the material with its strength and stiffness [1]. The use of carbon-fibre-reinforced polymer (CFRP) materials in multiple industrial applications is continuously increasing [2–7], leading to the use of a vast range of composite materials for the design and manufacture of high-performance composite products such as vehicle structures, sporting goods, etc. [2,8]. However, since the strength of the matrix is at least an order of magnitude lower than the strength of the reinforcement, the loads applied in a perpendicular direction to the reinforcement are almost exclusively carried by the low-strength matrix. This results in the onset of delamination, which can lead to the rapid degradation of the mechanical performance of the structure and cause premature failure [9–12]. Accordingly, multiple studies have investigated methods for adhesive layer modification [13,14] or composite laminate modification to delay delamination in a composite joint [15–17], such as the use of fibre metal laminates (FML), composite laminates with toughened layers [18,19], glass fabric reinforcement [20], the use of Z-pins [21,22], 3D weaving [23], stitching [23], braiding [24], or even the adoption of additional thermoplastic inter-plies [25]. However, the significant complexity of these techniques often restricts their usage. Furthermore, such methods normally require the implementation of at least one additional production step [26,27] and thus increase the costs associated to the production process.

Recent advancements in composite manufacturing techniques have led to the development of spread-tow technology [28], which results in flat and straight plies with a more homogeneous fibre distribution and smaller resin-rich regions [29,30]. In this case, a dry ply thickness as low as 0.02 mm can be achieved. Generally, plies with a thickness below 100 μm are known as thin plies [31]. By reducing the thickness of a single layer, the number of possible total layers and therefore the degrees of freedom in design are increased [32]. This also results in a larger number of interfaces in thin-ply laminates, lowering the shear stresses [32,33]. Moreover, thin-ply laminates are known for their ability to delay the onset of the matrix damage mechanisms and suppress transverse microcracking [31] and free edge delamination [32,34] for static, fatigue, and impact loadings. Due to their superior damage and delamination resistance, thin-ply laminates could exhibit higher interlaminar shear properties [35] and strain energy [36] compared to conventional plies. Therefore, thinner composite plies are acknowledged to have higher in situ transverse strength [33]. Nonetheless, the properties of the laminate can still rapidly deteriorate after the onset of damage, leading to premature structural failure [11]. Thin plies are now seen as a promising approach to improve the performance of adhesively bonded CFRP, mainly due to their ability to enhance the off-axis performance of composites and postpone delamination [36]. Moreover, studies have shown that through the use of thin plies in a structural joint, the damage location in the composite moves from the adhesive interface towards the mid-thickness of the composite adherends [36], mainly due to the in situ effect [37]. Through-thickness reinforcement can effectively provide improved interlaminar strength and delamination resistance while producing a more integrated composite structure [38].

A previous study by the authors [39] showed that replacing layers of a conventional composite in a unidirectional laminate with layers of thin ply can increase the composite's strength under transverse tensile loading. The authors postulate that this is due to the increase in laminate ductility, which could postpone the delamination under transverse tensile loads. Moreover, experimental observation clearly demonstrated that the presence of thin plies acts as a barrier against crack propagation. It was shown that the use of 25% (corresponding to the optimum amount) thin ply per total thickness of the laminate (12.5% on each top) increased the transverse tensile strength considerably. Figure 1a shows the studied configuration for a reference conventional composite: thin ply and the optimum hybrid laminate. Figure 1b illustrates the experimentally obtained failure loads for the mentioned configurations.

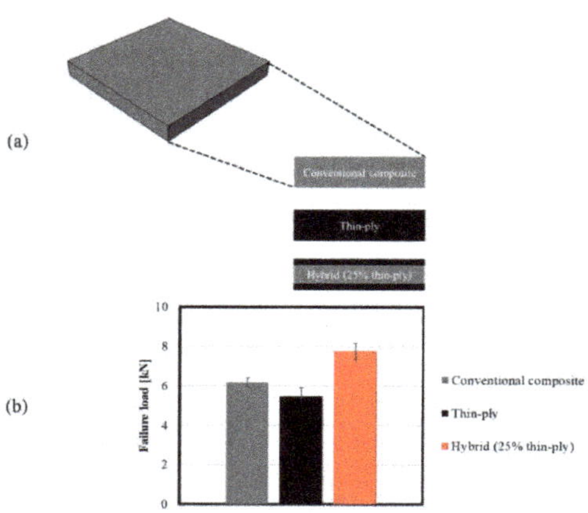

Figure 1. (**a**) Schematic design for conventional composite, thin-ply, and hybrid (25% thin ply) laminates and (**b**) summary of the experimentally obtained failure load for unidirectional reference conventional composite, thin-ply, and hybrid (25% thin ply) laminates. Adopted from [39].

The current study seeks to further investigate this topic, quantifying the performance of a hybrid composite single lap joint and analysing the effect of reinforcing the unidirectional conventional composite adherend using thin ply. In this work, "HS 160 T700" and "NTPT-TP415" are used as a conventional composite and thin-ply material, respectively. The tests were recorded using a high-speed camera to determine the location of damage initiation. Afterwards, the failure surface of specimens was analysed and measured via image analysis, allowing for an accurate estimation of the delaminated area. It was found that the use of hybrid single lap joints reinforced with thin-ply layers results in a considerable increase in the joint strength compared to the reference conventional composite joint. Numerical models were also created via cohesive zone modelling, allowing for the accurately replication and description of the experimentally determined failure processes.

2. Experimental Details

2.1. Adhesive

The adhesive used in this work was an epoxy structural adhesive, supplied in film form, with the commercial reference Scotch Weld AF 163-2k (3M, Saint Paul, MN, USA) [40]. The adhesive was cured following the manufacturer's recommendations at 130 °C for 2 h. The mechanical properties of the AF 163-2k adhesive are presented in Table 1.

Table 1. Main mechanical properties of "AF 163-2k" [41].

Mechanical Property	Value
Young's modulus [MPa]	1521.87
Shear modulus [MPa]	563.67
Tensile strength [MPa]	46.93
Shear strength [MPa]	46.93
G_{IC} [N/mm]	4.05
G_{IIC} [N/mm]	9.77

2.2. Adherend

2.2.1. Conventional Composite

The materials used in the studied configurations were chosen to be representative of a possible application within the aerospace sector. Accordingly, a unidirectional prepreg carbon–epoxy composite with a ply thickness of 0.15mm was selected, with the commercial reference Texipreg HS 160 T700 (Seal Spa, Legnano, Italy). This is an orthotropic material whose mechanical properties are presented in Table 2. The elastic mechanical properties of the conventional composite correspond to the orientation of a 0° composite ply (1 and 2 are defined as the fibre and transverse directions). Moreover, the cohesive properties of the conventional composite's resin are presented separately in Table 3.

Table 2. Conventional composite mechanical properties [42].

Mechanical Property	Value
E_1 [MPa]	109,000
E_2 [MPa]	8819
G_{12} [MPa]	4315
G_{23} [MPa]	3200
υ_{12}	0.34
υ_{23}	0.38

Table 3. Cohesive properties of conventional composite [43].

Property	Value
Tensile strength [MPa]	25
Shear strength [MPa]	13.5
G_{IC} [N/mm]	0.33
G_{IIC} [N/mm]	0.79

2.2.2. Thin-Ply

A unidirectional, 0° oriented carbon–epoxy prepreg composite with a ply thickness of 0.075 mm was selected for use in this work, serving as the thin-ply material. This material has the commercial reference NTPT-TP415 (North thin ply technology, Zory, Poland). The elastic orthotropic and cohesive properties for the thin ply, characterised by the authors in a previous work [44], are presented in Tables 4 and 5, respectively.

Table 4. Thin ply mechanical properties [44].

Mechanical Property	Value
E_1 [MPa]	101,720
E_2 [MPa]	5680
G_{12} [MPa]	3030
G_{23} [MPa]	3030
υ_{12}	0.38
υ_{23}	0.04

Table 5. Cohesive properties of the thin ply [44].

Property	Value
Tensile strength [MPa]	35
Shear strength [MPa]	32
G_{IC} [N/mm]	0.76
G_{IIC} [N/mm]	0.83

2.3. Single Lap Joint Manufacturing

Single lap joints (SLJs) were manufactured with the geometry shown in Figure 2. The width for all specimens under consideration was set at 15 mm.

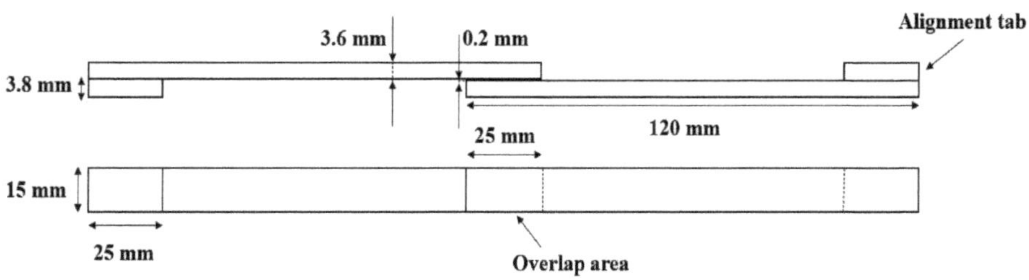

Figure 2. Single lap joint geometry.

The manufacturing process for the reference conventional composite and thin-ply adherends began with a layer-by-layer stacking of the conventional composite and thin-ply prepregs respectively, until the desired adherend thickness was attained (3.6 mm). In this case, 24 and 48 layers of conventional composite and thin-ply prepreg were used, respectively. For the hybrid (25% thin ply) adherends, 6 plies of conventional composite were replaced by 12 plies of thin ply on the adherend tops (6 layers of thin ply on each adherend top). The joints were then bonded by applying an additional layer of adhesive between the adherends. A mould was used to ensure the thickness of the adherends and adhesive. A mould-release agent was used to ensure easy debonding of the specimen from the mould after curing. It was observed that the curing sequence, i.e., curing the adhesive and substrate composite plies in one cure (co-curing) or in two separate cures, had no significant effect on the mechanical properties of the joint for the AF163-2k adhesive. Therefore, a one-step curing manufacturing method was preferred to simultaneous reducing manufacturing time and energy usage. Accordingly, the joint was co-cured at 130 °C for two hours, following the manufacturer's recommended procedure. A schematic design of the reference conventional composite, thin-ply, and hybrid (25% thin-ply) single lap joints are shown in Figure 3.

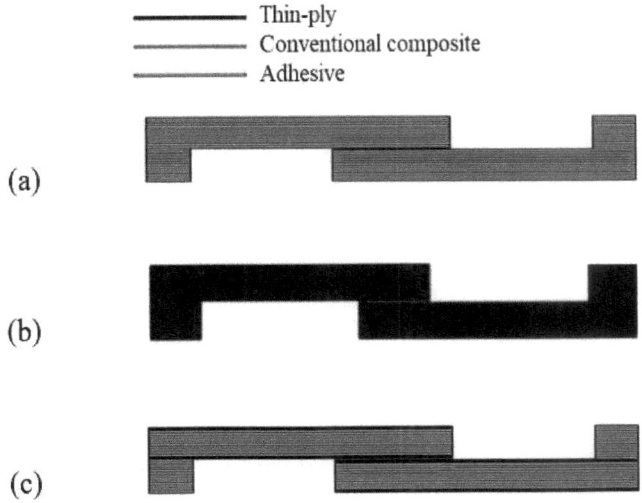

Figure 3. Schematic design of (**a**) conventional composite, (**b**) thin-ply, and (**c**) hybrid (25% thin-ply) joints.

2.4. Testing Condition

The SLJs were tested using an Instron 8801 servo hydraulic testing machine with a load cell of 100 kN and at a constant crosshead speed of 1 mm/min. All tests were performed under laboratory ambient conditions (room temperature of 24 °C, relative humidity of 55%). Four repetitions were performed for each configuration under analysis.

3. Experimental Results

3.1. Load–Displacement Curve

Figure 4 shows representative, experimentally obtained load–displacement curves for the studied configurations. The hybrid (25% thin ply) joint presented the highest failure load, with an increase in joint strength of approximately 90% compared to the reference conventional composite configuration.

3.2. Damage Initiation

A high-speed camera was used to record the specimens under load, seeking to determine whether the damage initiation occurred first in the adhesive layer or within the adherend. A Chronos 1.4 high speed camera was used, recording at 5000 frames per second. Figure 5 presents the images at damage initiation for each configuration. The adhesive and adherend boundaries were roughly defined by correlating the known specimen's dimensions and the equivalent image pixels. As can be seen in Figure 5, in the reference conventional composite and the thin-ply, damage initiation occurred in the composite adherend. In contrast, for the hybrid (25% thin ply) joint, the damage initiation occurred in the adhesive layer.

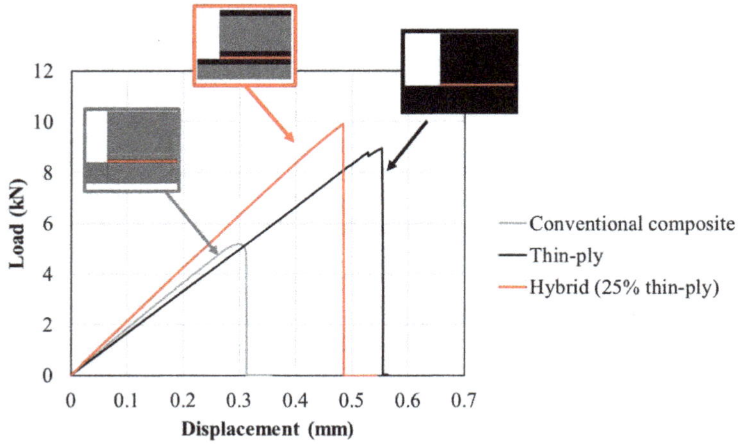

Figure 4. Representative load–displacement curves for reference conventional composite, thin-ply, and hybrid (25% thin ply) joints.

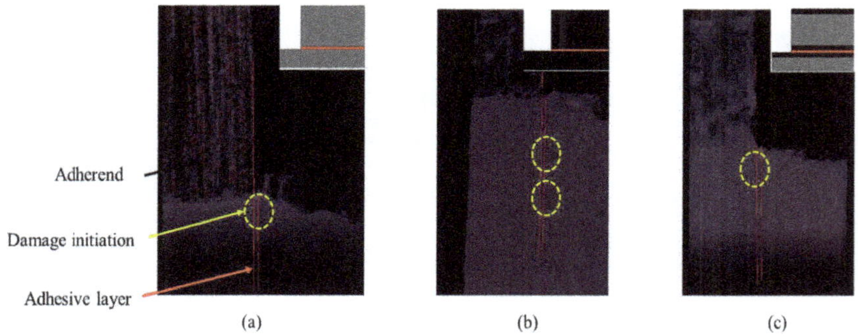

Figure 5. Damage initiation for (**a**) conventional composite, (**b**) thin-ply, and (**c**) hybrid (25% thin ply) joints.

3.3. Delamination

Digital images of the failure surface were analysed in order to obtain the delamination ratio for each configuration. The open-source software IC Measure was used to calculate the delamination area for each joint. The delamination ratio is defined as the delamination area divided by the total bonded area, as presented in Equation (1). It should be noted that the total bonded area was constant and equal to 375 mm^2 for all configurations. Figure 6 provides representative images of the failure surface for all configurations. The representative delamination area from Figure 6 and the average delamination area are presented in Table 6. The reference conventional composite joint shown in Figure 6 presents delamination of about 51%, while the hybrid (25% thin ply) configuration presents delamination of about 29%. In contrast, around 80% of the total area was observed to have suffered delamination in the reference thin-ply joint.

$$Delamination\ ratio\ (\%) = \frac{Delamination\ area}{Total\ bonded\ area} \quad (1)$$

(**a**)

(**b**)

(**c**)

Figure 6. Representative images of failure surface of (**a**) reference conventional composite, (**b**) thin-ply, and (**c**) hybrid (25% thin ply) joints.

Table 6. Representative and average delamination area.

Configuration	Representative [mm^2]	Average [mm^2]
Conventional composite	190.52	210.00 ± 32.75
Thin-ply	296.82	332.07 ± 39.65
Hybrid (25% thin ply)	110.83	128.20 ± 36.92

3.4. Microscopic Images

The failure surfaces of the reference conventional composite and hybrid (25% thin ply) joint were analysed using a ZEISS AXIOPHOT microscope. According to the microscopic images presented in Figure 7, multiple fibre breakages [45] and fibre pull-outs [46] were observed on the failure surface of the reference conventional composite failure surface. These failure mechanisms were not observed in the failure surface of the hybrid (25% thin ply) joint under analysis.

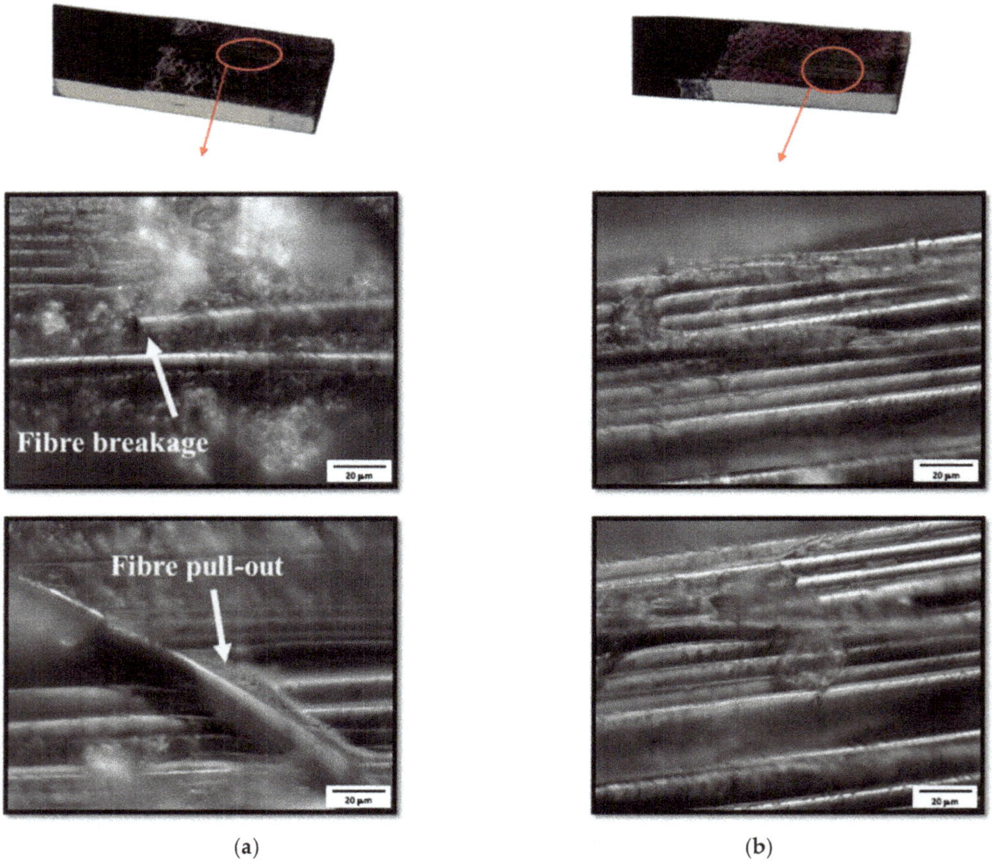

Figure 7. Microscopic images of (**a**) conventional composite and (**b**) hybrid (25% thin ply) joints.

4. Numerical Study

4.1. Load–Disploacement Curve

A two-dimensional statically loaded model was used to simplify the problem under analysis and reduce the computational time. Boundary conditions were defined as shown in Figure 8. The left end of the joint was fixed while a displacement was applied in the right end to replicate the testing fixtures. A cohesive zone model (CZM) was used to model the adhesive behaviour, employing four node elements: cohesive quadrilateral elements. Non-linear geometrical effects were included. Solid cohesive elements following triangular traction separation laws were applied to the adhesive layers of the model to simulate damage evolution (damage initiation and propagation). Cohesive behaviour was specified directly in terms of a traction–separation law, which has been shown to be suitable to represent delamination in composite laminates [47]. Therefore, a similar

CZM was introduced into the composite material (conventional composite or thin-ply) to model delamination due to the experimental failure mode obtained. These interlaminar cohesive element layers were placed in between elastic homogeneous sections (see Figure 9) and effectively simulated the possible debonding between the plies of composite. The CZM layers were placed at a distance of 0.15, 0.37, and 0.13 mm from the interface of the adherend and adhesive layer for the conventional composite, thin-ply, and hybrid (25% thin ply) joints, respectively. This distance roughly corresponded to the experimentally measured distance of the delamination plane from the adhesive layer. The thickness of the cohesive layer matched the thickness of one equivalent composite ply (0.075 mm for thin ply and 0.15 mm for the conventional composite).

Figure 8. Boundary condition of simulated single lap joint.

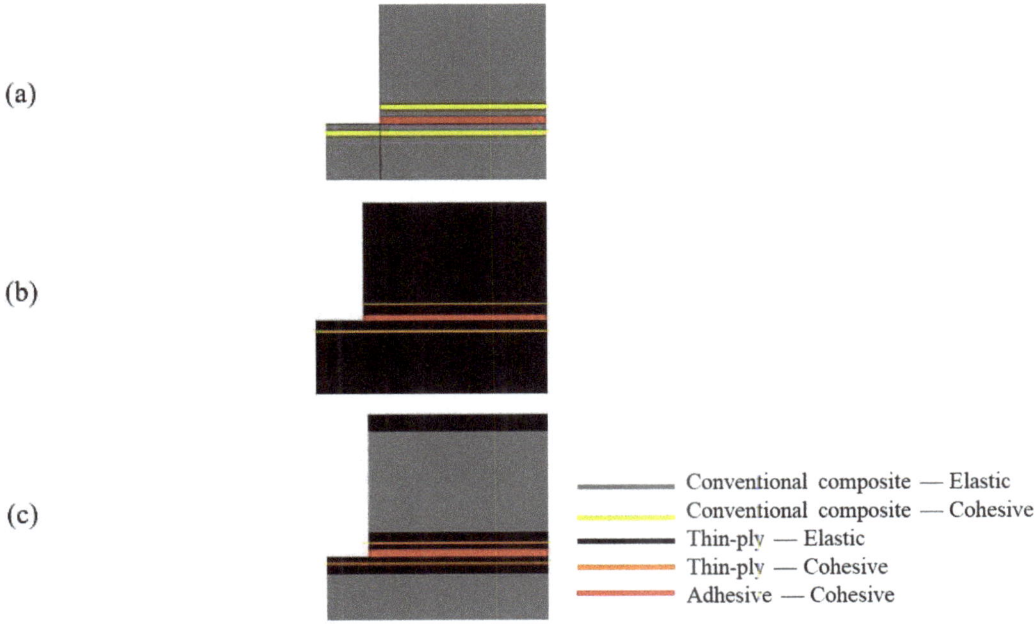

Figure 9. Assigned mechanical properties for (**a**) conventional composite, (**b**) thin-ply, and (**c**) hybrid (25% thin ply) joints.

Double and single biased mesh distributions were considered in the x direction (see Figure 10) for the bondline and the adherends, respectively. The minimum and maximum sizes for the mesh were considered 0.2 and 0.5 mm respectively. However, a uniform mesh distribution with the size of 0.5 mm was considered for the end tabs (in the x direction). Moreover, a uniform distribution through the mesh thickness (y direction) was considered for all models with a mesh size of 0.2 mm. Figure 10 illustrates the mesh distribution mentioned above. As a result, around 15,000 elements were generated for each numerical model. Figure 11 presents the numerical load–displacement curves obtained for all configurations. As shown, the numerical results are in good agreement with the experimentally obtained load–displacement curves.

4.2. Damage

The same model was used to determine the damage initiation and its propagation mode for each configuration under analysis. According to the numerical results presented in Figure 12, the damage for the reference conventional composite and thin-ply joint initiated in the composite (conventional composite and thin-ply joints, respectively). For the hybrid (25% thin ply) joint, damage initiation occurred in the adhesive layer. The loads at damage initiation for the mentioned configurations were 3.6, 6.9, and 7.3 kN, respectively. It should be noted that the damage generated in the configurations at the equivalent numerically obtained failure loads illustrates that delamination is expected to be the final failure mode for all configurations under analysis (see Figure 13).

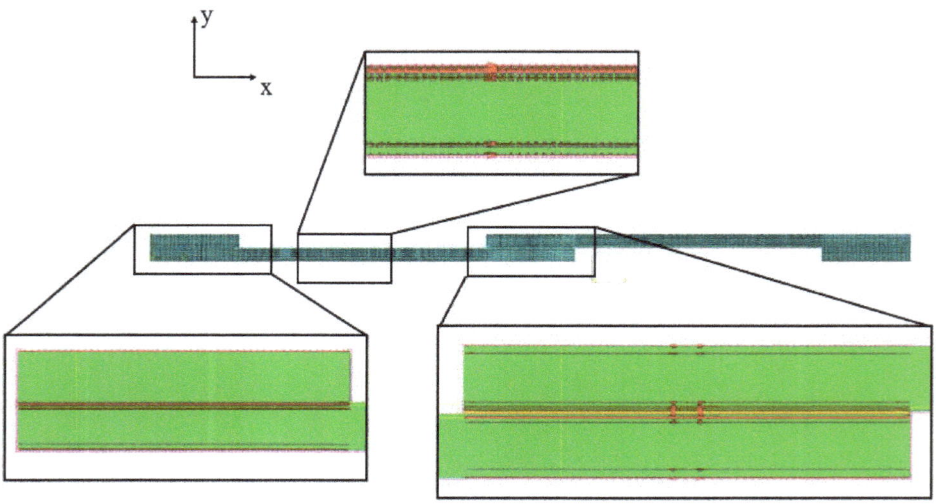

Figure 10. Mesh distribution for numerical models.

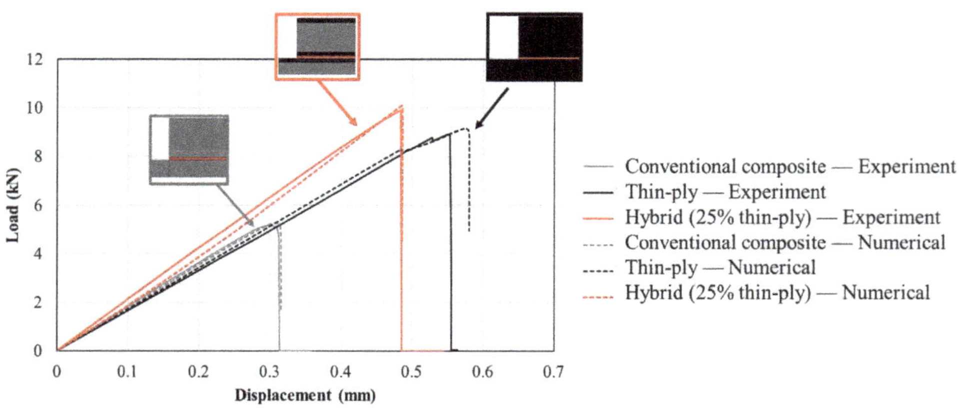

Figure 11. Comparison of numerically obtained load–displacement curves for conventional composite, thin-ply, and hybrid (25% thin ply) joints with the representative experimental results.

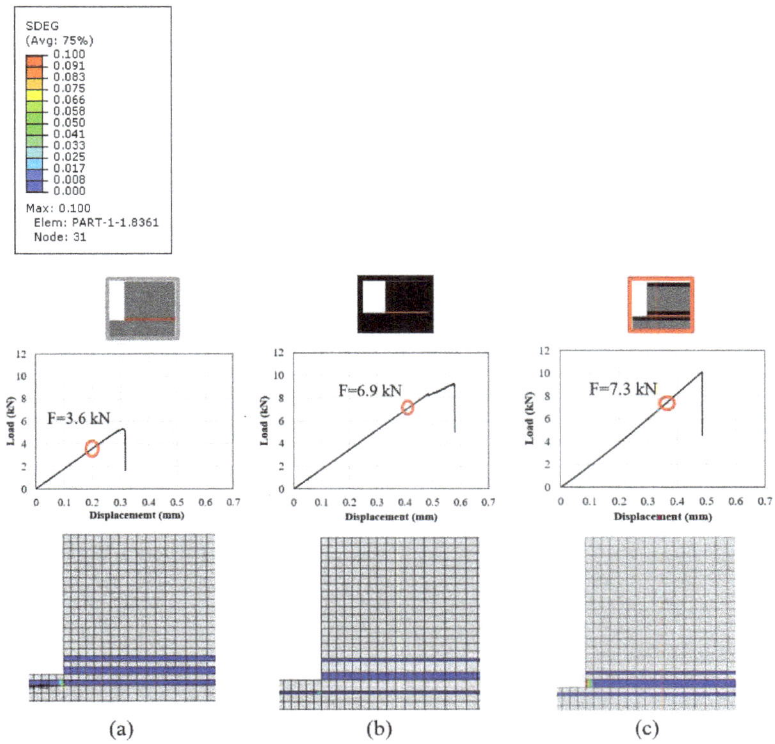

Figure 12. SDEG at damage initiation for (**a**) conventional composite, (**b**) thin-ply and (**c**) hybrid (25% thin ply) joints (equivalent loads for each configuration are 3.6, 6.9, and 7.3 kN, respectively).

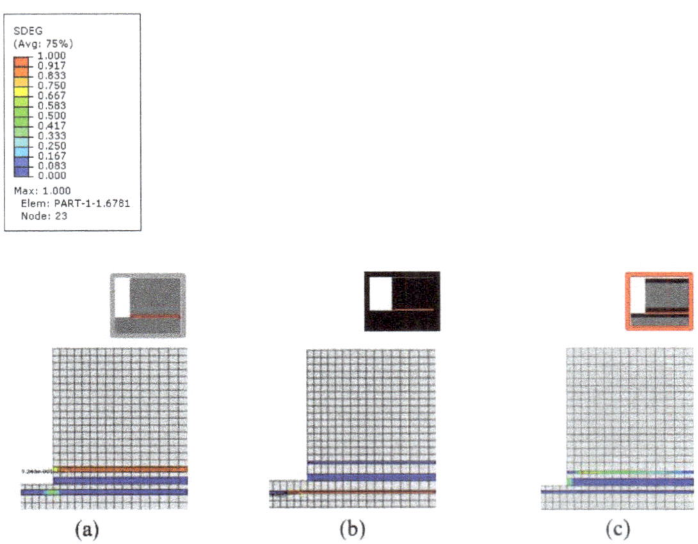

Figure 13. SDEG at failure load for (**a**) conventional composite, (**b**) thin-ply and (**c**) hybrid (25% thin ply) joints (equivalent loads for each configuration are 5.4, 9.2, and 10.0 kN, respectively).

5. Discussion

The use of hybrid, adhesively bonded composite joints reinforced with thin plies increases the tensile strength compared to that of reference joints manufactured using only a conventional composite. Figure 14 presents the average failure load obtained for the reference conventional composites, thin-ply, and hybrid (25% thin ply) single lap joints. An increase of about 90% in the failure load was obtained for the hybrid (25% thin ply) joints. Although similar failure loads were obtained for the hybrid (25% thin ply) and reference thin-ply joint, it must be mentioned that the manufacturing process for a hybrid (25% thin ply) joint costs less and is less time-consuming compared to the process of manufacturing the reference thin-ply joint. Moreover, experimental observation illustrates that the delamination ratio decreases considerably while using hybrid composite joints reinforced with thin- ply compared to the reference conventional composite and thin-ply single lap joint. The average delamination ratio obtained for this configuration can be found in Figure 15. According to the numerical and experimental study, the damage initiation location depends on the joint configuration. Damage first occurs in the adherend in the reference conventional composite and thin ply, but it initiates in the adhesive layer for the hybrid (25% thin ply) joint. The initiated damage propagates as a combination of delamination and cohesive failure for all configurations, but a lower level of delamination was obtained for the hybrid joint. A schematic representation of the mentioned failure mechanism could be found in Figure 16.

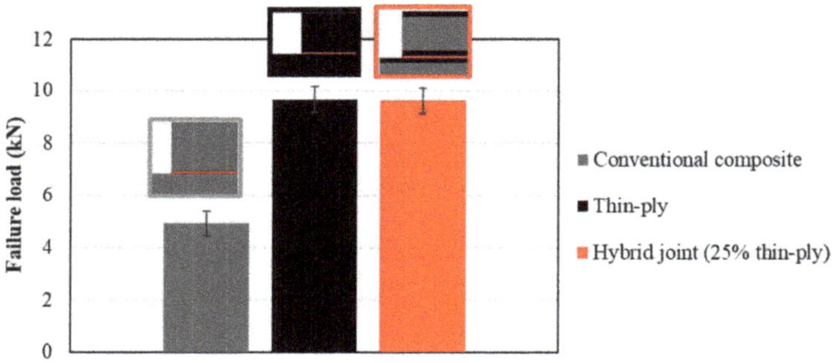

Figure 14. Average failure loads obtained experimentally for reference conventional composite, thin-ply, and hybrid (25% thin ply) joints.

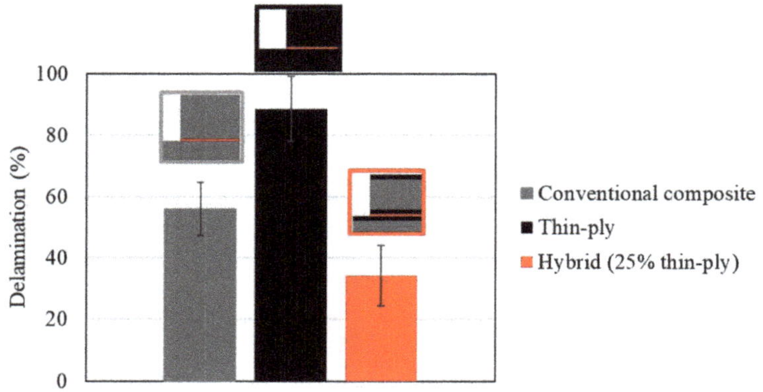

Figure 15. Average delamination ratios for reference conventional composite, thin-ply, and hybrid (25% thin ply) joints.

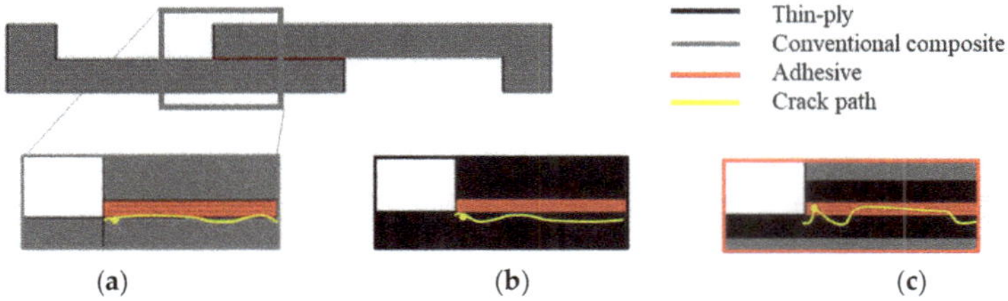

Figure 16. Schematic representation of the failure mechanism for (**a**) reference conventional composite, (**b**) thin-ply, and (**c**) hybrid (25% thin ply) joints.

6. Conclusions

This study investigated the mechanical performance of composite single lap joints using toughened adherends reinforced with thin plies. A numerical and experimental study was performed accordingly. The main conclusions drawn from this work are as follows:

- An increase of approximately 90% in the failure load was found for the hybrid joint reinforced with thin ply when compared to the reference conventional composite joint.
- According to the experimental observation, damage initiation occurs in the adherend for the reference conventional composite and thin-ply joint, while for the hybrid (25% thin ply) joint, damage initiation occurs in the adhesive layer.
- Damage propagates as a combination of delamination and cohesive failure for all configurations. However, a more limited amount of delamination was obtained for the hybrid joint.
- Microscopic images of the bond line allowed for the identification of multiple fibre breakages and fibre pull-outs on the failure surface of the reference conventional composite configuration. In contrast, the fibres were still intact and well-aligned in the failure surface of the hybrid joint.
- The configurations under analysis were modelled numerically, and a good agreement was obtained between the numerical and experimental results, allowing for a precise representation of the damage initiation and failure processes.

Author Contributions: Investigation, F.R.; Writing—original draft, F.R.; Writing—review & editing, R.J.C.C., E.A.S.M. and L.F.M.d.S.; Supervision, R.J.C.C., E.A.S.M. and L.F.M.d.S. All authors have read and agreed to the published version of the manuscript.

Funding: The authors gratefully acknowledge the Portuguese Foundation for Science and Technology (FCT) for supporting the work presented here through the individual grants CEECIND/03276/2018 and 2021.07943.BD and Project No. PTDC/EME-EME/2728/2021, "New approaches to improve the joint strength and reduce the delamination of composite adhesive joints".

Institutional Review Board Statement: Not applicable.

Informed Consent Statement: Not applicable.

Data Availability Statement: Not applicable.

Conflicts of Interest: The authors declare no conflict of interest.

References

1. Karataş, M.A.; Gökkaya, H. A review on machinability of carbon fiber reinforced polymer (CFRP) and glass fiber reinforced polymer (GFRP) composite materials. *Def. Technol.* **2018**, *14*, 318–326. [CrossRef]
2. Ashby, M.F.; Jones, D.R. *Engineering Materials 1: An Introduction to Properties, Applications and Design*; Elsevier: Amsterdam, The Netherlands, 2011; Volume 1.
3. Sahu, P.; Gupta, M.K. A review on the properties of natural fibres and its bio-composites: Effect of alkali treatment. *Proc. Inst. Mech. Eng. Part L J. Mater. Des. Appl.* **2020**, *234*, 198–217. [CrossRef]
4. Zhou, M.; Gu, W.; Wang, G.; Zheng, J.; Pei, C.; Fan, F.; Ji, G. Sustainable wood-based composites for microwave absorption and electromagnetic interference shielding. *J. Mater. Chem. A* **2020**, *8*, 24267–24283. [CrossRef]

5. Budzik, M.K.; Wolfahrt, M.; Reis, P.; Kozłowski, M.; Sena-Cruz, J.; Papadakis, L.; Nasr Saleh, M.; Machalicka, K.V.; Teixeira de Freitas, S.; Vassilopoulos, A.P. Testing mechanical performance of adhesively bonded composite joints in engineering applications: An overview. *J. Adhes.* **2022**, *98*, 2133–2209. [CrossRef]
6. Nassiraei, H.; Rezadoost, P. Static capacity of tubular X-joints reinforced with fiber reinforced polymer subjected to compressive load. *Eng. Struct.* **2021**, *236*, 112041. [CrossRef]
7. Nassiraei, H.; Rezadoost, P. Local joint flexibility of tubular T/Y-joints retrofitted with GFRP under in-plane bending moment. *Mar. Struct.* **2021**, *77*, 102936. [CrossRef]
8. Liu, B.; Zhang, Q.; Li, X.; Guo, Y.; Zhang, Z.; Yang, H.; Yuan, Y. Potential advantage of thin-ply on the composite bolster of a bogie for a high-speed electric multiple unit. *Polym. Compos.* **2021**, *42*, 3404–3417. [CrossRef]
9. Su, K.B. *Delamination Resistance of Stitched Thermoplastic Matrix Composite Laminates*; ASTM International: Conshohocken, PA, USA, 1989; pp. 279–300.
10. Zhou, X.; Li, J.; Qu, C.; Bu, W.; Liu, Z.; Fan, Y.; Bao, G. Bending behavior of hybrid sandwich composite structures containing 3D printed PLA lattice cores and magnesium alloy face sheets. *J. Adhes.* **2022**, *98*, 1713–1731. [CrossRef]
11. Guillamet, G.; Turon, A.; Costa, J.; Renart, J.; Linde, P.; Mayugo, J.A. Damage occurrence at edges of non-crimp-fabric thin-ply laminates under off-axis uniaxial loading. *Compos. Sci. Technol.* **2014**, *98*, 44–50. [CrossRef]
12. Tserpes, K.; Barroso-Caro, A.; Carraro, P.A.; Beber, V.C.; Floros, I.; Gamon, W.; Kozłowski, M.; Santandrea, F.; Shahverdi, M.; Skejić, D.; et al. A review on failure theories and simulation models for adhesive joints. *J. Adhes.* **2021**, *98*, 1855–1915. [CrossRef]
13. Sam-Daliri, O.; Farahani, M.; Araei, A. Condition monitoring of crack extension in the reinforced adhesive joint by carbon nanotubes. *Weld. Technol. Rev.* **2020**, *91*, 7–15. [CrossRef]
14. Ghabezi, P.; Farahani, M. Trapezoidal traction–separation laws in mode II fracture in nano-composite and nano-adhesive joints. *J. Reinf. Plast. Compos.* **2018**, *37*, 780–794. [CrossRef]
15. Ramezani, F.; Simões, B.D.; Carbas, R.J.; Marques, E.A.; da Silva, L.F. Developments in Laminate Modification of Adhesively Bonded Composite Joints. *Materials* **2023**, *16*, 568. [CrossRef]
16. Akhavan-Safar, A.; Ramezani, F.; Delzendehrooy, F.; Ayatollahi, M.R.; da Silva, L.F.M. A review on bi-adhesive joints: Benefits and challenges. *Int. J. Adhes. Adhes.* **2022**, *114*, 103098. [CrossRef]
17. Qin, Z.; Yang, K.; Wang, J.; Zhang, L.; Huang, J.; Peng, H.; Xu, J. The effects of geometrical dimensions on the failure of composite-to-composite adhesively bonded joints. *J. Adhes.* **2021**, *97*, 1024–1051. [CrossRef]
18. Ramezani, F.; Nunes, P.D.P.; Carbas, R.J.C.; Marques, E.A.S.; da Silva, L.F.M. The joint strength of hybrid composite joints reinforced with different laminates materials. *J. Adv. Join. Process.* **2022**, *5*, 100103. [CrossRef]
19. Simões, B.D.; Nunes, P.D.; Ramezani, F.; Carbas, R.J.; Marques, E.A.; da Silva, L.F. Experimental and Numerical Study of Thermal Residual Stresses on Multimaterial Adherends in Single-Lap Joints. *Materials* **2022**, *15*, 8541. [CrossRef]
20. Shang, X.; Marques, E.A.S.; Machado, J.J.M.; Carbas, R.J.C.; Jiang, D.; Da Silva, L.F.M. A strategy to reduce delamination of adhesive joints with composite substrates. *Proc. Inst. Mech. Eng. Part L J. Mater. Des. Appl.* **2019**, *233*, 521–530. [CrossRef]
21. Potter, K.D.; Guild, F.J.; Harvey, H.J.; Wisnom, M.R.; Adams, R.D. Understanding and control of adhesive crack propagation in bonded joints between carbon fibre composite adherends I. Experimental. *Int. J. Adhes. Adhes.* **2001**, *21*, 435–443. [CrossRef]
22. Mouritz, A.P. Review of z-pinned composite laminates. *Compos. Part A Appl. Sci. Manuf.* **2007**, *38*, 2383–2397. [CrossRef]
23. Ko, F.K.; Chou, T.W. *Textile Structural Composites*, 6th ed.; Composite Materials Series; Elsevier Science: Amsterdam, The Netherlands, 1989; Volume 3.
24. Sawyer, J.W. Effect of stitching on the strength of bonded composite single lap joints. *AIAA J.* **1985**, *23*, 1744–1748. [CrossRef]
25. Hader-Kregl, L.; Wallner, G.M.; Kralovec, C.; Eyßell, C. Effect of inter-plies on the short beam shear delamination of steel/composite hybrid laminates. *J. Adhes.* **2019**, *95*, 1088–1100. [CrossRef]
26. Verpoest, I.; Wevers, M.; De Meester, P.; Declercq, P. 2.5 D-fabrics and 3D-fabrics for delamination resistant composite laminates and sandwich structures. *Sampe J.* **1989**, *25*, 51–56.
27. Dransfield, K.; Baillie, C.; Mai, Y.W. Improving the delamination resistance of CFRP by stitching—A review. *Compos. Sci. Technol.* **1994**, *50*, 305–317. [CrossRef]
28. Sihn, S.; Kim, R.Y.; Kawabe, K.; Tsai, S.W. Experimental studies of thin-ply laminated composites. *Compos. Sci. Technol.* **2007**, *67*, 996–1008. [CrossRef]
29. Amacher, R.; Cugnoni, J.; Botsis, J.; Sorensen, L.; Smith, W.; Dransfeld, C. Thin ply composites: Experimental characterization and modeling of size-effects. *Compos. Sci. Technol.* **2014**, *101*, 121–132. [CrossRef]
30. Roure, T. C-PLY™, a new structural approach to multiaxials in composites: BI-ANGLE NCF. *JEC Compos.* **2011**, *68*, 53–55.
31. Arteiro, A.; Catalanotti, G.; Xavier, J.; Linde, P.; Camanho, P.P. A strategy to improve the structural performance of non-crimp fabric thin-ply laminates. *Compos. Struct.* **2018**, *188*, 438–449. [CrossRef]
32. Kötter, B.; Karsten, J.; Körbelin, J.; Fiedler, B. CFRP thin-ply fibre metal laminates: Influences of ply thickness and metal layers on open hole tension and compression properties. *Materials* **2020**, *13*, 910. [CrossRef]
33. Wisnom, M.R.; Khan, B.; Hallett, S.R. Size effects in unnotched tensile strength of unidirectional and quasi-isotropic carbon/epoxy composites. *Compos. Struct.* **2008**, *84*, 21–28. [CrossRef]
34. Kim, R.Y.; Soni, S.R. Experimental and analytical studies on the onset of delamination in laminated composites. *J. Compos. Mater.* **1984**, *18*, 70–80. [CrossRef]

35. Huang, C.; He, M.; He, Y.; Xiao, J.; Zhang, J.; Ju, S.; Jiang, D. Exploration relation between interlaminar shear properties of thin-ply laminates under short-beam bending and meso-structures. *J. Compos. Mater.* **2018**, *52*, 2375–2386. [CrossRef]
36. Kupski, J.; Zarouchas, D.; de Freitas, S.T. Thin-plies in adhesively bonded carbon fiber reinforced polymers. *Compos. Part B Eng.* **2020**, *184*, 107627. [CrossRef]
37. Camanho, P.P.; Dávila, C.G.; Pinho, S.T.; Iannucci, L.; Robinson, P. Prediction of in situ strengths and matrix cracking in composites under transverse tension and in-plane shear. *Compos. Part A Appl. Sci. Manuf.* **2006**, *37*, 165–176. [CrossRef]
38. Chan, W.S. Design approaches for edge delamination resistance in laminated composites. *J. Compos. Technol. Res.* **1991**, *13*, 91–96.
39. Ramezani, F.; Carbas, R.J.; Marques, E.A.; Ferreira, A.M.; da Silva, L.F. A study of the fracture mechanisms of hybrid carbon fiber reinforced polymer laminates reinforced by thin-ply. *Polym. Compos.* **2023**, *44*, 1672–1683. [CrossRef]
40. 3M. *3m Scotch-Weld Structural Adhesive Lm Af 163-2k Technical Datasheet*; Technical Report; 3M: St. Paul, MN, USA, 2009.
41. Morgado, M.A.; Carbas, R.J.C.; Dos Santos, D.G.; Da Silva, L.F.M. Strength of CFRP joints reinforced with adhesive layers. *Int. J. Adhes. Adhes.* **2020**, *97*, 102475. [CrossRef]
42. Campilho, R.D.; De Moura, M.F.S.F.; Domingues, J.J.M.S. Modelling single and double-lap repairs on composite materials. *Compos. Sci. Technol.* **2005**, *65*, 1948–1958. [CrossRef]
43. Machado, J.J.M.; Marques, E.A.S.; Campilho, R.D.S.G.; da Silva, L.F. Mode I fracture toughness of CFRP as a function of temperature and strain rate. *J. Compos. Mater.* **2017**, *51*, 3315–3326. [CrossRef]
44. Ramezani, F.; Carbas, R.; Marques, E.A.S.; Ferreira, A.M.; da Silva, L.F.M. Study on out-of-plane tensile strength of angle-plied reinforced hybrid CFRP laminates using thin-ply. *Mech. Adv. Mater. Struct.* **2023**, 1–14. [CrossRef]
45. Liu, P.F.; Chu, J.K.; Liu, Y.L.; Zheng, J.Y. A study on the failure mechanisms of carbon fiber/epoxy composite laminates using acoustic emission. *Mater. Des.* **2012**, *37*, 228–235. [CrossRef]
46. Okoli, O.I.; Smith, G.F. Failure modes of fibre reinforced composites: The effects of strain rate and fibre content. *J. Mater. Sci.* **1998**, *33*, 5415–5422. [CrossRef]
47. Ghabezi, P.; Farahani, M. A cohesive model with a multi-stage softening behavior to predict fracture in nano composite joints. *Eng. Fract. Mech.* **2019**, *219*, 106611. [CrossRef]

materials

MDPI

Article

Effect of Al_2O_3 (x = 0, 1, 2, and 3 vol.%) in CrFeCuMnNi-x High-Entropy Alloy Matrix Composites on their Microstructure and Mechanical and Wear Performance

S. Sivasankaran [1,*], Hany R. Ammar [1,*], El-Sayed M. Sherif [2], Abdulaziz S. Alaboodi [1] and Abdel-baset H. Mekky [3]

[1] Department of Mechanical Engineering, College of Engineering, Qassim University, Buraydah 51452, Saudi Arabia
[2] Center of Excellence for Research in Engineering Materials (CEREM), Deanship of Scientific Research, King Saud University, Riyadh 11421, Saudi Arabia
[3] Department of Physics, College of Science and Arts El-Meznab, Qassim University, Buraydah 51931, Saudi Arabia
* Correspondence: sivasankaran@qec.edu.sa (S.S.); hanyammar@qec.edu.sa (H.R.A.)

Abstract: This work aims to study the influence of Al_2O_3 in CrFeCuMnNi high-entropy alloy matrix composites (HEMCs) on their microstructure, phase changes, and mechanical and wear performances. CrFeCuMnNi-Al_2O_3 HEMCs were synthesized via mechanical alloying (MA) followed by hot compaction (550 °C at 550 MPa), medium frequency sintering (1200 °C), and hot forging (1000 °C at 50 MPa). The XRD results demonstrate the formation of both FCC and BCC phases in the synthesized powders, which were transformed into major stable FCC and minor ordered B2-BCC phases, as confirmed by HRSEM. The microstructural variation of HRSEM-EBSD, in terms of the coloured grain map (inverse pole figures), grain size distribution, and misorientation angle, was analysed and reported. The grain size of the matrix decreased with the increase in Al_2O_3 particles owing to the higher structural refinement by MA and zener pinning of the incorporated Al_2O_3 particles. The hot-forged CrFeCuMnNi-3 vol.% Al_2O_3 sample exhibited an ultimate compressive strength of 1.058 GPa, which was 21% higher than that of the unreinforced HEA matrix. Both the mechanical and wear performance of the bulk samples increased with an increase in Al_2O_3 content due to solid solution formation, high configurational mixing entropy, structural refinement, and the effective dispersion of the incorporated Al_2O_3 particles. The wear rate and coefficient of friction values decreased with the increase in Al_2O_3 content, indicating an improvement in wear resistance owing to the lower domination of abrasive and adhesive mechanisms, as evidenced by the SEM worn surface morphology.

Keywords: high-entropy alloys; Al_2O_3; X-ray diffraction; microstructures; mechanical; wear

1. Introduction

Recently, high-entropy alloys (HEAs) have attracted the attention of the research community owing to their excellent mechanical, wear, thermal, fatigue, and surface properties. HEAs are a multicomponent solid solution category that usually possesses complicated microstructures due to the lack of restrictions on the alloy design. HEAs retain mechanical properties (strength and hardness) at elevated temperatures and are less susceptible to corrosion [1–3]. According to Yeh et al. [4], HEAs can be designed and developed by mixing five or more metallic elements in equiatomic or near equiatomic ratios, which introduces a high-configurational entropy effect in materials. High configurational entropy (promoting stable solid solution), sluggish diffusion (controlling microstructure and enhancing creep resistance), severe lattice distortion (promoting solid solution hardening and thermal resistance), and cocktail effects (controlling phase formation and promoting hardness) are the four effects that alter the microstructure and mechanical properties of HEAs [5–8].

The incorporation of ceramic particles (oxides, carbides, and nitrides) into HEAs can further enhance their mechanical properties and be applied to industrial and structural parts [9,10]. Generally, HEAs exhibit single-phase FCC or BCC or a combination of FCC/BCC/HCP phases (multi-phase alloys) [11–14]. The addition of ceramic particles to HEAs causes the production of more interfaces, dislocation loops, and voids in the grain boundaries, which are expected to promote a hardening effect in the structure [15]. Furthermore, maintaining a stable phase is a major challenge in HEAs because of grain growth at elevated temperatures, which can be eliminated by incorporating ceramic particles [16]. Hadraba et al. [17] developed CoCrFeNiMn-0.3 wt.% Y_2O_3 high-entropy alloy matrix composites (HEAMCs) by an in situ reaction technique. The $Al_{0.3}CoCrFeMnNi$-0.3 Y_2O_3 HEAMC was recently developed and investigated by Gwalani et al. [18]. The HEAMC produced 1800 MPa of mechanical strength, which was 80% higher than the $Al_{0.3}CoCrFeMnNi$ HEA matrix. Prasad et al. [19] synthesised FeNiCoCrAlMn reinforced with 2 wt.% of Al_2O_3 HEAMC by the mechanical alloying (MA) route and studied the thermal behaviour. Their results indicated that the developed sample exhibited a higher resistance to temperature and oxygen. CrMnFeCoNi-0.25 wt.% Y_2O_3 HEAMC was synthesized by Liu et al. [20], who studied the tensile and wear properties. Jia et al. [21] developed an FeCrCoNi-5 wt.% Y_2O_3 HEAMC through the MA route, and studied the microstructure and mechanical behaviour. Yang et al. [22] developed $Al_{0.4}FeCrCo_{1.5}NiTi_{0.3}$-8 vol.% Al_2O_3 HEAMC and achieved a single-phase FCC structure.

Based on the literature, it can be observed that improved strength and ductility can be achieved by the addition of ceramic particles into the HEA matrix, which produces high-entropy alloy matrix composites (HEAMCs) [23]. HEAMCs can be manufactured via vacuum arc melting (VAM [24]), mechanical alloying and spark plasma sintering (MA-SPS [25,26]), induction melting (IM [27]), MA and hot-isostatic pressing (MA-HIP [9,28]), vacuum hot-pressing (VHP [29]), fusion deposition (FD [30]), and laser surface injection (LSI [31]). Al_2O_3, Y_2O_3, WC, TiC, TiN, and SiC [32] are commonly used ceramic particles in HEA that extensively alter the matrix properties. Based on the literature, it is evident that the addition of oxide particles successfully alters the mechanical and structural properties of HEAs [23,28,31,32]. Furthermore, there has been no study on the development and investigation of CrFeCuMnNi HEA matrices reinforced with different vol.% of ultra-fine Al_2O_3 particles (105 nm).

Therefore, the present study is conducted because of the limited number of studies on the synthesis of HEAMC powders of CrFeCuMnNi-x vol.% Al_2O_3 (x = 0, 1, 2, and 3) by MA. The synthesized nanostructured powders were consolidated using hot compaction, followed by medium-frequency sintering and hot pressing. The synthesized powders and bulk samples were characterized by XRD, HRSEM with EDAX, and EBSD analyses. The relative density, mechanical properties (in terms of compressive stress–strain curves, ultimate compressive strength, and strain at the ultimate point), and wear behaviour by the sliding wear test of the prepared bulk samples with the function of Al_2O_3 reinforcements are investigated and reported. Finally, this study recommends the developed HEAMCs for use in aerospace and automotive parts.

2. Materials and Methods

2.1. Synthesis of CrFeCuMnNi-Al_2O_3 HEAMCs

To synthesise CrFeCuMnNi- x vol.% Al_2O_3 (x = 0, 1, 2, and 3) HEAMCs, pure elemental powders related to Cr, Fe, Cu, Mn, and Ni, and ultra-fine α-Al_2O_3 (105 nm) were procured from Messrs (M/s) Nanografi, Jena, Germany. The purity of as-received powder was ensured to be more than 99.5%. The atomic and weight percentages of the synthesized HEAMCs are listed in Table 1. An electronic weighing balance with four digit accuracy was used to weigh the powders, as shown in Table 1, and charged into a four-station planetary high-energy ball mill (HEBM). M/s. TENCON, XQM-4A, Chin, HEBM was used and run in auto-mode according to the parameters. The set parameters during high-energy ball milling (milling media: hardened stainless-steel balls and vials) were 50 h of milling time,

15:1 ball-to-powder ratio, 350 rpm milling speed, 15 min forward, 15 min pause, and 15 min reverse. Ethanol was used as a process control agent (PCA) to minimize cold welding and oxidization during the MA process. The mechanically alloyed powders were dried and the stress was relieved inside a vacuum tube furnace at 393 ± 30 K (M/s. Nabatherm, Luxembourg, Germany).

Table 1. Chemical composition of CrFeCuMnNi-Al_2O_3 high-entropy alloy matrix composites (HEAMCs).

Sample (ID)	Cr		Fe		Cu		Mn		Ni		Al_2O_3	
	at.%	**wt.%**	**at.%**	**wt.%**	**at.%**	**wt.%**	**at.%**	**wt.%**	**at.%**	**wt.%**	**at.%**	**wt.%**
CrFeCuMnNi-0 A_2O_3 (HEAMC-0A)	0.200	0.196	0.200	0.182	0.200	0.223	0.200	0.193	0.200	0.206	0.000	0.000
CrFeCuMnNi-1A_2O_3 (HEAMC-1A)	0.198	0.193	0.198	0.179	0.198	0.219	0.198	0.189	0.198	0.202	0.010	0.018
CrFeCuMnNi-2A_2O_3 (HEAMC-2A)	0.196	0.189	0.196	0.176	0.196	0.215	0.196	0.186	0.196	0.199	0.020	0.035
CrFeCuMnNi-3A_2O_3 (HEAMC-3A)	0.194	0.186	0.194	0.173	0.194	0.211	0.194	0.186	0.194	0.195	0.030	0.052

2.2. Hot Compaction, Medium-Frequency Sintering, and Hot Forging

To prepare the bulk samples, stress-relieved powders were charged into the H13 die-set, heated to 823 ± 20 K, held for 0.75 h, and then, hot-compacted at 550 MPa in a universal testing machine (M/s MTS, USA: 250 kN capacity) and held for 10 min. An H13 steel die-set (heat-treated, 65 HRC) with a bore of 15 mm and a punch diameter of 15 mm was used [33]. A hot pellet with a diameter of 15 mm and a height of 25 ± 1 mm was produced by hot compaction. A graphite lubricant was applied to the die-wall surface before charging the powder. At least five pellets were prepared to study the mechanical behaviour, five samples with the same dimensions were produced to examine the wear behaviour, and three hot pellet samples with a height of 10 mm were used for X-ray diffraction and microstructural characterizations.

The hot-compacted samples were then subjected to medium-frequency sintering in an argon atmosphere. A medium-frequency electric induction furnace manufactured by M/s Zhengzhou Yuanjie Chemical co., Ltd., Zhengzhou, China was used. The maximum heat input capacity of the furnace was 4000 W and 2000 W was used with a heating rate of 120°/min. Sintering was conducted at 1473 ± 20 K and held at this temperature for approximately 15 min; the sintered hot-pellets were transferred to a graphite-coated pre-heated die of 16 mm in diameter (H13 steel, preheated to 750 °C), and forged at 1273 ± 40 K with 50 MPa forging pressure (forging ratio: 1.13:1) and then air-cooled. Figure 1 shows a schematic diagram representing the unit cell of the as-received powders, MA process, hot compaction, medium-frequency sintering, and hot forging.

2.3. Characterization Using HRSEM, XRD, and EBSD

HRSEM (M/s Apreo FEG) was used to examine the surface morphologies of the as-received powder, milled powders, and forged samples. This HRSEM was operated at 30 keV with an accuracy of 1.3 nm resolution at 1 keV. An X-ray diffraction machine (M/s Empyrean, Malven Panalytical) was used to characterize the as-received, milled, and forged samples. The XRD test was performed at a scanning speed of 0.6°/min and step size of 0.01°. Various phase formations were analysed using the X'Pert high score plus software. HRSEM with EBSD (electron backscatter diffraction) detector was used to examine the detailed microstructural analyses in terms of the grain coloured map of inverse pole figures (IPF), grain size distribution, and misorientation angle distribution. For EBSD analyses, the hot-forged samples were polished with electropolishing in an electrolyte, which consisted of 80% methanol and 20% perchloric acid.

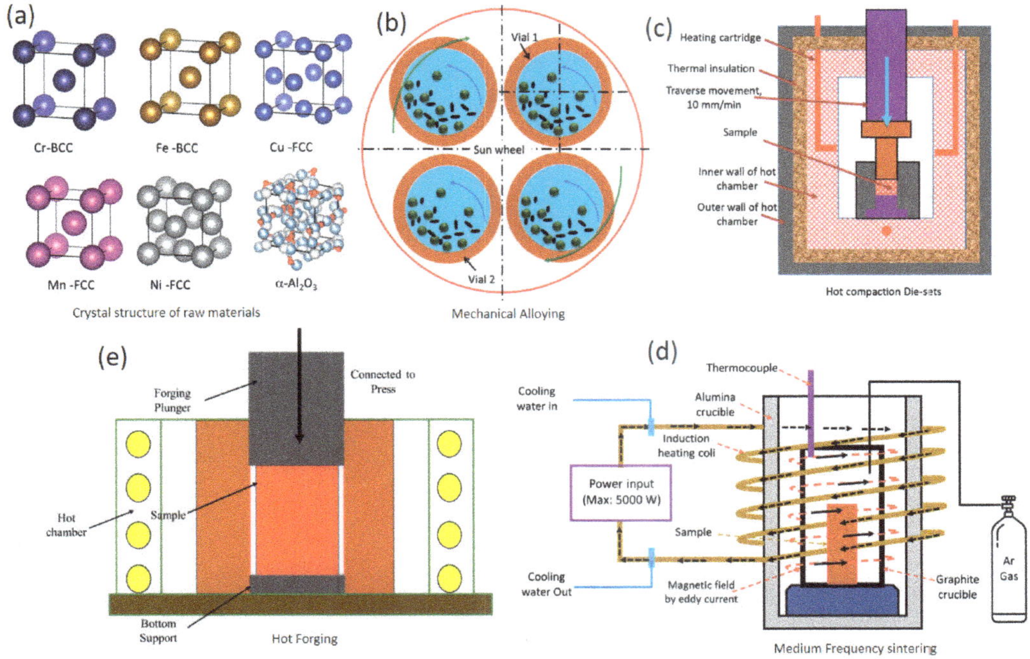

Figure 1. Schematic representing the development of CrFeCuMnNi-x vol.% γ-Al_2O_3 (0, 1, 2, and 3) HEAMCs: (**a**) crystal structure of the raw materials; (**b**) mechanical alloying process; (**c**) hot compaction; (**d**) medium-frequency sintering; and (**e**) hot forging.

2.4. Density Measurement and Mechanical Testing (Hardness, Compression, and Wear Test) of HEAMCs

The density of the hot-forged CrFeCuMnNi-x vol.% Al_2O_3 HEAMC samples was determined by the Archimedes principle, as shown in Equation (1):

$$Density\ of\ forged\ sample,\ \rho_{act} = Relative\ density\ (R) \times density\ of\ water\ (\rho_{water}) \quad (1)$$

The relative density (*R*) was calculated using Equation (2):

$$R = \frac{Mass\ of\ sample\ in\ air\ (m_{air})}{[mass\ of\ sample\ in\ air\ (m_{air}) - mass\ of\ sample\ in\ water\ (m_{water})]} \quad (2)$$

The percentage theoretical density ($\%TD$) was determined by Equation (3):

$$\%TD = \frac{Actual\ density\ (\rho_{act})}{Theoretical\ density\ (\rho_{th})} \times 100 \quad (3)$$

where ρ_{th} is calculated based on the rule of mixtures as per Equation (4):

$$\rho_{th} = \sum_{i=1}^{n} f_i \rho_i \quad (4)$$

where f_i is the volume fraction of the incorporated metallic elements and ρ_i is the theoretical density of the added powders. At least five measurements were performed in each trial, and the average was used for interpretation. Mechanical testing was performed using the Vickers microhardness and compression tests. Before executing the Vickers hardness test, the forged bulk samples were ground using SiC papers of different grit sizes and polished using a ceramic lapping paste to ensure a scratch-free surface. A load (F) of 100 kgf/mm^2 was applied and held for 15 s. The indentation size (d) was measured and, then, the Vickers hardness number (VHN) was determined using Equation (5) [34]:

$$VHN = 1.854\frac{F}{d^2} \tag{5}$$

Compression tests on the forged samples were performed using a 250 kN capacity universal testing machine made by M/s MTS Corporation, USA. The samples were prepared according to the ASTM E 9 standard (ϕ13 × 18 mm). The tests were performed at a loading rate of 1 mm/min. A data acquisition system with test-work software was used to capture the data. Applied compressive load (P) and compression (deformation, δ) were used to determine the mechanical properties. The engineering compressive stress (σ_c) was determined using Equation (6):

$$\sigma_c = \frac{Compressive\ load\ (P)}{Initial\ cross\ sectional\ area\ (A)} \tag{6}$$

The engineering compressive strain (ε) was determined using Equation (7):

$$\varepsilon_c = \frac{Change\ in\ deformation\ (\delta)}{Initial\ sample\ height\ (h)} \tag{7}$$

2.5. Sliding Wear Test

A tribometer (M/s DUCOM, Banglore, India) was used to conduct a sliding wear test on the developed HEAMCs at room temperature. Previously, the samples were prepared with 10 mm diameter and 16 mm height (by wire-cut EDM) and the test was conducted according to the ASTM G99 standard. The samples acted as pins in the sample holder, which was attached to an external load. A counter disc made of EN-32-hardened chromium steel was used. This counter disc has a Rockwell hardness of 63 HRC with 0.55 ± 2 µm surface roughness. A load of 10 N, sliding distance of 2000 m, sliding speed of 3 m/s, track distance of 130 mm, and sliding time of 11.11 min were used during the wear tests [35–38]. A load cell was used to measure the induced frictional force (F) and a linear variable differential transducer was used to measure the log wear depth. Five trials were used during the wear test and the average was used for the investigation. Mass loss was calculated before and after the wear tests. Figure 2 shows a schematic of the wear test apparatus used in this study (Figure 2a) and the sliding wear test using the pin-on-disc approach (Figure 2b). Figure 2c,d shows photographs of the sliding wear test apparatus used in this study. The wear performance was investigated using Equations (8)–(10):

$$Specific\ wear\ rate,\ W_R = \frac{Worn\ surface\ volume\ (WSV)}{Applied\ load\ (P) \times Sliding\ distance\ (D)},\ \frac{m^3}{Nm} \tag{8}$$

$$W_R = \frac{Mass\ loss\ (\Delta m)}{Applied\ load\ (P) \times Theoretical\ density\ (\rho) \times Sliding\ distance\ (D)},\ \frac{m^3}{Nm} \tag{9}$$

$$Coefficient\ of\ friction\ (CoF) = \frac{Friction\ force\ (F)}{Thrust\ force\ (P)} \tag{10}$$

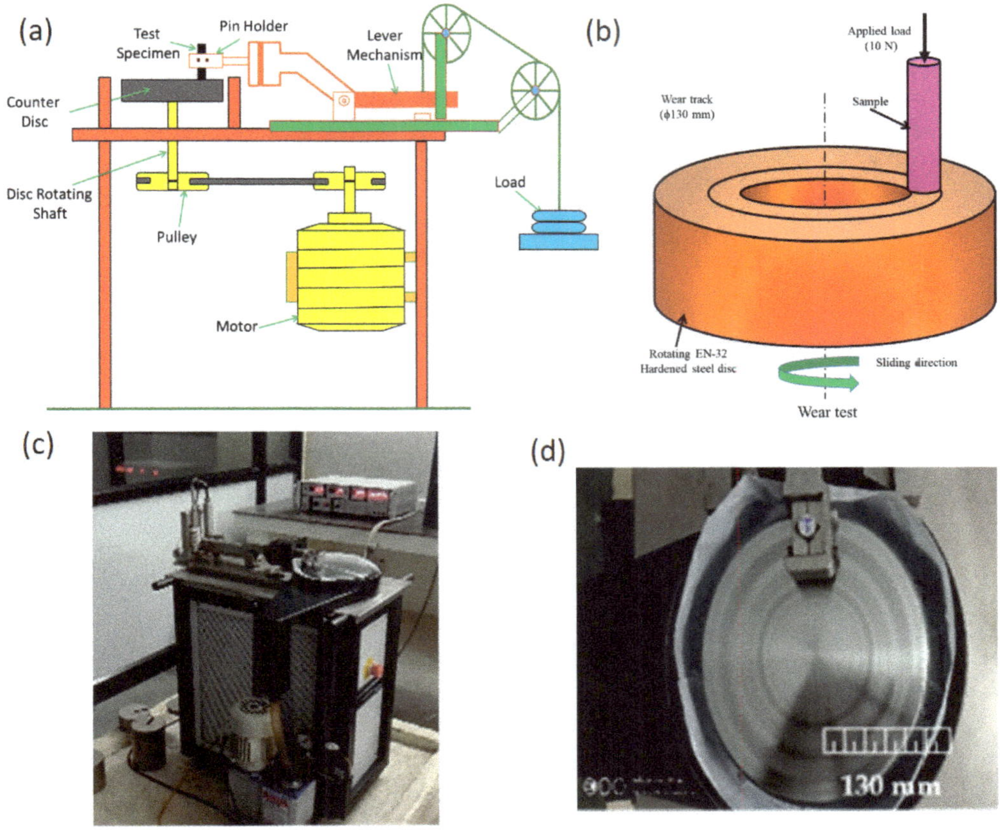

Figure 2. Schematic diagram representing: (**a**) wear test apparatus used in this work; (**b**) sliding wear principle by pin-on-disc approach; (**c**) photograph of the wear test apparatus used in this work; and (**d**) photograph of the pin and counter disc.

3. Results and Discussion

3.1. Examination of the Powder Surface Morphology of HEAMCs

Figure 3 shows the HRSEM-backscattered electron (BSE) images of the as-received elemental powder surface morphology. It was observed that irregular flake-like shapes with an average particle size of 48 ± 3.4 µm (Figure 3a), irregular and equiaxed shapes with an average size of 45 ± 2.8 µm (Figure 3b), regular spherical shapes with an average size of 26 ± 6.45 µm (Figure 3c), irregular polygonal shapes with an average size of 27 ± 5.78 µm (Figure 3d), and spherical shapes attached to some satellite particles with an average size of 20.5 ± 3.89 µm (Figure 3e) were noticed for Cr, Fe, Cu, Mn, and Ni, respectively. The HRSEM BSE image of the as-received α-Al_2O_3 powder particles is shown in Figure 4a. From Figure 4a, it is clear that almost spherical particles in the agglomerated condition were observed in the as-received condition. Based on several BSE images and ImageJ software, the average particle size of the as-received α-Al_2O_3 was 105 ± 3.5 nm. The formation of agglomeration is attributed to van der Waals forces of attraction and a higher surface energy, which is a common phenomenon in nanoparticles [39]. The X-ray diffraction (XRD) peak profile of the as-received α-Al_2O_3 powder is shown in Figure 4b, in which the peaks related to Al_2O_3 at different planes were observed and the crystal structure of the as-received Al_2O_3 was rhombohedral (space group: R-3C, and space group no: 167). Figure 4b does not show any other unusual peaks, which confirms the purity of the as-received Al_2O_3 powders.

Figure 3. HRSEM images of the as-received elemental powders: (**a**) Cr; (**b**) Fe; (**c**) Cu; (**d**) Mn; and (**e**) Ni.

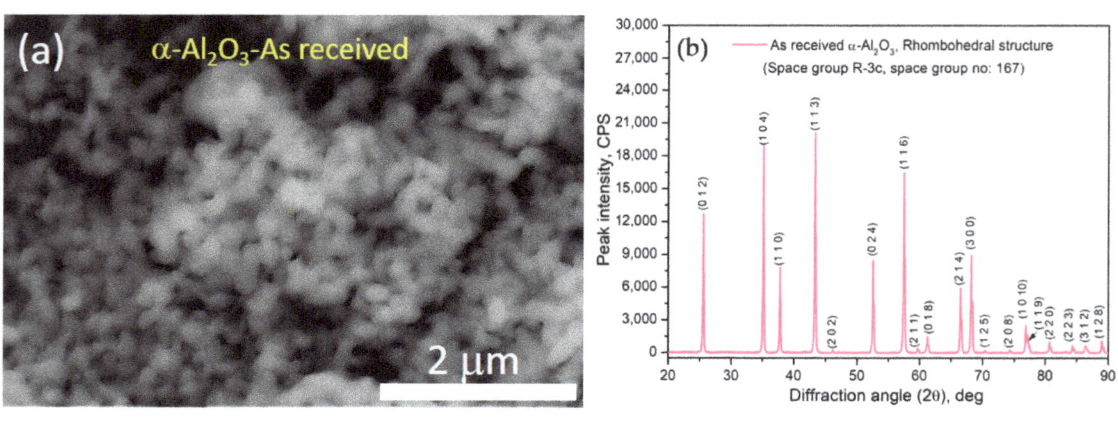

Figure 4. (**a**) HRSEM image of the as-received α-Al_2O_3 powders; (**b**) corresponding X-ray diffraction profile of α-Al_2O_3.

To confirm the solid solution formation and dispersion of the incorporated Al_2O_3 ceramic particles, HRSEM BSE images were captured for the CrFeCuMnNi alloy matrix (Figure 5a,b) and CrFeCuMnNi-2 vol.% Al_2O_3 HEAMC (Figure 5c,d) samples after 50 h of the MA process. Almost equiaxed powder particles were observed in both samples, indicating the attainment of a steady state [40]. Most of the powder particles were very fine in size, and a few particles exhibited a flake-like shape with a large size. More flake-like large powder particles were observed in the CrFeCuMnNi of the unreinforced alloy (Figure 5a), whereas fewer flake-like large powder particles were observed in the CrFeCuMnNi-2 vol.% Al_2O_3 HEAMCs sample (Figure 5b). Based on several images and using ImageJ software, the average sizes of the CrFeCuMnNi and CrFeCuMnNi-2 vol.% Al_2O_3 samples were 2.1 μm, and 1.1 μm, respectively. The formation of a very fine matrix (2.1 μm size of unreinforced HEA and 1.1 μm size of HEMCs) and composite particles was attributed to the high energy imported into charged materials by HEBM and the high-configurational entropy effect. These results demonstrate that the matrix powder particle size decreased by 47% in the CrFeCuMnNi-2 vol.% Al_2O_3 sample compared to that of the unreinforced sample. This was attributed to the incorporation of Al_2O_3 particles that acted as the milling medium. Repeated cold welding and fracturing are the main mechanisms involved in the MA process owing to the kinetic mechanical collisions produced by hardened stainless-steel balls and vials. Cold welding produces agglomerated large powder particles, whereas the fracturing mechanism helps to decrease the particle size. The insets at the bottom of Figure 5a,c show the attainment of a solid solution and the homogenization of the powder particles after 50 h of MA. The yellow arrow in the bottom inset of Figure 5c represents the dispersion of Al_2O_3 particles in the matrix powders, which confirms the successful production of HEAMCs. The HRSEM EDS spectra of both samples are shown in Figure 5b,c, which show all elements related to the incorporated one. The results confirm the composite nature of the milled powders as the presence of O and Al peaks in Figure 5d was observed in addition to Cr, Fe, Cu, Mn, and Ni peaks. Figure 6 shows the elemental overlay maps of the CrFeCuMnNi and CrFeCuMnNi-2 vol.% Al_2O_3 samples. These results also indicate the formation of a solid solution, homogeneous state, and dispersion of Al_2O_3 particles in the CrFeCuMnNi-2 vol.% Al_2O_3 sample (Figure 6b).

3.2. *X-ray Diffraction Phase Analyses on HEAMCs*

Figure 7 shows the X-ray diffractometer results of the blended (0 h) samples of CrFeCuMnNi and CrFeCuMnNi-3 vol.% Al_2O_3, exhibiting high intensity with narrow peaks owing to the lack of kinetic energy given to the charged powders. The X-ray peak profiles of the 50 h mechanically alloyed powders of the CrFeCuMnNi, CrFeCuMnNi-1 vol.% Al_2O_3, CrFeCuMnNi-2 vol.% Al_2O_3, and CrFeCuMnNi-3 vol.% Al_2O_3 HEAMC samples are shown in Figure 8. The results demonstrate that the observed peak profiles decreased extensively and peak broadening was observed. In addition, several peaks vanished and merged into single peaks, indicating the formation of a solid solution. Major body-centred cubic (BCC), face-centred cubic (FCC), and Al_2O_3 phases were observed, which confirms the successful fabrication of HEAMCs. Table 2 lists the atomic radii, crystal structures, and physical properties of the incorporated metallic elements and the ceramic particles. As expected, a supersaturated solid solution (Figures 5 and 6) was easily obtained because the atomic radii of the incorporated elements were similar (Table 2). A severe reduction in peak intensity and a drastic increase in peak broadening were expected owing to the 50 h MA process, which introduced the kinetic energy of mechanical collisions obtained from the milling medium (stainless-steel balls and vials) [40].

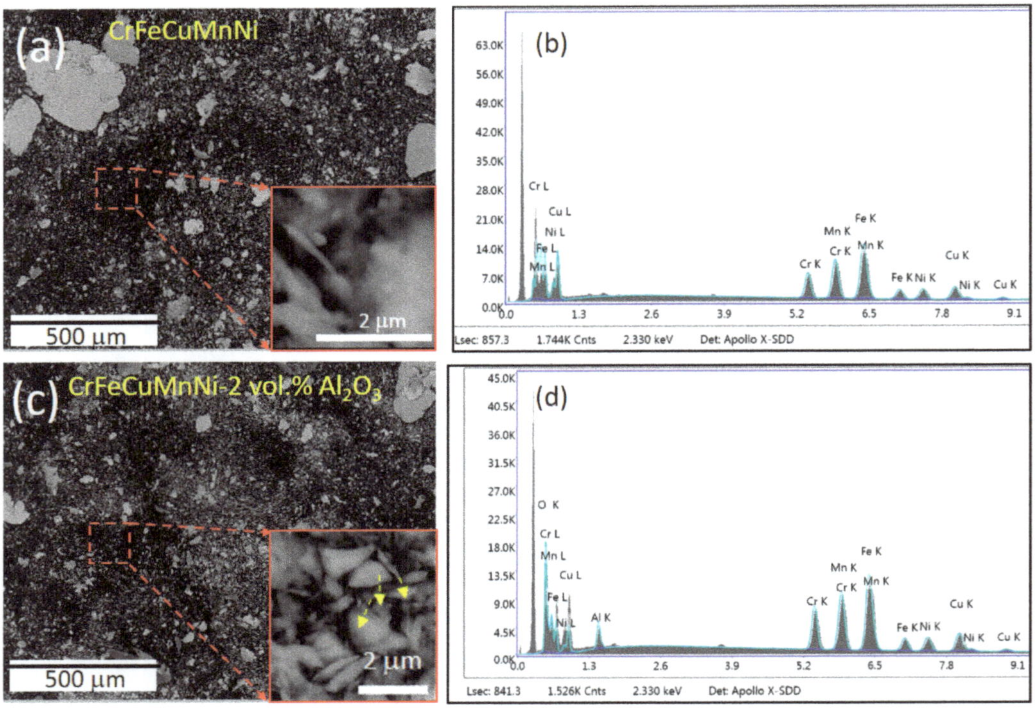

Figure 5. HRSEM powder surface morphologies and EDAX spectrum for the developed HEAMCs: (**a**) powder surface morphology of CrFeCuMnNi (inset shows the solid solution formation); (**b**) the corresponding EDS of (**a**); (**c**) powder surface morphology of CrFeCuMnNi-2 vol.% Al_2O_3 (inset shows the solid solution formation and dispersion of Al_2O_3 particles); and (**d**) the corresponding EDS of (**c**).

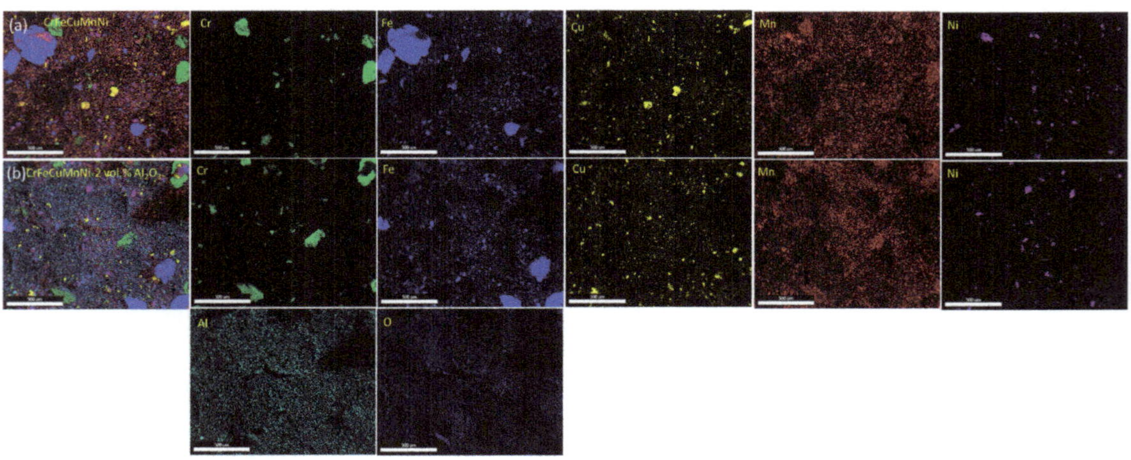

Figure 6. HRSEM elemental overlay map analyses on the developed high-entropy alloy matrix composites (HEAMCs): (**a**) CrFeCuMnNi; and (**b**) CrFeCuMnNi-2 vol.% Al_2O_3.

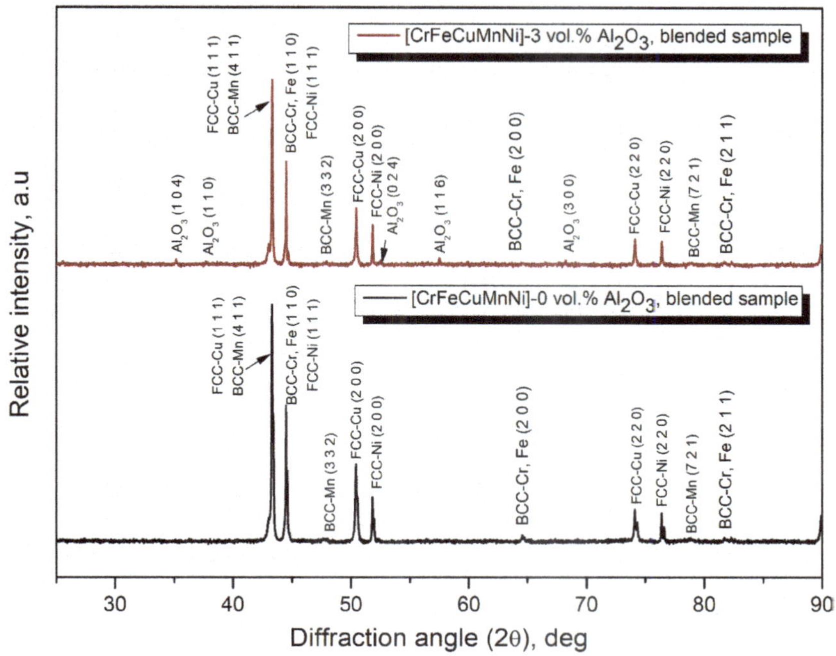

Figure 7. X-ray diffraction peak profile of the blended (0 h) samples of CrFeCuMnNi and CrFeCuMnNi-3 vol.% Al_2O_3 HEAMCs.

Table 2. Crystal structure, atomic radius, and physical properties of the developed HEAMCs.

Name of Element	Atomic Radius, pm	Crystal Structure	Melting Point (K)	Theoretical Density, kg/m^3
Cr	140	BCC	1811	7140
Fe	140	BCC	2180	7874
Cu	135	FCC	1357	8920
Mn	140	FCC	1519	7260
Ni	135	FCC	1728	8908
Al_2O_3	-	Rhombohedral	2345	3987

The X-ray diffractometer results of the hot-forged CrFeCuMnNi, CrFeCuMnNi-1 vol.% Al_2O_3, CrFeCuMnNi-2 vol.% Al_2O_3, and CrFeCuMnNi-3 vol.% Al_2O_3 HEAMCs are shown in Figure 9a, showing a major FCC phase and a minor B2-BCC phase. The major BCC phase observed in the 50 h milled powders almost disappeared after hot forging due to high-temperature sintering followed by hot forging and air cooling. Some authors [41,42] also observed the disappearance of the major BCC phase and the appearance of a major FCC phase after consolidation. Figure 9b shows the magnified view representing the major peak shift occurs with the variation of Al_2O_3 content due to more structural refinement. Further, Table 3 list the calculated internal structural parameters of hot-forged samples. The presence of a major FCC and minor B2-BCC phases along with the incorporated Al_2O_3 particles was expected to improve the mechanical and wear properties of the fabricated HEAMCs [43,44]. In addition, the Fe_2O_3 phase was observed in all hot-forged HEAMCs samples owing to contamination from the milling medium and high-temperature sintering followed by forging and air cooling.

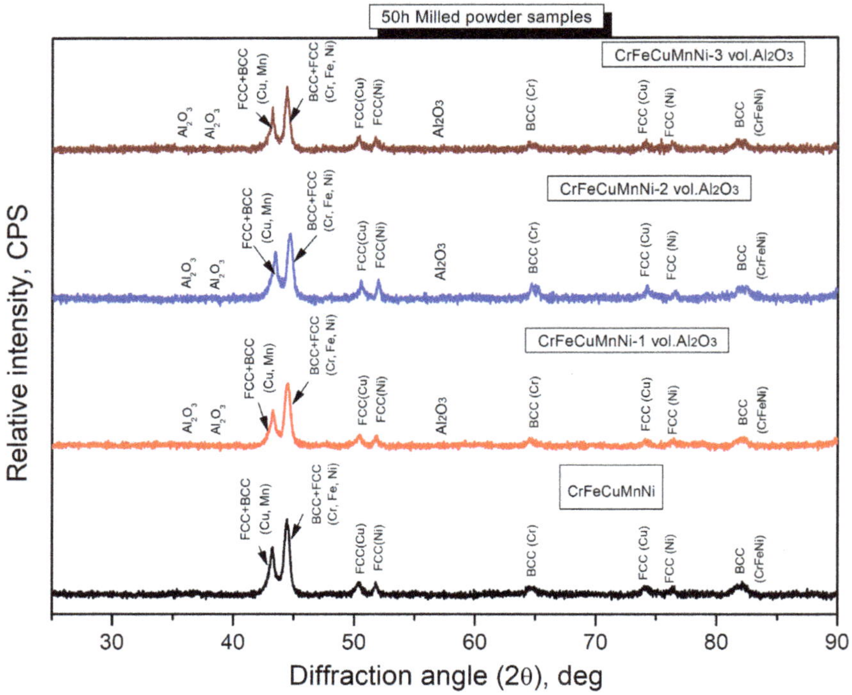

Figure 8. X-ray diffraction peak profile of the 50 h milled HEAMCs of CrFeCuMnNi, CrFeCuMnNi-1 vol.% Al_2O_3, CrFeCuMnNi-2 vol.% Al_2O_3, and CrFeCuMnNi-3 vol.% Al_2O_3.

Table 3. Structural parameters (peak position, FWHM, peak area, d-spacing, and peak height) of the hot-forged HEAMCs obtained from the XRD results.

Sample	Peak Position, 2θ (deg)	FWHM, deg	Peak Area	d-Spacing [Å]	Maximum Peak Height [cts]
CrFeCuMnNi	44.973	0.34158	379.325	2.01570	965
CrFeCuMnNi-1 vol.% Al_2O_3	44.1652	0.3657	246.28	2.05068	788.36
CrFeCuMnNi-2 vol.% Al_2O_3	43.5311	0.4093	307.34	2.07907	868.3256
CrFeCuMnNi-3 vol.% Al_2O_3	43.4662	0.4196	248.36	2.08202	1014

3.3. Physicochemical and Thermodynamic Parameters of the Developed HEMCs

Various parameters, namely, Gibbs free energy (ΔG_{mix}), enthalpy of mixing (ΔH_{mix}), entropy of mixing (ΔS_{mix}), thermodynamic parameter (Ω), atomic size deviation (δ), valence electron concentration (VEC), and electro-negativity (ΔX), can be used for predicting the phase formation and the phase stability of the developed HEAMCs. All parameters were calculated based on Equations (11)–(19) [45,46].

$$\Delta G_{mix} = \Delta H_{mix} - T\Delta S_{mix} \tag{11}$$

$$\Delta S_{mix} = -R\sum_{i=1}^{N}(C_i lnC_i) \tag{12}$$

where R is the gas constant (8.314 J/mol·K) and C_i represents the atomic fraction of the ith element.

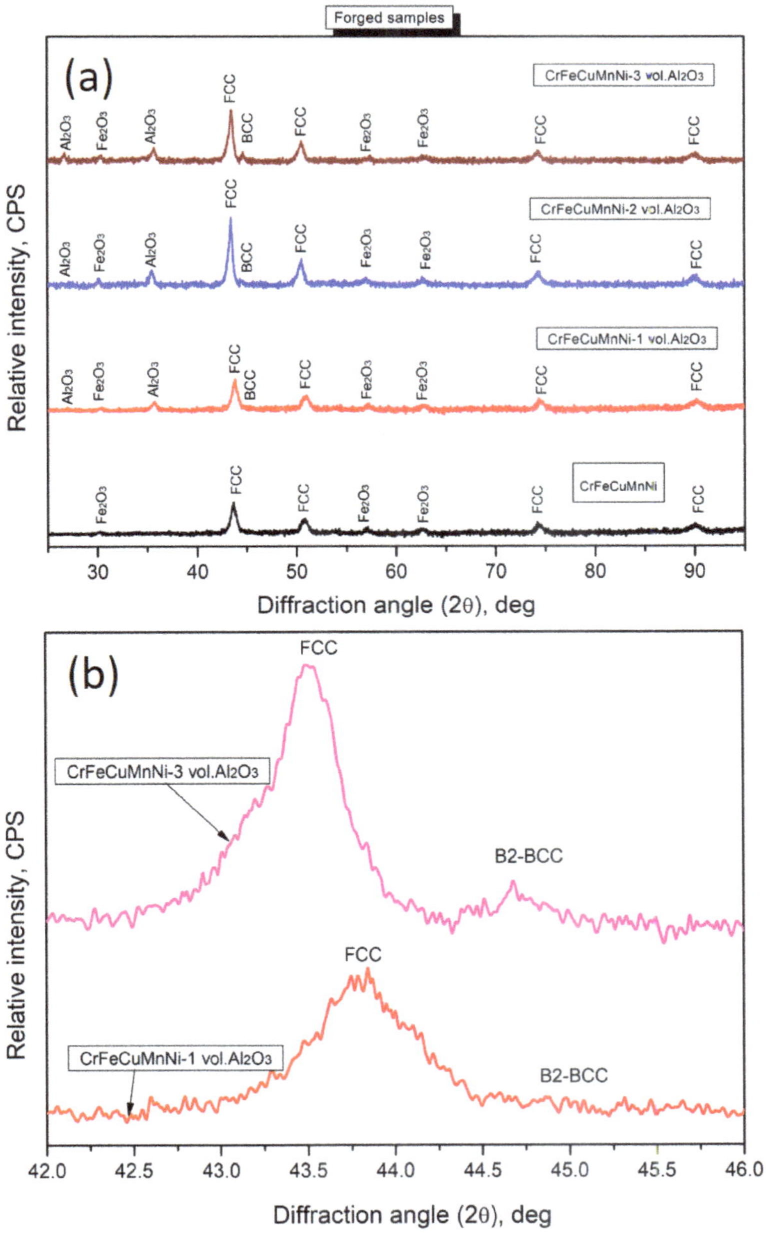

Figure 9. X-ray diffraction peak profile of the hot-forged HEAMCs of: (**a**) CrFeCuMnNi, CrFeCuMnNi-1 vol.% Al_2O_3, CrFeCuMnNi-2 vol.% Al_2O_3, and CrFeCuMnNi-3 vol.% Al_2O_3, (**b**) magnified view (**a**) showing the peak shift.

$$\Delta H_{mix} = \sum_{i=1,\ i\neq j}^{N} (4\Delta H_{mix}^{AB} C_i C_j) \quad (13)$$

where ΔH_{mix}^{AB} is the mixing enthalpy between the *i*th and *j*th elements (Table 4 [47]).

$$\Omega = \frac{T_m \Delta S_{mix}}{\Delta H_{mix}} \quad (14)$$

where T_m is the melting point in K (see Table 2).

$$\delta = \sqrt{\sum_{i=1}^{N} C_i \left(1 - \frac{r_i}{\bar{r}}\right)^2} \quad (15)$$

$$\bar{r} = \sum_{i=1}^{N} C_i r_i \quad (16)$$

where r_i is the atomic radius in m (see Table 2).

$$VEC = \sum_{i=1}^{N} C_i (VEC)_i \quad (17)$$

$$\Delta X = \sqrt{\sum_{i=1}^{N} C_i (X_i - \bar{X})^2} \quad (18)$$

$$\bar{X} = \sum_{i=1}^{N} C_i X_i \quad (19)$$

where X_i represents the Pauling electronegativity of the *i*th element.

Table 4. Magnitude of the enthalpy mix (H_{mix}^{AB}, kJ/mol) calculated using Miedema's model for the incorporated metallic elements.

	Fe	Cr	Cu	Mn	Ni
Fe	Fe	−1	13	0	−2
Cr	−1	Cr	12	2	−7
Cu	13	12	Cu	4	4
Mn	0	2	4	Mn	−8
Ni	−2	−7	4	−8	Ni

The values of δH_{mix}, δS_{mix}, δG_{mix}, Ω, δ, VEC, and δX for the developed HEAMCs are listed in Table 5. The criteria for the formation of a solid solution were $\Omega > 1.1$ and $\delta < 6.6\%$. From Table 5, it is clear that a solid solution was possible in the developed HEAMCs. Furthermore, the mixing entropy increased with the addition of Al_2O_3 and decreased with the formation of intermetallic phases after hot forging at elevated temperatures. VEC can be used to predict the phases and phase stabilities of the developed HEAMCs. The conditions for phase formation were: stable FCC phase formation when VEC > 8; single BCC phase formation when VEC < 6.87; and the mixing of FCC and BCC phase formation when 6.87 < VEC < 8 [48]. Based on the VEC values, the developed HEAMCs exhibited values greater than 8, indicating that the stable FCC phase was expected. The predicted phases matched the XRD results of the hot-forged samples (Figure 9). The minor B2-BCC phase formation was attributed to the Cr- and Fe-rich precipitates formed after the high-temperature forging process. The formation of the minor B2-BCC phase was attributed to the increase in the Pauling electronegativity difference (Table 5) with increasing the Al_2O_3 content in the CrFeCuMnNi HEA matrix [49]. Elements possessing high electronegativity differences are prone to losing or gaining electrons, which is expected to promote the formation of the B2-BCC phase in the developed HEAMCs.

Table 5. Calculated thermodynamic parameters values (δH_{mix}, δS_{mix}, δG_{mix}, Ω, r, VEC, and δX) for the developed ODS-HECs.

	δH_{mix}, kJ/mol	δS_{mix} J/mol.k	δG_{mix} J/mol	Ω	r, nm	δ, %	VEC	δX	Expected Phases
CrFeCuMnNi	2.72	13.38	−1267.5	8.4565	0.138	1.77	8.400	0.1418	FCC
CrFeCuMnNi−1 vol.%Al_2O_3	2.67	13.71	−1420.5	8.8743	0.136	2.05	8.316	0.1422	FCC
CrFeCuMnNi−2 vol.%Al_2O_3	2.61	13.93	−1538.36	9.2322	0.135	2.70	8.232	0.1430	FCC
CrFeCuMnNi−3 vol.%Al_2O_3	2.56	14.10	−1642.46	9.5739	0.133	3.54	8.148	0.15	FCC

3.4. HRSEM Microstructural Examination with EBSD Analyses on Hot-Forged HEAMCs

Figure 10a–d shows the HRSEM microstructures of the hot-forged HEAMCs that exhibited both the FCC phase (large area) and a small amount of the B2-BCC phase (dark region in a small area). These results correlate well with the XRD peak profiles of the hot-forged samples (Figure 9). It can be seen that the observed major FCC grains and minor BCC grains were refined with the addition of Al_2O_3, indicating that the incorporated Al_2O_3 particles decreased the CrFeCuMnNi HEA matrix grain growth. Figure 10e,f show the magnified HRSEM microstructures of the CrFeCuMnNi-2 vol.% Al_2O_3 and CrFeCuMnNi-3 vol.% Al_2O_3 samples, respectively. The dispersion of the incorporated Al_2O_3 particles was observed in both samples; however, the little agglomeration of the Al_2O_3 particles over the HEA matrix grain boundaries was observed in the CrFeCuMnNi-3 vol.% Al_2O_3, indicating the limit of the addition of Al_2O_3 particles in the developed HEAMCs. Furthermore, the formation of the partial dissolution and overlapping of the B2-BCC phase over the FCC phase was observed because of the hot-forging process; consequently, improved mechanical properties were expected.

HRSEM with EBSD analyses was conducted for the developed HEAMCs to examine the grain size of the matrix by inverse pole figures (IPFs) and grain refinements by grain size and misorientation angle distribution. An FEI Quanta FEG SEM with TSL-OIM software was used to examine the EBSD data and the linear intercept method was used to measure the grain size. Figure 11 shows the EBSD inverse pole figures (IPFs) of the developed HEAMCs, which clearly demonstrate the variation in the CrFeCuMnNi matrix grain size with the addition of Al_2O_3 particles. The observed grain size started to decrease with the increase in Al_2O_3 particles owing to the solid solution formation, dislocation pinning by Al_2O_3 particles, high-configurational entropy, and structural refinement of MA and hot-forging processes. In addition, the observed matrix grain size was almost equiaxed in shape, indicating an isotropic nature [50]. The grain size distribution of the CrFeCuMnNi matrix based on the EBSD microstructures over the area fraction was determined and is shown in Figure 12. The results demonstrate that the CrFeCuMnNi matrix grain size decreased with the increase in Al_2O_3 particles, confirming the grain refinement process by MA followed by the hot-forging process [51]. The average matrix grain sizes were 4.96, 4.40, 2.39, and 2.07 μm for CrFeCuMnNi, CrFeCuMnNi-1 vol.% Al_2O_3, CrFeCuMnNi-2 vol.% Al_2O_3, and CrFeCuMnNi-3 vol.% Al_2O_3, respectively. The misorientation angle distribution over the number fraction of developed HEAMCs is shown in Figure 13. The misorientation angle of the observed grains represents structural refinement. Low-angle grain boundaries (LAGBs, <15°) indicate a coarse grain nature, whereas high-angle grain boundaries (HAGBs, >15°) denote the fine-grained nature of the samples. In this study, the observed percentages of LAGBs and HAGBs were 52.83% and 47.17%, 49.38% and 50.62%, 40.12% and 59.88%, and 33.81% and 66.19% for CrFeCuMnNi, CrFeCuMnNi-1 vol.% Al_2O_3, CrFeCuMnNi-2 vol.% Al_2O_3, and CrFeCuMnNi- 3 vol.% Al_2O_3, respectively. The results indicate that the value of HAGBs started to increase with the increase in Al_2O_3 particles, ensuring an increase in structural refinement owing to the high-configurational mixing entropy effect, solid solution formation, dislocation pinning, and hot-forging process [50,52].

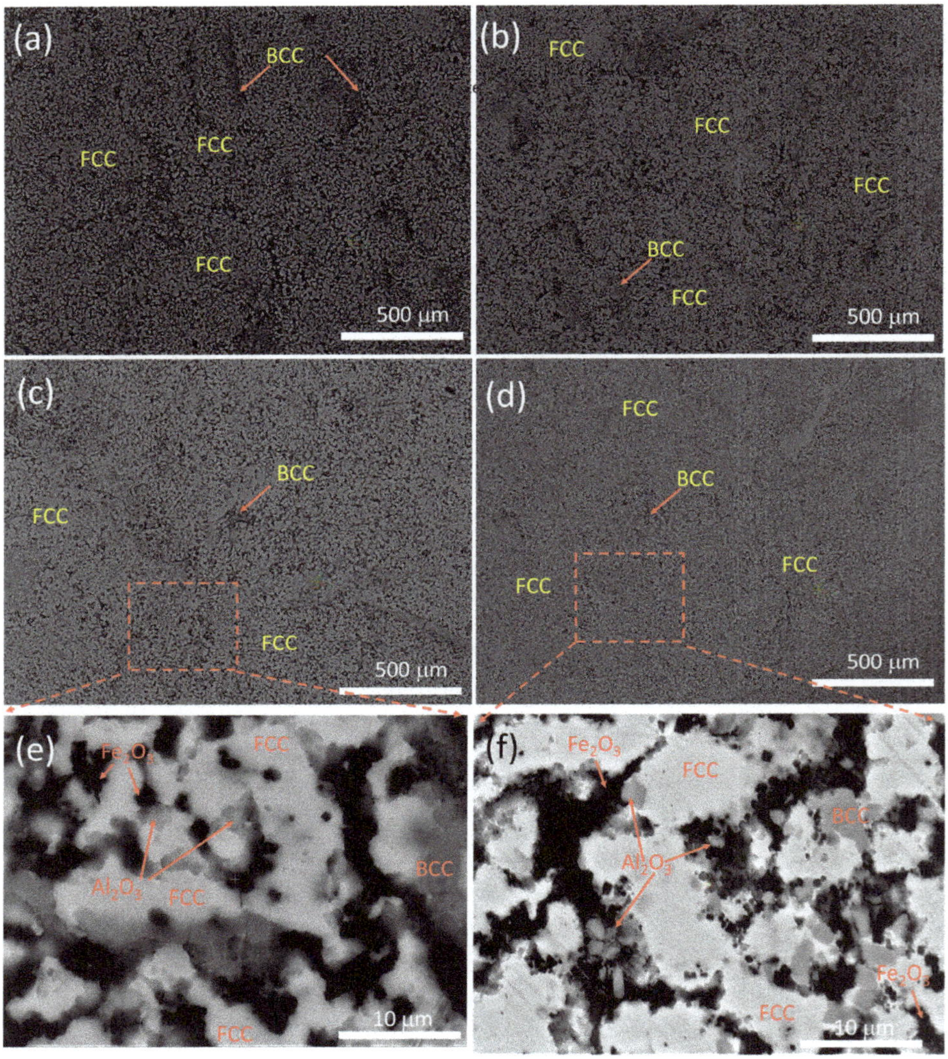

Figure 10. HRSEM microstructures of the hot-forged HEAMCs of: (**a**) CrFeCuMnNi; (**b**) CrFeCuMnNi-1 vol.% Al_2O_3; (**c**) CrFeCuMnNi-2 vol.% Al_2O_3; and (**d**) CrFeCuMnNi-3 vol.% Al_2O_3. (**e**,**f**) shows the magnified view of (**c**,**d**), respectively illustrating the dispersion of Al_2O_3 over matrix.

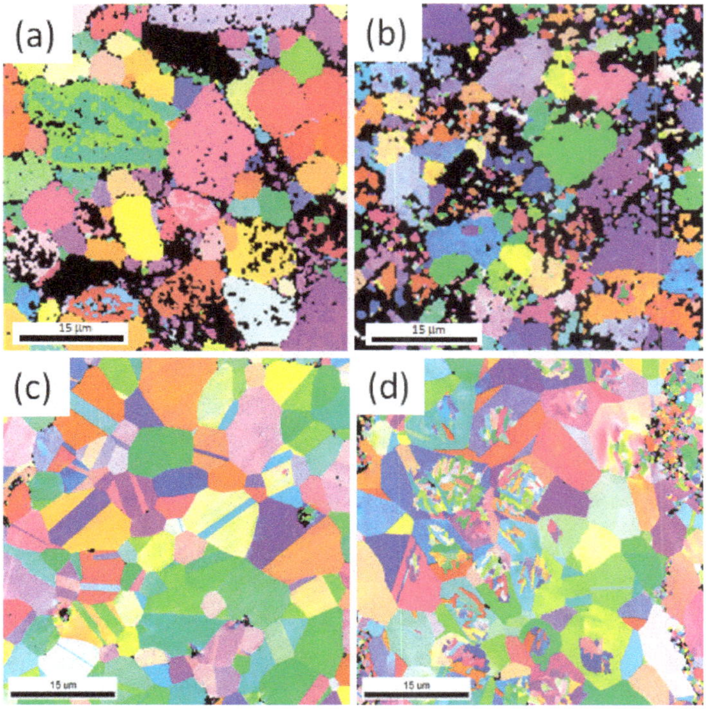

Figure 11. HRSEM with EBSD inverse pole figures (IPFs) of the developed HEAMCs of: (**a**) CrFeCuMnNi; (**b**) CrFeCuMnNi-1 vol.%Al_2O_3; (**c**) CrFeCuMnNi-2 vol.% Al_2O_3; and (**d**) CrFeCuMnNi-3 vol.% Al_2O_3.

3.5. Examination of the Relative Density and Vickers Hardness of the Developed HEAMCs

The relative density of the hot-compacted pellets and hot-forged samples were measured using Archimedes' principle and the values are listed in Table 6. The variation in relative density with the volume percentage of Al_2O_3 in the CrFeCuMnNi HEA matrix is shown in Figure 14. The results show that the measured relative density decreased slightly with the increase in the amount of Al_2O_3 in the CrFeCiMnNi HEA matrix of both hot-compacted and hot-forged samples owing to the dislocation pinning of the Al_2O_3 particles over the grain boundaries (Figure 10e,f), increasing the structural refinement (Figures 10–13), and a high-configurational mixing entropy effect (Table 5). The presence of fine Al_2O_3 particles over the grain boundaries, more structural refinement, and high-configurational entropy effect led to an increased resistance to packing during hot compaction and decreased dislocation motion during hot forging, which resulted in a decrease in the relative density. However, the relative density of the hot-forged samples were higher than those of the hot-compacted pellets. This was attributed to the decrease in the yield strength of the bulk samples during the high-temperature forging process, which led to a decrease in the height of the samples and consequently increased the density. For example, the relative density of the unreinforced CrFeCuMnNi matrix after hot compaction and hot forging were 0.6902 ± 0.0057 and 0.8295 ± 0.0124, respectively. The relative density of the hot-forged CrFeCuMnNi HEA matrix was 20.18% higher than that obtained under hot-compacted conditions. Similarly, the relative density of the CrFeCuMnNi-3 vol.% Al_2O_3 HEAMC sample exhibited 0.6525 ± 0.0019 and 0.7957 ± 0.0115 after hot compaction and hot forging, respectively. An increase of almost 21.95% in density was achieved in the hot-forged CrFeCuMnNi-3 vol.% Al_2O_3 HEAMC sample compared to hot-compacted one. This is attributed to grain growth during the high-temperature medium-frequency sintering followed by hot forging and a decrease in the yield strength at elevated temperatures. These results clearly demonstrate that the hot-forging process helps to enhance the relative density, which is a low-cost and easily applicable process in industries.

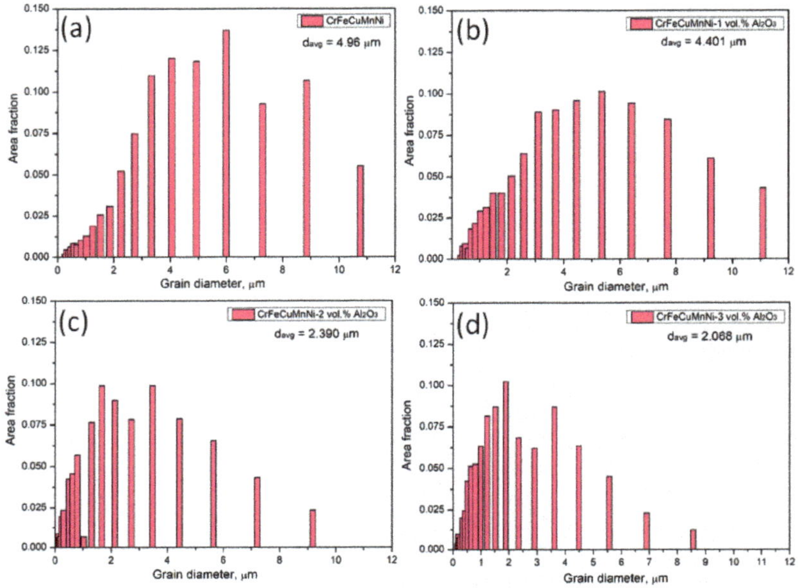

Figure 12. CrFeCuMnNi matrix grain size distribution of the developed HEAMCs over the area fraction: (**a**) CrFeCuMnNi alloy; (**b**) CrFeCuMnNi-1 vol.% Al_2O_3; (**c**) CrFeCuMnNi-2 vol.% Al_2O_3; and (**d**) CrFeCuMnNi-3 vol.% Al_2O_3.

Table 6. Mechanical properties in terms of relative density, ultimate compressive strength, strain at ultimate point, and the Vickers hardness strength of the developed HEAMCs.

Name of Sample	Relative Density at 550 Mpa Compaction Pressure	Relative Density at 1100 Followd Forging	Ultimate Compressive Stress, Mpa	Strain at Ultimate Point	Vickers Hardness Strength, Mpa
CrFeCuMnNi	0.6902 ± 0.0057	0.8295 ± 0.0124	872.3256 ± 5.3689	0.1388 ± 0.009	2648.3650 ± 7.3659
CrFeCuMnNi-1 vol.% Al2O3	0.6786 ± 0.0024	0.8137 ± 0.0066	954.0280 ± 3.5698	0.1072 ± 0.0057	2895.4560 ± 5.369
CrFeCuMnNi-2 vol.% Al2O3	0.6643 ± 0.0036	0.8027 ± 0.0014	1025.5140 ± 4.897	0.0857 ± 0.0065	3108.2350 ± 7.236
CrFeCuMnNi-3 vol.% Al2O3	0.6525 ± 0.0019	0.7957 ± 0.0115	1058.0670 ± 3.4780	0.0843 ± 0.0037	3203.5600 ± 5.698

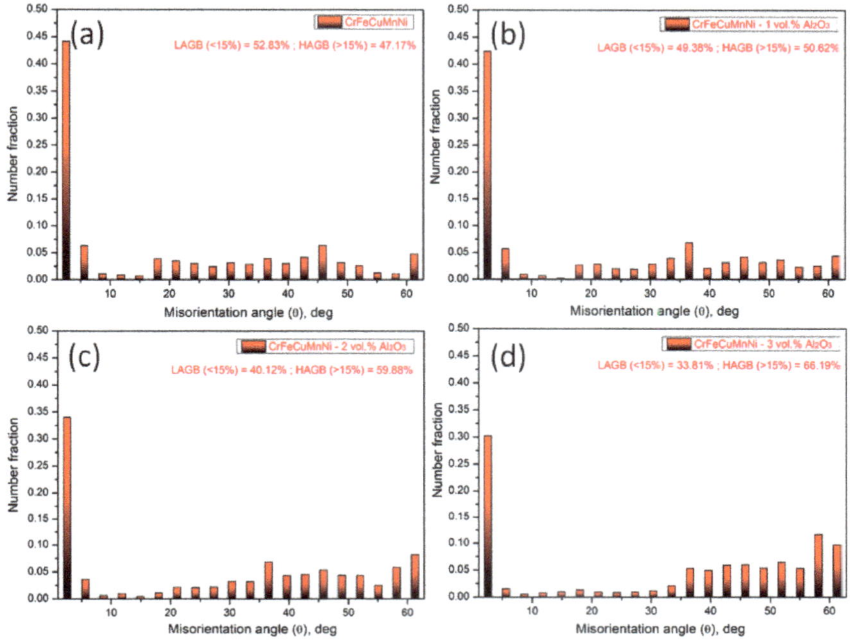

Figure 13. Misorientation angle distribution over the number fraction of the developed HEAMCs of: (**a**) CrFeCuMnNi alloy; (**b**) CrFeCuMnNi-1 vol.% Al_2O_3; (**c**) CrFeCuMnNi-2 vol.% Al_2O_3; (**d**) and CrFeCuMnNi-3 vol.% Al_2O_3.

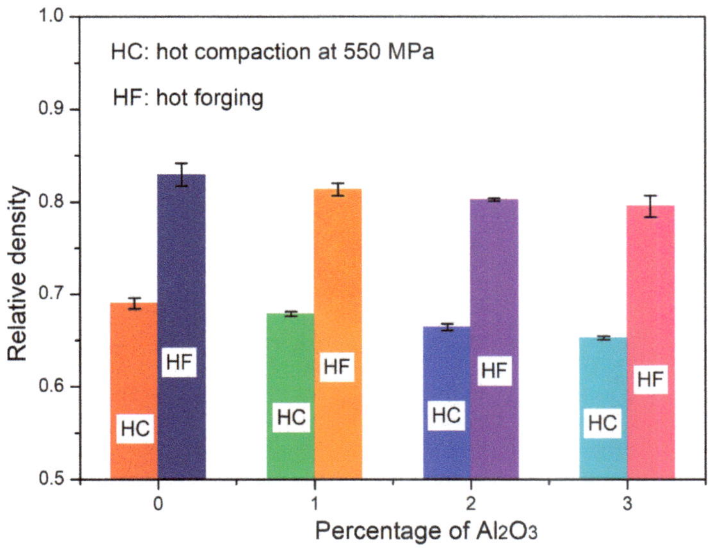

Figure 14. Variation of the relative density of the developed HEAMCs.

Figure 15 shows the increase in the Vickers hardness strength (VHS) as a function of Al_2O_3 in the CrFeCuMnNi HEA matrix. The VHS of the CrFeCuMnNi, CrFeCuMnNi-1 vol.% Al_2O_3, CrFeCuMnNi-2 vol.% Al_2O_3, and CrFeCuMnNi-3 vol.% Al_2O_3 HEAMC samples were 2.648, 2.895, 3.108, and 3.203 GPa, respectively. The Vickers hardness strength of the CrFeCuMnNi-3 vol.% Al_2O_3 HEAMC sample was 1.21 times higher than that of the unreinforced CrFeCuMnNi HEA matrix. The increase in the hardness with the Al_2O_3 addition was attributed to the dislocation pinning of the Al_2O_3 particles over the grain boundaries (Figure 10e,f), increasing the structural refinement (Figures 10–13) and high-configurational mixing entropy effect (Table 5). Based on the XRD results in Figure 9 and HRSEM micrographs in Figure 10, it is clear that the formation of the B2-BCC phase increased with the addition of Al_2O_3 and the hardness of the B2-BCC phase was expected to be higher than that of the FCC phase owing to the refined B2-BCC aggregates (dark region in Figure 10). Therefore, the applied load during the Vickers test was expected to be absorbed by the incorporated Al_2O_3 particles followed by the FCC phase of the CrFeCuMnNi matrix. The same observation was reported during the hot forging of the AlxCoCrFeNi alloys [50,53].

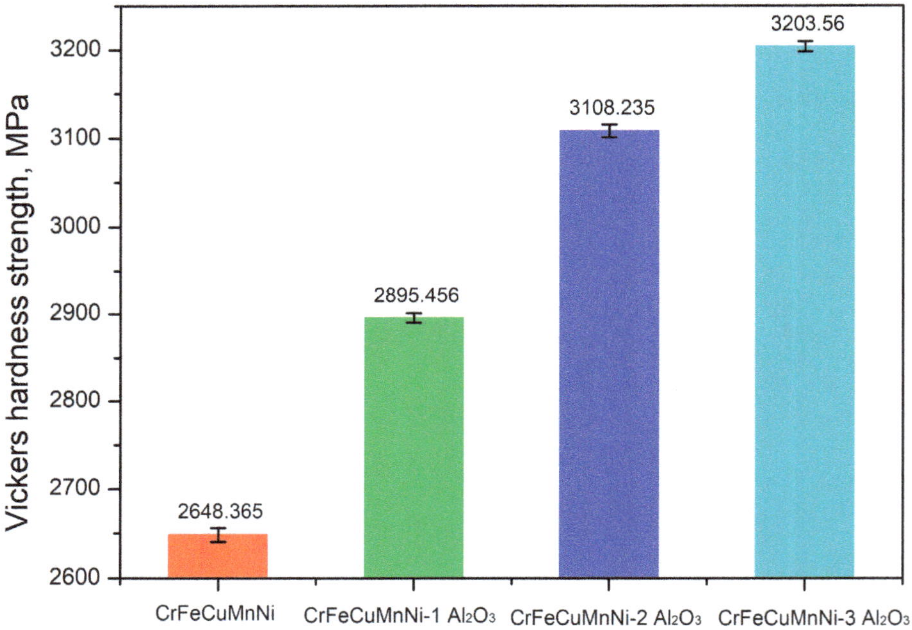

Figure 15. Variation of the Vickers hardness strength for the developed HEAMCs.

3.6. Mechanical Performance of the HEAMCs

Figure 16 shows the compressive engineering stress–strain curves of the CrFeCuMnNi-x vol.% Al_2O_3 HEAMCs. The ultimate compressive strength (UCS) and the corresponding compressive strain at the ultimate point are shown in Figure 17. The results of Figures 16 and 17 demonstrate that the compressive strength of the developed HEAMCs increased steadily with the increase in Al_2O_3 in the CrFeCuMnNi matrix. However, the corresponding strain at the ultimate point decreased, indicating a decrease in plasticity. The microstructure development after hot foring decreased the grain size of the CrFeCuMnNi matrix (Figures 10–12). In general, grain refinement can enhance the strength and suppress ductility, owing to the formation of more grain boundaries (producing more surface defects) and leading to a diminished dislocation motion [54]. The UCS values of 872, 954, 1025, and 1058 MPa were obtained for the CrFeCuMnNi, CrFeCuMnNi-1 vol.% Al_2O_3, CrFeCuMnNi-2 vol.% Al_2O_3, and CrFeCuMnNi-3 vol.% Al_2O_3, respectively. CrFeCuMnNi-3 vol.% Al_2O_3 exhibited the highest UCS, which was around 1.21% times higher than that of the unreinforced CrFeCuMnNi matrix. The increase in the UCS with Al_2O_3 in the CrFeCuMnNi HEA matrix is attributed to the embedding of Al_2O_3 particles (Figure 10e,f), dislocation pinning produced by Al_2O_3 particles (Figure 10e,f), solid solution of incorporated metallic elements (Figure 10), structural refinement (Figure 10), and high-configurational mixing entropy effect (Table 5). However, the strain at the ultimate point was 0.1388, 0.1072, 0.0857, and 0.0843 for the CrFeCuMnNi, CrFeCuMnNi-1 vol.% Al_2O_3, CrFeCuMnNi-2 vol.% Al_2O_3, and CrFeCuMnNi-3 vol.% Al_2O_3 samples, respectively, which decreased drastically (Figure 17). The strain at the ultimate point of the unreinforced CrFeCuMnNi HEA was 1.64 times higher than that of the CrFeCuMnNi-3 vol.% Al_2O_3 HEAMC sample. The decrease in strain as a function of Al_2O_3 in the CrFeCuMnNi matrix was attributed to the increase in resistance against the dislocation motion produced by the Al_2O_3 particles, the B2-BCC phase, structural refinement, and high-configurational entropy [55,56]. Figure 18 shows the HRSEM microstructures of the deformed samples observed perpendicularly to the loading direction near the circumference of the sample. The results clearly show that more deformed grains in the FCC phase were observed in the CrFeCuMnNi-1 vol.% Al_2O_3 sample (Figure 18a) in comparison with those of the CrFeCuMnNi-3 vol.% Al_2O_3 sample (Figure 18c). It can be concluded that the developed HEAMCs can be recommended for use in industrial applications, particularly in aerospace, defence, and automotive industries.

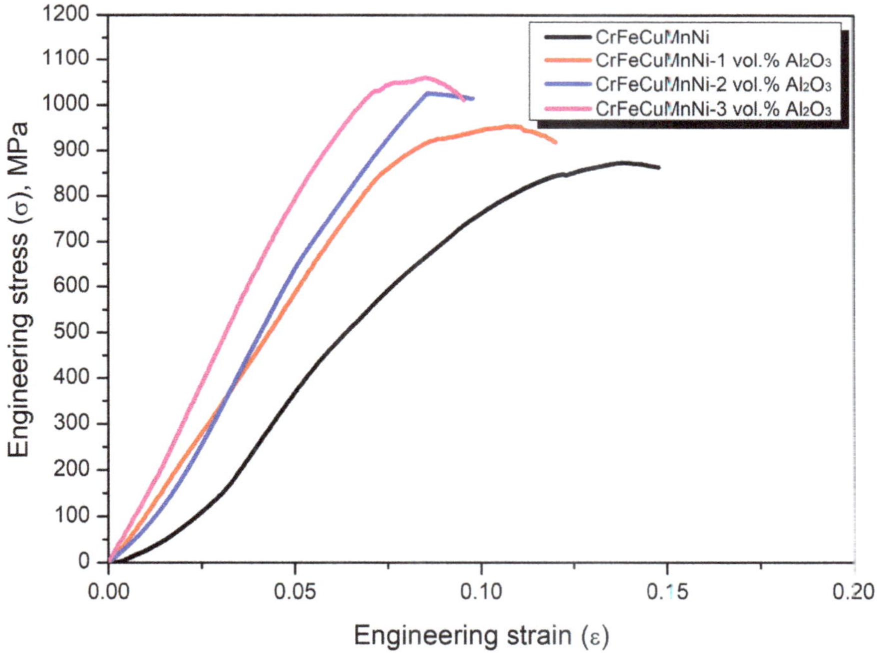

Figure 16. Compressive engineering stress–strain curves of the hot-forged CrFeCuMnNi-x vol.% Al_2O_3 HEAMCs.

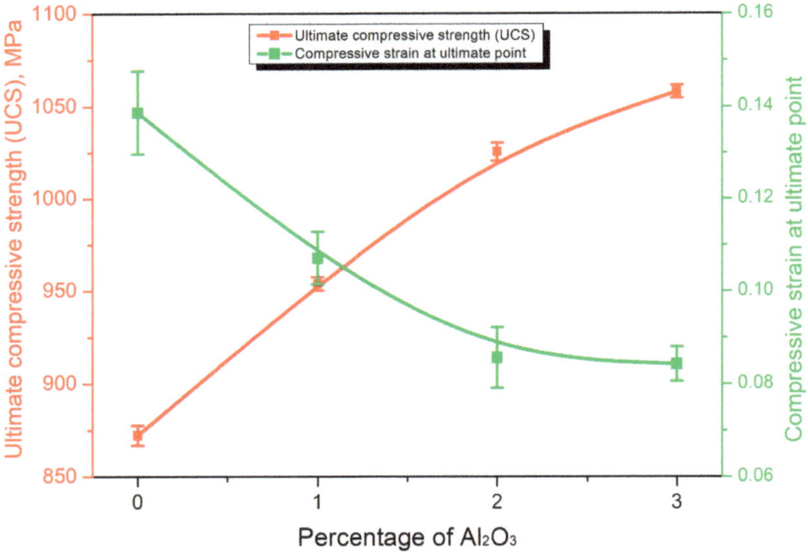

Figure 17. Variation of the ultimate compressive strength (UCS) and compressive strain at the ultimate point with the function of Al_2O_3 in CrFeCuMnNi HEAMCs.

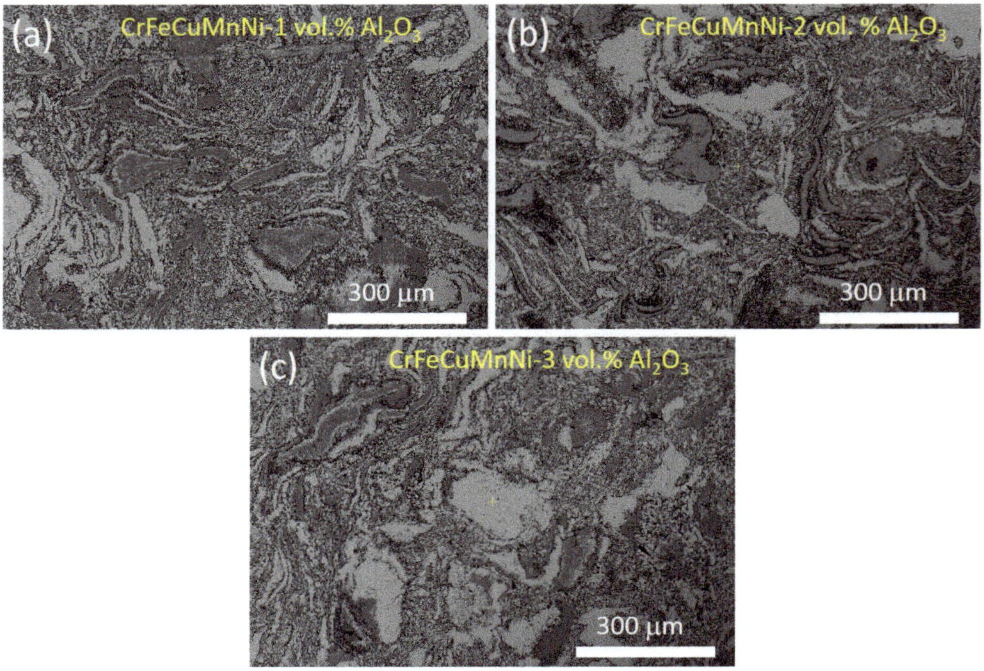

Figure 18. HRSEM microstructures of the deformed samples after the compression test of: (**a**) CrFeCuMnNi-1 vol.% Al_2O_3; (**b**) CrFeCuMnNi-2 vol.% Al_2O_3; and (**c**) CrFeCuMnNi-3 vol.% Al_2O_3 (microstructures captured at the outer periphery of the sample, which is perpendicular to the loading direction).

3.7. Examination of the Wear Behavior of the HEAMCs

The influence of the Al_2O_3 content in the CrFeCuMnNi HEA matrix on the wear rate (WR) and coefficient of friction (CoF) of the developed hot-forged HEAMCs is illustrated in Figures 19 and 20, respectively. The dry sliding wear rate at room temperature demonstrates that the wear rate decreased significantly with the increase in the content of Al_2O_3 in the CrFeCuMnNi HEA matrix. This is attributed to the solid solution of the incorporated metallic elements (Figures 6 and 9), the effective dispersion of the Al_2O_3 particles in the matrix (Figure 10), structural refinement (Figures 11–13), and high-configurational mixing entropy (Table 5). The WR of 7.12×10^{-12}, 6.46×10^{-12}, 5.32×10^{-12}, and 4.66×10^{-12} m^3/Nm for CrFeCuMnNi, CrFeCuMnNi-1 vol.% Al_2O_3, CrFeCuMnNi-2 vol.% Al_2O_3, and CrFeCuMnNi-3 vol.% Al_2O_3, respectively. The CrFeCuMnNi-3 vol.% Al_2O_3 HEAMC sample exhibited a 52.9% lower wear rate compared to the CrFeCuMnNi matrix. In addition, the increasing amount of the minor B2-BCC phase might have diminished the wear rate owing to the refined B2-BCC aggregates (dark region in Figure 10). These results clearly indicate that the wear resistance of the developed HEAMCs can be improved by the addition of Al_2O_3 particles, which suggests their potential for industrial applications. The experimental CoF value started to decrease considerably with the addition of Al_2O_3 to the CrFeCuMnNi matrix, indicating a decrease in friction, which can enhance the wear resistance of the developed HEAC samples. CoF values of 0.205, 0.183, 0.165, and 0.156 were observed for CrFeCuMnNi, CrFeCuMnNi-1 vol.% Al_2O_3, CrFeCuMnNi-2 vol.% Al_2O_3, and CrFeCuMnNi-3 vol.% Al_2O_3, respectively. The CrFeCuMnNi-3 vol.% Al_2O_3 HEAMC sample had around 31.41% lower CoF value compared to the CrFeCuMnNi matrix. The incorporation of Al_2O_3 particles and the formed Fe_2O_3 might have acted as a protective layer, consequently decreasing the metal contact between the sample (pin) and counter disc (hardened one), which leads to a decrease in CoF.

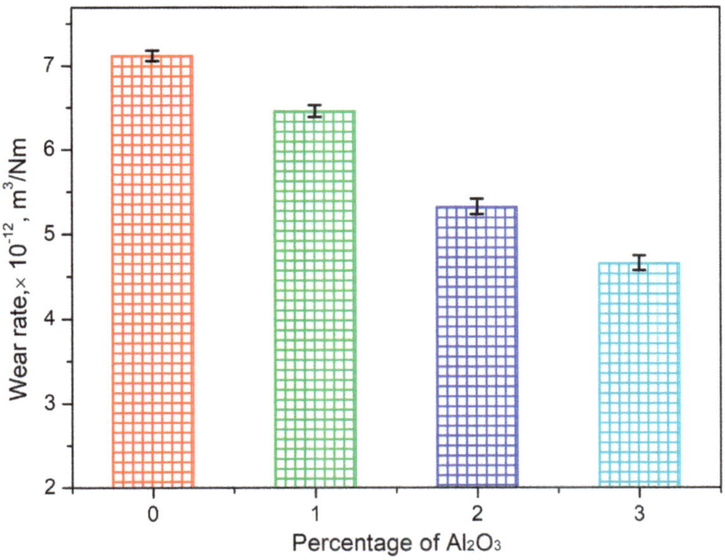

Figure 19. Variation in the wear rate (WR) with Al_2O_3 content in the CrFeCuMnNi of the hot-forged samples HEAMCs.

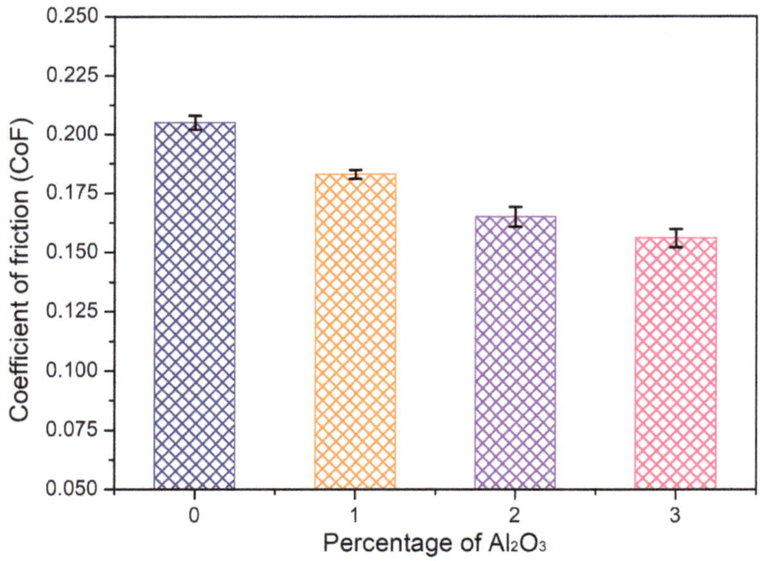

Figure 20. Variation in the coefficient of friction (CoF) with Al_2O_3 content in the CrFeCuMnNi of the hot-forged samples HEAMCs.

Figure 21 shows the SEM worn surface morphologies of the hot-forged samples after the sliding wear test at room temperature. It is clear that the incorporation of Al_2O_3 in the CrFeCuMnNi HEA matrix significantly influenced the wear performance. Deep grooves (more ploughing), severe delamination, and mild cracks were observed in the CrFeCuMnNi matrix (Figure 21a). In addition, more oxides on the outer surface of the wear track were formed in this sample, which clearly explains that the CrFeCuMnNi HEA matrix possesses a poor wear resistance owing to the domination of more abrasive and adhesive wear mechanisms [57]. This was attributed to a lower structural refinement (Figure 11a, i.e., large grain size), the absence of Al_2O_3 content (Figure 9), lower hardness (Table 6, Figure 15), and lower mechanical strength (Table 6 and Figure 17). However, with the increase in the content of Al_2O_3 in the CrFeCuMnNi matrix, the formation of deep grooves caused by ploughing, severe delamination, amount of debris, and the presence of oxides on the outer surface of the wear track decreased (Figure 21b–d). In addition, the formation of oxide particles inside the wear track, leading to the formation of an oxide protective layer, increased with the increase in the Al_2O_3 content in the CrFeCuMnNi matrix. These results demonstrate that the wear resistance increased with the addition of Al_2O_3 particles. In other words, the dominance of the adhesive and abrasive wear mechanisms decreased. A very smooth surface along with a tribo-layer was observed in the CrFeCuMnNi-3 vol.% Al_2O_3 sample (Figure 21d), indicating a high wear resistance compared to the other samples. A high-magnification SEM image of the worn surface morphology of CrFeCuMnNi-2 vol.% Al_2O_3 as an example is shown in Figure 21e, which shows the smooth surface and wear track with tribo-layer and oxide protective layer formation. The high wear resistance due to the presence of a smooth surface and wear track with a tribo-layer was attributed to the high-configurational mixing entropy effect (Table 5), the effective dispersion of Al_2O_3 particles over the matrix, structural refinement (Figure 10), and high mechanical strength (both compressive strength and hardness, Table 6 [49]). Overall, the addition of Al_2O_3 particles enhanced the wear resistance of the CrFeCuMnNi HEA matrix, which is suitable for both structural and defensive applications.

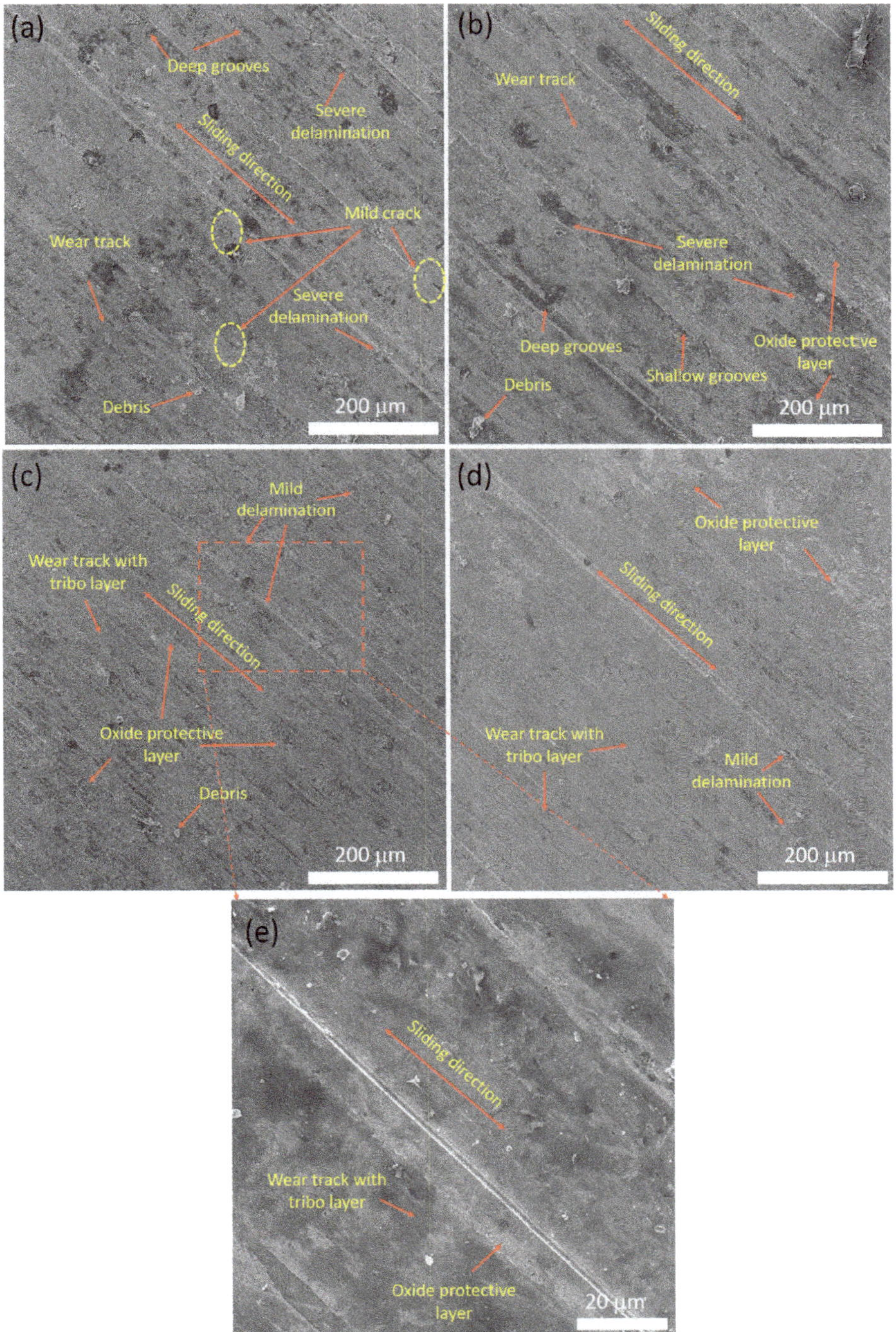

Figure 21. SEM worn surface morphology of the developed hot-forged samples after the dry sliding wear test of: (**a**) CrFeCuMnNi; (**b**) CrFeCuMnNi-1 vol.% Al_2O_3; (**c**) CrFeCuMnNi-2 vol.% Al_2O_3, (**d**) CrFeCuMnNi-3 vol.% Al_2O_3; and (**e**) magnified view of (**c**).

4. Conclusions

The major findings of this study are presented below:

- ❖ The developed hot-forged HEMCs exhibited a major face-centred cubic (FCC) structure that matched the predicted structure using physio- and thermo-mechanical parameters (ΔG_{mix}, ΔH_{mix}, ΔS_{mix}, Ω, VEC, and ΔX) and a minor body-centred cubic (B2-BCC) phase owing to the high electronegativity difference and high temperature used during the hot-forging process.
- ❖ The addition of Al_2O_3 to the CrFeCuMnNi matrix enhances the high-configurational mixing entropy effect and decreases the mixing enthalpy effect, leading to the formation of a stable FCC structure. The grain size reduction and structural refinement of Al_2O_3 were studied using EBSD analysis.
- ❖ The hot-forging process helped to increase the relative density by approximately 21% compared to the hot-compacted samples owing to grain growth and decrease in yield strength at elevated temperatures.
- ❖ With the increase in the Al_2O_3 content, the hardness and mechanical strength gradually increased and the VHS and UCS of the CrFeCuMnNi-3 vol.% Al_2O_3 HEAMC sample were around 1.21 times higher than those of the unreinforced CrFeCuMnNi HEA matrix.
- ❖ Moreover, the wear resistance of the CrFeCuMnNi-3 vol.% Al_2O_3 HEAMC sample steadily improved by 52.9% compared to the unreinforced one due to the lower domination of adhesive and abrasive mechanisms and increase in the oxide protective layer.

Author Contributions: Conceptualization, S.S. and H.R.A.; methodology, S.S., H.R.A. and A.S.A.; formal analysis, S.S., H.R.A., E.-S.M.S., A.S.A. and A.-b.H.M.; investigation, S.S., H.R.A., A.S.A., E.-S.M.S. and A.-b.H.M.; data curation, S.S., H.R.A., E.-S.M.S. and A.-b.H.M.; writing—original draft preparation, S.S. and H.R.A.; writing—review and editing, S.S., H.R.A. and A.S.A. All authors have read and agreed to the published version of the manuscript.

Funding: This research was funded by Qassim University, represented by the Deanship of Scientific Research, under the number (10257-qec-2020-1-3-I).

Institutional Review Board Statement: Not Applicable.

Informed Consent Statement: Not Applicable.

Data Availability Statement: The experimental datasets obtained from this research work and the analysed results during the current study are available from the corresponding author upon reasonable request.

Acknowledgments: The authors gratefully acknowledge Qassim University, represented by the Deanship of Scientific Research, for the financial support for this research under the number (10257-qec-2020-1-3-I) during the academic year 1441 AH/2020 AD.

Conflicts of Interest: The authors declare no conflict of interest.

References

1. Kim, J.H.; Lim, K.R.; Won, J.W.; Na, Y.S.; Kim, H.-S. Mechanical properties and deformation twinning behavior of as-cast CoCrFeMnNi high-entropy alloy at low and high temperatures. *Mater. Sci. Eng. A* **2018**, *712*, 108–113. [CrossRef]
2. Zhang, Y.; Ting, T.; Tang, Z.; Gao, M.C.; Dahmen, K.A.; Liaw, P.K.; Ping, Z. Microstructures and properties of high-entropy alloys. *Prog. Mater. Sci.* **2014**, *61*, 1–93. [CrossRef]
3. Zhang, W.; Ma, Z.; Zhao, H.; Ren, L. Breakthrough the strength-ductility trade-off in a high-entropy alloy at room temperature via cold rolling and annealing. *Mater. Sci. Eng. A* **2021**, *800*, 140264. [CrossRef]
4. Yeh, J.-W. Alloy Design Strategies and Future Trends in High-Entropy Alloys. *JOM* **2013**, *65*, 1759–1771. [CrossRef]
5. Gao, M.C.; Yeh, J.-W.; Liaw, P.K.; Zhang, Y. *High-Entropy Alloys*; Springer: Cham, Switzerland, 2016.
6. Tong, C.-J.; Chen, Y.-L.; Yeh, J.-W.; Lin, S.-J.; Chen, S.-K.; Shun, T.-T.; Tsau, C.-H.; Chang, S.-Y. Microstructure characterization of Al x CoCrCuFeNi high-entropy alloy system with multiprincipal elements. *Metall. Mater. Trans. A* **2005**, *36*, 881–893. [CrossRef]
7. Tsai, K.-Y.; Tsai, M.-H.; Yeh, J.-W. Sluggish diffusion in Co–Cr–Fe–Mn–Ni high-entropy alloys. *Acta Mater.* **2013**, *61*, 4887–4897. [CrossRef]
8. Yeh, J.-W. Physical Metallurgy of High-Entropy Alloys. *JOM* **2015**, *67*, 2254–2261. [CrossRef]

9. Rogal, Ł.; Kalita, D.; Litynska-Dobrzynska, L. CoCrFeMnNi high entropy alloy matrix nanocomposite with addition of Al_2O_3. *Intermetallics* **2017**, *86*, 104–109. [CrossRef]
10. Chen, C.-S.; Yang, C.-C.; Chai, H.-Y.; Yeh, J.-W.; Chau, J.L.H. Novel cermet material of WC/multi-element alloy. *Int. J. Refract. Met. Hard Mater.* **2014**, *43*, 200–204. [CrossRef]
11. Liu, W.H.; Wu, Y.; He, J.Y.; Nieh, T.G.; Lu, Z.P. Grain growth and the Hall–Petch relationship in a high-entropy FeCrNiCoMn alloy. *Scr. Mater.* **2013**, *68*, 526–529. [CrossRef]
12. Huang, Y.-S.; Chen, L.; Lui, H.-W.; Cai, M.-H.; Yeh, J.-W. Microstructure, hardness, resistivity and thermal stability of sputtered oxide films of AlCoCrCu0.5NiFe high-entropy alloy. *Mater. Sci. Eng. A* **2007**, *457*, 77–83. [CrossRef]
13. Rogal, Ł.; Morgiel, J.; Świątek, Z.; Czerwiński, F. Microstructure and mechanical properties of the new Nb25Sc25Ti25Zr25 eutectic high entropy alloy. *Mater. Sci. Eng. A* **2016**, *651*, 590–597. [CrossRef]
14. Czerwinski, F.; Jochym, P.T.; Litynska-Dobrzynska, L. Microstructure and mechanical properties of the novel Hf25Sc25Ti25Zr25 equiatomic alloy with hexagonal solid solutions. *Mater. Des.* **2016**, *92*, 8–17. [CrossRef]
15. Oono, N.; Ukai, S.; Kondo, S.; Hashitomi, O.; Kimura, A. Irradiation effects in oxide dispersion strengthened (ODS) Ni-base alloys for Gen. IV nuclear reactors. *J. Nucl. Mater.* **2015**, *465*, 835–839. [CrossRef]
16. Praveen, S.; Basu, J.; Kashyap, S.; Kottada, R.S. Exceptional resistance to grain growth in nanocrystalline CoCrFeNi high entropy alloy at high homologous temperatures. *J. Alloys Compd.* **2016**, *662*, 361–367. [CrossRef]
17. Hadraba, H.; Chlup, Z.; Dlouhy, A.; Dobes, F.; Roupcova, P.; Vilemova, M.; Matejicek, J. Oxide dispersion strengthened CoCrFeNiMn high-entropy alloy. *Mater. Sci. Eng. A* **2017**, *689*, 252–256. [CrossRef]
18. Gwalani, B.; Pohan, R.M.; Lee, J.; Lee, B.; Banerjee, R.; Ryu, H.J.; Hong, S.H. High-entropy alloy strengthened by in situ formation of entropy-stabilized nano-dispersoids. *Sci. Rep.* **2018**, *8*, 14085. [CrossRef]
19. Prasad, H.; Singh, S.; Panigrahi, B.B. Mechanical activated synthesis of alumina dispersed FeNiCoCrAlMn high entropy alloy. *J. Alloys Compd.* **2017**, *692*, 720–726. [CrossRef]
20. Liu, X.; Yin, H.; Xu, Y. Microstructure, Mechanical and Tribological Properties of Oxide Dispersion Strengthened High-Entropy Alloys. *Materials* **2017**, *10*, 1312. [CrossRef] [PubMed]
21. Jia, B.; Liu, X.; Wang, H.; Wu, Y.; Lu, Z. Microstructure and mechanical properties of FeCoNiCr high-entropy alloy strengthened by nano-Y2O3 dispersion. *Sci. China Technol. Sci.* **2017**, *61*, 179–183. [CrossRef]
22. Yang, S.; Zhang, Y.; Yan, X.; Zhou, H.; Pi, J.; Zhu, D. Deformation twins and interface characteristics of nano-Al_2O_3 reinforced Al0.4FeCrCo1.5NiTi0.3 high entropy alloy composites. *Mater. Chem. Phys.* **2018**, *210*, 240–244. [CrossRef]
23. Zhu, G.; Liu, Y.; Ye, J. Early high-temperature oxidation behavior of Ti(C,N)-based cermets with multi-component AlCoCrFeNi high-entropy alloy binder. *Int. J. Refract. Met. Hard Mater.* **2014**, *44*, 35–41. [CrossRef]
24. Liu, W.H.; Lu, Z.P.; He, J.Y.; Luan, J.H.; Wang, Z.J.; Liu, B.; Liu, Y.; Chen, M.W.; Liu, C.T. Ductile CoCrFeNiMox high entropy alloys strengthened by hard intermetallic phases. *Acta Mater.* **2016**, *116*, 332–342. [CrossRef]
25. Shun, T.-T.; Chang, L.-Y.; Shiu, M.-H. Microstructure and mechanical properties of multiprincipal component CoCrFeNiMox alloys. *Mater. Charact.* **2012**, *70*, 63–67. [CrossRef]
26. Liu, X.; Zhang, L.; Xu, Y. Microstructure and mechanical properties of graphene reinforced Fe50Mn30Co10Cr10 high-entropy alloy composites synthesized by MA and SPS. *Appl. Phys. A* **2017**, *123*, 567. [CrossRef]
27. Riva, S.; Tudball, A.; Mehraban, S.; Lavery, N.P.; Brown, S.G.R.; Yusenko, K. V A novel High-Entropy Alloy-based composite material. *J. Alloys Compd.* **2018**, *730*, 544–551. [CrossRef]
28. Hussain, S.W.; Mehmood, M.A.; Karim, M.R.A.; Godfrey, A.; Yaqoob, K. Microstructural evolution and mechanical characterization of a WC-reinforced CoCrFeNi HEA matrix composite. *Sci. Rep.* **2022**, *12*, 9822. [CrossRef]
29. Liu, J.; Guo, X.; Lin, Q.; He, Z.; An, X.; Li, L.; Liaw, P.K.; Liao, X.; Yu, L.; Lin, J. Excellent ductility and serration feature of metastable CoCrFeNi high-entropy alloy at extremely low temperatures. *Sci. China Mater.* **2019**, *62*, 853–863. [CrossRef]
30. Karthik, G.; Panikar, S.; Ram, G.J.; Kottada, R.S. Additive manufacturing of an aluminum matrix composite reinforced with nanocrystalline high-entropy alloy particles. *Mater. Sci. Eng. A* **2017**, *679*, 193–203. [CrossRef]
31. Meng, G.; Yue, T.; Lin, X.; Yang, H.; Xie, H.; Ding, X. Laser surface forming of AlCoCrCuFeNi particle reinforced AZ91D matrix composites. *Opt. Laser Technol.* **2015**, *70*, 119–127. [CrossRef]
32. Mehmood, M.A.; Shehzad, K.; Mujahid, M.; Bin Yaqub, T.; Godfrey, A.; Fernandes, F.; Muhammad, F.Z.; Yaqoob, K. Ceramic-reinforced HEA matrix composites exhibiting an excellent combination of mechanical properties. *Sci. Rep.* **2022**, *12*, 21486. [CrossRef]
33. Sivasankaran, S.; Ammar, H.R.; Al-Mufadi, F.A. Continuous hot-compaction behavior of nanostructured FeCrCuMnTi-(V, Zn) high-entropy alloys. *Mater. Manuf. Process.* **2022**, *37*, 1122–1131. [CrossRef]
34. Gong, J.; Wu, J.; Guan, Z. Examination of the indentation size effect in low-load vickers hardness testing of ceramics. *J. Eur. Ceram. Soc.* **1999**, *19*, 2625–2631. [CrossRef]
35. Basavarajappa, S.; Chandramohan, G.; Mahadevan, A.; Thangavelu, M.; Subramanian, R.; Gopalakrishnan, P. Influence of sliding speed on the dry sliding wear behaviour and the subsurface deformation on hybrid metal matrix composite. *Wear* **2007**, *262*, 1007–1012. [CrossRef]
36. Mondal, D.P.; Das, S.; Jha, N. Dry sliding wear behaviour of aluminum syntactic foam. *Mater. Des.* **2009**, *30*, 2563–2568. [CrossRef]
37. Baskaran, S.; Anandakrishnan, V.; Duraiselvam, M. Investigations on dry sliding wear behavior of in situ casted AA7075–TiC metal matrix composites by using Taguchi technique. *Mater. Des.* **2014**, *60*, 184–192. [CrossRef]

38. Basavarajappa, S.; Chandramohan, G.; Mukund, K.; Ashwin, M.; Prabu, M. Dry sliding wear behavior of Al 2219/SiCp-Gr hybrid metal matrix composites. *J. Mater. Eng. Perform.* **2006**, *15*, 668. [CrossRef]
39. Chen, X.-H.; Yan, H. Fabrication of nanosized Al_2O_3 reinforced aluminum matrix composites by subtype multifrequency ultrasonic vibration. *J. Mater. Res.* **2015**, *30*, 2197–2209. [CrossRef]
40. Sivasankaran, S.; Al-Mufadi, F.A.; Ammar, H.R. Influence of V and Zn in FeCrCuMnTi High-Entropy Alloys on Microstructures and Uniaxial Compaction Behavior Prepared by Mechanical Alloying. *Crystals* **2021**, *11*, 1413. [CrossRef]
41. Zhao, Q.; Pan, Z.; Wang, X.; Luo, H.; Liu, Y.; Li, X. Corrosion and passive behavior of AlxCrFeNi3−x (x = 0.6, 0.8, 1.0) eutectic high entropy alloys in chloride environment. *Corros. Sci.* **2022**, *208*, 110666. [CrossRef]
42. Ma, Y.; Jiang, B.; Li, C.; Wang, Q.; Dong, C.; Liaw, P.K.; Xu, F.; Sun, L. The BCC/B2 morphologies in Al x NiCoFeCr high-entropy alloys. *Metals* **2017**, *7*, 57. [CrossRef]
43. Huo, W.; Zhou, H.; Fang, F.; Xie, Z.; Jiang, J. Microstructure and mechanical properties of CoCrFeNiZrx eutectic high-entropy alloys. *Mater. Des.* **2017**, *134*, 226–233. [CrossRef]
44. Chen, X.; Qi, J.Q.; Sui, Y.W.; He, Y.Z.; Wei, F.X.; Meng, Q.K.; Sun, Z. Effects of aluminum on microstructure and compressive properties of Al-Cr-Fe-Ni eutectic multi-component alloys. *Mater. Sci. Eng. A* **2017**, *681*, 25–31. [CrossRef]
45. Alshataif, Y.A.; Sivasankaran, S.; Al-Mufadi, F.A.; Alaboodi, A.S.; Ammar, H.R. Synthesis, Microstructures and Mechanical Behaviour of Cr0.21Fe0.20Al0.41Cu0.18 and Cr0.14Fe0.13Al0.26Cu0.11Si0.25Zn0.11 Nanocrystallite Entropy Alloys Prepared by Mechanical Alloying and Hot-Pressing. *Met. Mater. Int.* **2021**, *27*, 139–155. [CrossRef]
46. Yang, X.; Zhang, Y. Prediction of high-entropy stabilized solid-solution in multi-component alloys. *Mater. Chem. Phys.* **2012**, *132*, 233–238. [CrossRef]
47. Takeuchi, A.; Inoue, A. Classification of Bulk Metallic Glasses by Atomic Size Difference, Heat of Mixing and Period of Constituent Elements and Its Application to Characterization of the Main Alloying Element. *Mater. Trans.* **2005**, *46*, 2817–2829. [CrossRef]
48. Guo, S.; Ng, C.; Lu, J.; Liu, C.T. Effect of valence electron concentration on stability of fcc or bcc phase in high entropy alloys. *J. Appl. Phys.* **2011**, *109*, 103505. [CrossRef]
49. Shu, C.; Yao, Z.; Li, X.; Du, W.; Tao, X.; Yang, H. Microstructure and wear mechanism of CoCrCuFeNiVx high entropy alloy by sintering and electron beam remelting. *Phys. B Condens. Matter* **2022**, *638*, 413834. [CrossRef]
50. Sourav, A.; Yebaji, S.; Thangaraju, S. Structure-property relationships in hot forged AlxCoCrFeNi high entropy alloys. *Mater. Sci. Eng. A* **2020**, *793*, 139877. [CrossRef]
51. Li, G.; Liu, M.; Lyu, S.; Nakatani, M.; Zheng, R.; Ma, C.; Li, Q.; Ameyama, K. Simultaneously enhanced strength and strain hardening capacity in FeMnCoCr high-entropy alloy via harmonic structure design. *Scr. Mater.* **2021**, *191*, 196–201. [CrossRef]
52. Wu, X.; Yang, M.; Yuan, F.; Wu, G.; Wei, Y.; Huang, X.; Zhu, Y. Heterogeneous lamella structure unites ultrafine-grain strength with coarse-grain ductility. *Proc. Natl. Acad. Sci. USA* **2015**, *112*, 14501–14505. [CrossRef] [PubMed]
53. Annasamy, M.; Haghdadi, N.; Taylor, A.; Hodgson, P.; Fabijanic, D. Dynamic recrystallization behaviour of AlxCoCrFeNi high entropy alloys during high-temperature plane strain compression. *Mater. Sci. Eng. A* **2019**, *745*, 90–106. [CrossRef]
54. He, J.Y.; Wang, H.; Huang, H.L.; Xu, X.D.; Chen, M.W.; Wu, Y.; Liu, X.J.; Nieh, T.G.; An, K.; Lu, Z.P. A precipitation-hardened high-entropy alloy with outstanding tensile properties. *Acta Mater.* **2016**, *102*, 187–196. [CrossRef]
55. Zhou, J.; Liao, H.; Chen, H.; Huang, A. Effects of hot-forging and subsequent annealing on microstructure and mechanical behaviors of Fe35Ni35Cr20Mn10 high-entropy alloy. *Mater. Charact.* **2021**, *178*, 111251. [CrossRef]
56. Shahmir, H.; Mousavi, T.; He, J.; Lu, Z.; Kawasaki, M.; Langdon, T.G. Microstructure and properties of a CoCrFeNiMn high-entropy alloy processed by equal-channel angular pressing. *Mater. Sci. Eng. A* **2017**, *705*, 411–419. [CrossRef]
57. Xing, Q.; Feltrin, A.C.; Akhtar, F. Processing, microstructure and high temperature dry sliding wear of a Cr-Fe-Hf-Mn-Ti-Ta-V high-entropy alloy based composite. *Mater. Today Commun.* **2021**, *28*, 102657. [CrossRef]

materials

MDPI

Article

Chirality-Dependent and Intrinsic Auxeticity for Single-Walled Carbon Nanotubes

Hai-Ning Zhang, Yin Fan * and Hui-Shen Shen

School of Aeronautics and Astronautics, Shanghai Jiao Tong University, Shanghai 200240, China
* Correspondence: yfan1987@sjtu.edu.cn

Abstract: Single-walled carbon nanotubes (SWCNTs) have superior mechanical properties which originate from a strong C-C covalent bond and unique nanostructure. Chirality, one of the helical structural parameters of SWCNTs, leads to differences in mechanical performance. In this work, molecular dynamics (MD) simulation was performed to analyze engineering Poisson's ratio (EPR) and incremental Poisson's ratio (IPR) of SWCNTs with different chiral angles, respectively, under tensile and compressive load, as well as the chiral effect on rigidity. We reported the minimum EPR for (4, 1) SWCNT and obtained the distribution and trend of EPR which is dependent on chiral index m. In addition, a new observation showed two exactly opposite trends of EPR existing not only in tension and compression but also in the longitudinal and radial directions. Furthermore, we found that the critical strain, over which SWCNT would be auxetic, ranged from 6% to 18% and was also chirality-dependent. Three representative SWCNTs with chiral angle of 0° (zigzag), 10.89° (chiral), and 30° (armchair) were selected for the mechanism study of auxeticity. Finally, a method of the contribution to radial strain for two main deformation modes proposed in this paper could well explain the negative IPR phenomenon.

Keywords: carbon nanotubes; chirality; Poisson's ratio; molecular dynamics; auxetic; rigidity

1. Introduction

The discovery [1] of carbon nanotubes (CNTs) has accelerated exploration of the nano area and nowadays nanotechnologies are changing our world. Due to excellent mechanical properties, CNTs, especially for single-walled carbon nanotubes (SWCNTs), are regarded as an ideal candidate for reinforcement of next-generation advanced composite materials [2]. Hence, a better understanding of the mechanical performance of CNTs is key issue of future applications in engineering and industrial materials. Among all mechanical parameters, Poisson's ratio is an important mechanical indicator, reflecting the response of a material subjected to external loads. Conventionally, the value equal to the negative ratio of radial strain to axial strain is normally defined as engineering Poisson's ratio (EPR, υ) of CNT, while incremental Poisson's ratio (IPR, υ^{in}) describing variation of material EPR at some time is the value of the negative ratio of radial differential strain to axial differential strain. In addition, a reasonable and accurate characterization of rigidity (C), which reflects the resist deformation ability, is necessary for application of CNTs as a candidate for reinforcement.

Molecular mechanics (MM) paved an effective way to study EPR of SWCNTs. In the MM model, the C-C bond in CNTs is always treated as a beam of nonlinear spring structure [3]. However, the predicted results of SWCNT EPR from different research teams seem inconsistent, even though they were all based on MM, and without considering chirality. For example, Zhu and Wang [4] estimated that the EPRs of the zigzag and armchair SWCNTs with a big enough diameter would both reach about 0.27, which is approaching to the reported results [3,5–8]. Meanwhile, the EPRs of the SWCNTs (with diameter > 2 nm) calculated by Xiao et al. [9] were close to 0.21, which is very close to a lattice-dynamics result [10]. Other researches [11–13] indicated that the EPR of large-diameter SWCNT would go down to 0.16, that is also the EPR of graphite sheet [14]. The main reason for these differences is because of various interaction between carbon atoms. As is well known, chiral SWCNTs are those excepting zigzag and armchair SWCNTs, and chirality is represented by indices (n, m) which determines both size and chiral angle of a SWCNT. Some of above studies also reported that ERPs of chiral SWCNTs were between those of armchair SWCNTs and zigzag SWCNTs. On the other hand, EPRs of CNTs can be expressed in analytical formulas which are independent on deformation through MM modelling, therefore, IPRs of SWCNTs seem impossible to be studied.

On this problem, molecular dynamics (MD) can effectively predict historical EPRs which is adaptive to a further analysis of IPRs [15,16]. Based on a second-generation reactive empirical bond order (REBO) potential [17], Ni et al. [18] obtained variations of EPRs of both zigzag and armchair SWCNTs during both tensile and compressive processes. The compressive and tensile simulation of chiral SWCNTs was performed by Shintani and Narita [19], and they found chiral-dependence could be in proportion to the magnitude of strain. Especially, both Ni et al. [18] and Shintani and Narita [19] reported that EPRs show thoroughly reversed trends, respectively, under tensile and compressive load. Recently, Canadija [20] proved that a deep learning neural network developed on the basis of MD results could accurately predict mechanical properties (including EPR) of SWCNTs. Although IPR can be calculated from MD simulation, there have not been any SWCNT IPR results found from the existing literature, as far as the authors can find. Whereas another allotrope of carbon, graphene, was reported to have negative IPR at the tensile strain of about 6% [21]. So, we guess the IPR of SWCNTs would also turn out to be negative in the tension process and the turning point is chirality-dependent.

In mechanics, there is a controversy on how to model SWCNT following continuum theory. For instance, Bian et al. [3] treated CNTs as an isotropic material in their MM model, as they thought effective wall thickness (h) of CNTs was too small and thus neglected the radial stress. This assumption was similar to that in the study of Zhu and Wang [4]. Nevertheless, some other researchers [11,22] have noted that the effective elastic properties of a SWCNTs were permanently transversely isotropic. Compared with work of Chang et al. [22] where only two force constants were used, Li and Guo [23] adopted four force constants to describe C-C bonds and claimed that SWCNTs were prone to degenerate from transversely isotropic to isotropic if the diameter of SWCNTs turned out to be larger. A similar conclusion was also reported by Refs. [13,24]. More specifically, Li and Chou [25] conducted MM method on SWNTs under hydrostatic pressure and found that radial Young's modulus (E_r) was not equal to circumferential Young's modulus (E_θ). Furthermore, studies employing the finite element (FE) model also gave rise to variable conclusions. Pereira et al. [26] considered that it was advisable to calculate Poisson's ratio assuming isotropy. Jalan et al. [27], through FE analysis, obtained the five transversely isotropic properties of SWCNTs, while Ghavamian et al. [28] thought that CNTs behave rather as an anisotropic material.

On the other hand, determination of the effective thickness of the SWCNT "wall" in mechanics is still a problem. To calculate Young's modulus of SWCNTs, one usually needs to define the effective wall thickness based on continuum elasticity theory. However, the structure of SWCNTs is inherently discrete [29] and the "wall" of a SWCNT consisting of several carbon atoms practically do not form the continuous spatial distribution [30]. Hence, the assumption of the effective wall thickness whose values range from 0.617 Å [31] to 6.9 Å [32] in part results in the dispersion of Young's moduli [30,33–35]. Many researchers [4,6–9,20,33,36,37] took the interplanar spacing of graphite layers 3.4 Å as the effective wall thickness, while some other researchers [38,39] considered that the effective wall thickness is equal to the extent of π orbitals in the out-of-plane direction. Through the calculation of the flexural rigidity $Eh^3/12(1-v^2)$ and axial rigidity $C = Eh$, the thickness of 0.66 Å and Young's modulus about 5.5 TPa were reported by Yakobson et al. [38].In addition to choosing the value that possesses specific physical meaning, researchers [31,32,40,41] also tried many different ways to obtain the effective wall thickness. For instance, Vodenitcharova and Zhang [31] used the ring theory continuum mechanics model to predict the Young's modulus about 4.88 TPa and thickness of 0.617 Å, which was cited by Mylvaganam and Zhang [42]. Huang et al. [41] linked the thickness directly to the interatomic potential and found that the thickness depends on the choice of potential, type of loading, the diameter and chirality of CNTs. Odegard et al. [32] obtained the thickness of 6.9 Å by equating the molecular potential energy with the strain energy stored in the equivalent truss and continuum model.

In fact, Fan and Shen [43] have already claimed that hypothesis of continuum mechanics cannot exactly predict mechanical characteristics of nanomaterials, although some concepts from continuum mechanics we have employed to analyze their mechanical behaviors. Thus, to avoid the confused assumption of h, we use the ratio of rigidity as a substitute for analysis of the ratio of Young's moduli. Obviously, ratio of rigidity in the longitudinal direction (C_l) to rigidity in the radial direction (C_r) equals to the ratio of the longitudinal Young's modulus (E_l) to radial Young's modulus (E_r), namely $C_l/C_r = E_l\,h/E_r\,h = E_l/E_r$. To the best of the authors' knowledge, this will be the first time to address this problem specially.

To fully understand the chiral effect on Poisson's ratios and ratios of moduli of SWCNT, we implemented an MD simulation to analyze EPRs, IPRs, and ratio of moduli of SWCNTs subjected to tensile or compressive load. Particularly, we engaged in two types of comparison: longitudinal vs. radial properties; tensile vs. compressive results. The adaptive intermolecular reactive empirical bond order (AIREBO) [44] potential is used for the force field between carbon atoms. The auxetic phenomena as negative IPRs in chiral SWCNTs will be discussed in detail. The present work not only reveals chiral-dependent insights of mechanical properties of SWCNTs, but also first discovered the auxetic phenomenon of perfect SWCNT. It is believed that with these new findings, our work will contribute to future study and application of SWCNTs [45,46].

2. Models and Methodology

As shown in Figure 1a, a cylindrical SWCNT can be imagined as a rolled graphene sheet where the two crystallographically equivalent points, O and A, coincide. The chiral vector $\boldsymbol{C_h}$ that starts from point O and ends on point A, can be expressed in terms of the basic vectors $\boldsymbol{a}_1$ and $\boldsymbol{a}_2$ as:

$$C_h = n\boldsymbol{a}_1 + m\boldsymbol{a}_2 \equiv (n,\ m), \tag{1}$$

where (n, m) are chiral indices that define helical structural parameters of a SWCNT, i.e., diameter d and chiral angle θ. Chiral angle can be theoretically calculated by

$$\theta = \cos^{-1}\left(\frac{2n+m}{2\sqrt{n^2+nm+m^2}}\right), \tag{2}$$

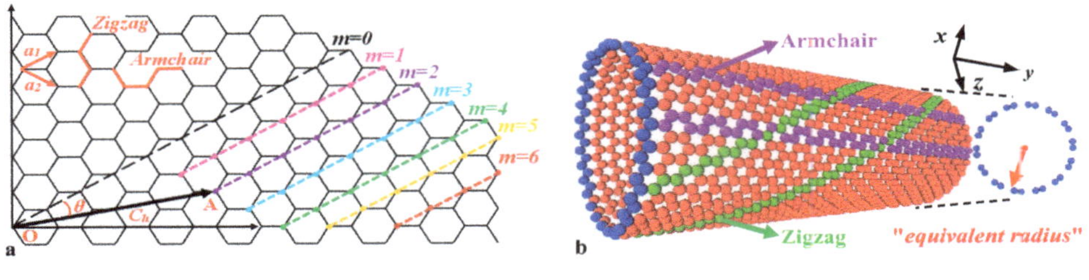

Figure 1. Schematic of a chiral SWCNT. (**a**) Schematic of the chiral parameters in graphene. Chiral vector $\mathbf{C_h}$ for different chiral indices m is shown. (**b**) Schematic of equivalent radius in the x-z plane for a chiral SWCNT.

It should be noted that θ is the angle between the chiral vector $\boldsymbol{C_h}$ and the unit vector $\boldsymbol{a_1}$, which varies from 0° to 30° when $n \geq m$. Specially, chiral angle 0° refers to "zigzag" SWCNT (m equals to zero), while a chiral angle of 30° refers to "armchair" SWCNT (*m and n* are equal). The rest of SWCNT are regarded as "chiral", as shown in Figure 1b.

Due to its tubular structure, transverse deformation of a SWCNT can be calculated by variation of tube radius. When a SWCNT is stretched along the longitudinal (or radial) direction, the EPR and IPR of the SWCNT are, respectively, calculated as:

$$v_{12} = -\frac{\varepsilon_r}{\varepsilon_l};\ v_{21} = -\frac{\varepsilon_l}{\varepsilon_r}, \tag{3}$$

$$v_{12}^{in} = -\frac{d\varepsilon_r}{d\varepsilon_l};\ v_{21}^{in} = -\frac{d\varepsilon_r}{d\varepsilon_r}, \tag{4}$$

where ε_l and ε_r are strains along the longitudinal and radial directions, respectively. Likewise, subscripts 1 and 2 refer to longitudinal and radial directions, respectively. In the MD simulation, however, the circles constructed by carbon atoms in any arbitrary section of a SWCNT can never be perfect in the simulation because of molecular vibration. If we do not consider the vibration effect, any atom in a section can be found its symmetric atom about circle center. Assuming that there are N atom pairs in a section, the equivalent radius $\bar{r}$ can be seen as the half of the mean value of the distances for these atom pair and expressed as:

$$\bar{r} = \frac{1}{2N}\sum_{1}^{N} d_{ij}, \tag{5}$$

in which d_{ij} denotes the distance between the atom i and the atom j, that are in a pair.

The rigidity of SWCNTs subjected to tension or compression is derived from force approach as:

$$C_{11}^{\mathrm{T}} = E_{11}^{\mathrm{T}}h = -\frac{\sigma_l^{\mathrm{T}}}{\varepsilon_l^{\mathrm{T}}}h,\ C_{22}^{\mathrm{C}} = E_{22}^{\mathrm{C}}h = -\frac{\sigma_r^{\mathrm{C}}}{\varepsilon_r^{\mathrm{C}}}h, \tag{6}$$

where superscripts T and C stand for tension and compression, respectively. σ_l is stress along the longitudinal direction evaluated by $\sigma_l = -S_l/V_0$. Here, V_0 is the volume of a SWCNT determined as $V_0 = 2\pi\bar{r}hL$ where L is the length of a SWCNT. S_l is the stress tensor [47] composed of kinetic energy component and virial component and defined as follows:

$$S_{ab} = -mv_a v_b - W_{ab}, \tag{7}$$

$$W_{ab} = \frac{1}{2}\sum_{k=1}^{N_p}(\boldsymbol{r}_{I_a}\cdot\boldsymbol{F}_{I_b}), \tag{8}$$

in which subscripts a and b represent the direction of x, y, z in Cartesian coordinates, and I refers to individual atom. v is the velocity component of atom I and hence $v_a \cdot v_b$ represents kinetic energy contribution. The second term W_{ab} is the virial contribution due to intra and intermolecular interactions, where k loops over the N_p neighbors of specified atom I. Term $\boldsymbol{r}\cdot\boldsymbol{F}$ is the dot product between position vector $\boldsymbol{r}$ and force vector $\boldsymbol{F}$ resulting from the pairwise interaction on specified atom, i.e., projection of $\boldsymbol{F}$ on the position vector $\boldsymbol{r}$ direction. Therefore, radial stress tensor (if decomposed in x-y plane perpendicular to the CNT axis) can be expressed as:

$$S_r \approx -m(v_x^2 + v_y^2) - \frac{1}{2}\sum_{k=1}^{N_p}\boldsymbol{r}_I\cdot(\boldsymbol{F}_{I_x} + \boldsymbol{F}_{I_y}), \tag{9}$$

The choice of the potential is known to have crucial consequences for the results of calculated properties because each potential has its drawback. In the present case, if there is no other statement, the AIREBO [44] potential with a cutoff radius (3 Å) is selected to describe carbon–carbon interactions, which is widely used to study SWCNTs [20,37,48–51]. The definition of AIREBO potential consists of three terms [52,53]:

$$E^{AIREBO} = \frac{1}{2}\sum_{i}\sum_{j\neq i}\left[E_{ij}^{REBO} + E_{ij}^{LJ} + \sum_{k\neq i,j}\sum_{g\neq i,j,k} E_{kijg}^{TORSION}\right], \quad (10)$$

where subscripts i, j, k, and g refer to individual atom. It is clear that the AIREBO potential considers non-boned interactions E^{LJ} and torsional interactions $E^{TORSION}$ besides REBO terms, which is a short-range model describing a C–C atomic interaction with a fixed cut-off range of 2.0 Å. In Equation (10), E_{ij}^{REBO} is given as:

$$E_{ij}^{REBO} = f(r_{ij})(V_{ij}^{R} + B_{ij}V_{ij}^{A}), \quad (11)$$

where B_{ij} is a bond-dependent parameter which weighs the bond order, V_{ij}^{R} and V_{ij}^{A} are repulsive and attractive pair energy terms, respectively, $f(r_{ij})$ is the cut-off function of the distance r_{ij} between atoms i and j [17,54].

The simulations were carried out using open-source molecular dynamics software LAMMPS [55] (large-scale atomic/molecular massively parallel simulator). Periodic boundary conditions are only set in the longitudinal direction, so that the length of the simulated SWCNT can be seen as infinitely long. In the meantime, we make it possible to remove the effect of discrepancy of length with respect to a varied chiral angle [56]. The simulation box boundaries, apart from longitudinal direction, are free to shrink. The duration of each timestep is 0.5 fs [42] and the temperature of system is set and maintained 4.3 K by employing the Nosé–Hoover thermostat to minimize the impact of randomization of C atoms thermal fluctuations. The minimized structure is then equilibrated for a period of 10^5 timesteps. Following equilibration, stretching of SWCNTs follows a specified constant engineering strain rate of 2×10^{-5}/ps with 10^6 timesteps in the canonical ensemble (NVT). Additionally, to obtain the transverse properties, the movement of each atom in a radial direction is controlled by the command *addforce* with 10^6 timesteps, which applied a time-varying force increment based on each atom coordinate [57]. When performing the compression test, engineering strain rate and *addforce* are set to be negative.

3. Results and Discussion

3.1. Engineering Poisson's Ratio (EPR) under Tension

The longitudinal strain of SWCNTs for each timestep can be directly obtained from the simulation, while the corresponding transverse strain can be calculated by means of equivalent radius. Then, an estimation from the linear variation of the transverse strain within 0.5% longitudinal strain is performed for EPR v_{12}. In contrast, EPR v_{21} is calculated through a linear fitting procedure within 0.5% transverse strain when SWCNTs expand along the radial direction. We simulated a series of 240 SWCNTs with diameter d ranging from 3.59 Å for the (4, 1) SWCNT [58] to 119.85 Å for the (150, 6) SWCNT and their EPRs v_{12}^{T} and v_{21}^{T} with different chiral indices m (between one and six) are shown in Figure 2a and Figure 2b, respectively. It can be observed that both v_{12}^{T} and v_{21}^{T} are varied regularly with chiral angle when m is fixed, in other words, m decides the trend of EPR. As regards v_{12}^{T}, however, this variation is not so remarkable when m is over three. From another perspective, the increase of index n also declines the insensibility of EPR to m. Furthermore, another important finding from Figure 2a is that the EPR curves with the same m would converge to one point at zero chiral angle, that is to say, EPR of zigzag SWCNT is chiral-independent and its value is fixed at about 0.29, which is very close to a reported MD result [59]. Figure 2b reveals that v_{21}^{T} also converge to a value near 0.33, surpassing v_{12}^{T}. Comparatively, EPRs v_{21}^{T} of each group of same chiral index m reduce more pronouncedly and keep sensible to m.

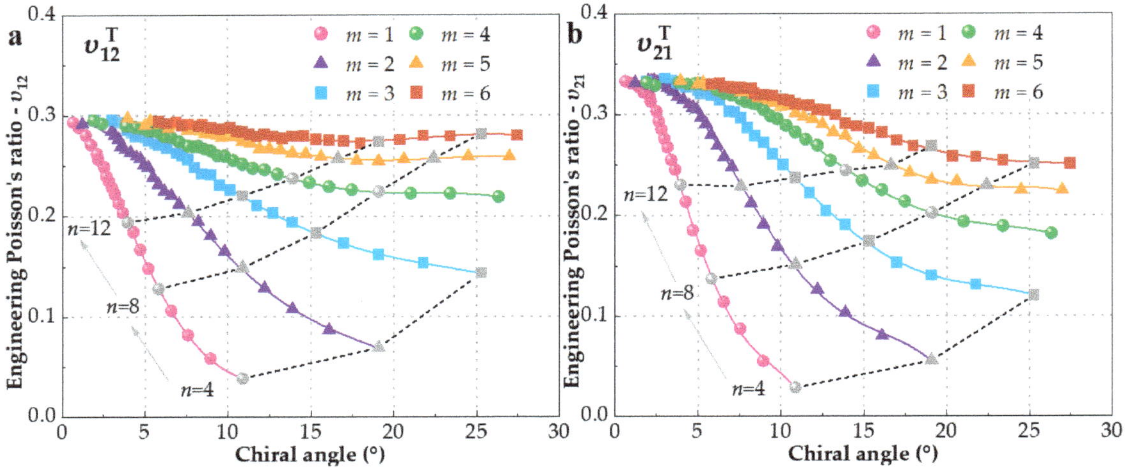

Figure 2. EPRs variation along with chiral angle. SWCNTs are grouped by chiral index m and their diameters are not restricted. (**a**) v_{12}^T obtained from tension in longitudinal direction (**b**) v_{21}^T obtained from tension in radial direction.

EPRs variation with respect to the diameter is illustrated in Figure 3 with the aim of comparing the influence of chiral angle and diameter. Figure 3 uncovers that EPRs are positively associated with the diameter and more approaching SWCNTs with larger diameters [18]. As in conjunction with Figure 2, it can be inferred that SWCNTs are influenced by the combined function of chiral angle and diameter when the diameter is less than 20 Å. Chiral index m seems to affect more v_{12}^T than v_{21}^T. For diameters above 20 Å, EPRs are almost constant and chiral angle becomes the prevailing factor, in close agreement with what was stated in Ref. [20]. Based on the above observations, we selected four groups with diameters of 10 Å, 15 Å, 20 Å and 25 Å to study the effect of the chiral angle separately in the following section. For convenience, we decide to represent them as G25, G20, G15 and G10, respectively.

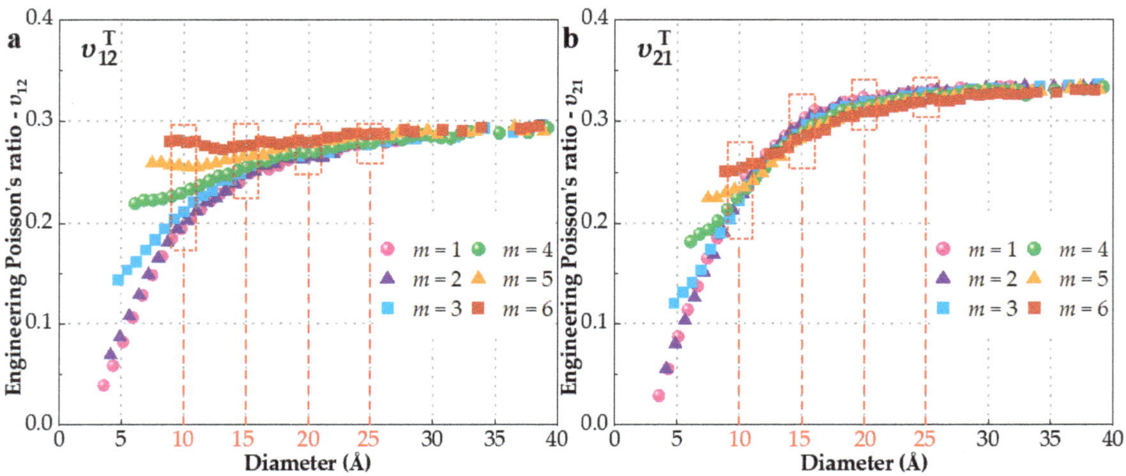

Figure 3. EPRs variation with respect to the diameter. SWCNTs are grouped by chiral index m and SWCNTs with the diameter of 10 Å, 15 Å, 20 Å, 25 Å are labeled in red. (**a**) v_{12}^T (**b**) v_{21}^T.

To explain these phenomena, the corresponding initial potential energy (IPE) of 240 SWCNTs is depicted in Figure 4. It is evident that the distribution and trend of IPE is similar to those of EPR in Figures 2 and 3. The IPE reflects the SWCNT capacity on resistance of axial deformation. In other words, the higher IPE means the stronger the resistance capacity to deformation; as a result, EPR is smaller. Although IPE accounts for the overall changes of EPR with chiral angle and diameter satisfactorily, IPE is not the cause of the inconsistency between υ_{12} and υ_{21}, given that each configuration (n, m) possesses only one IPE. Actually, IPE just reflects the stationary state of SWCNTs and cannot describes dynamic processes. The discrepancy has essentially arisen from the performance variations among zigzag and armchair structure, which has been reported not only in SWCNTs [18–20,37] but also in graphene [16,60]. It also shows that chirality-dependence is not negligible [14,49,61].

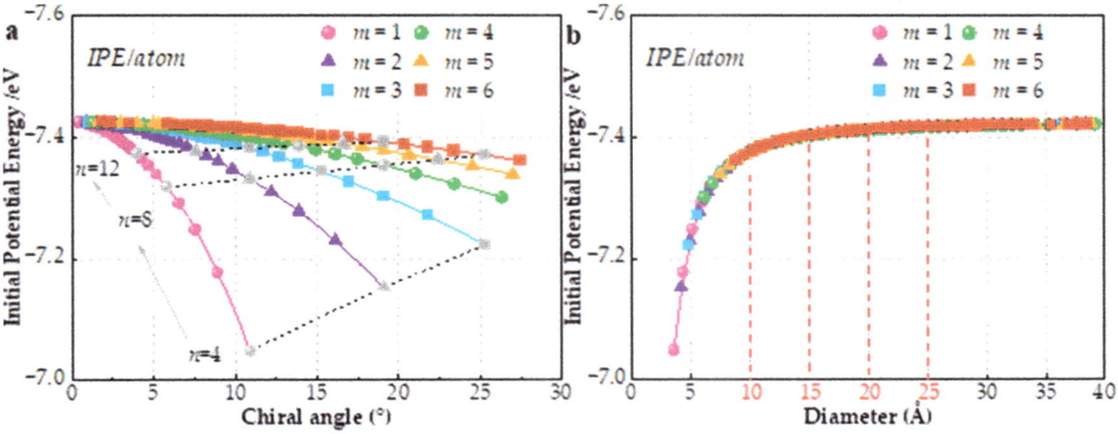

Figure 4. The initial potential energy of SWCNTs corresponding to Figure 2 (after relaxation and before stretch) (**a**) IPE differs according to chiral angle (**b**) IPEs differs according to diameter.

Theoretically, we can construct various SWCNTs with minimum diameter according to different chiral indices m. The EPRs of the SWCNTs, whose diameter is the smallest for a certain m, are listed in Table 1. It is worth noting that (4, 1) SWCNT possesses the smallest diameter in theory and its EPR reaches the value of 0.0385 for $\upsilon_{12}^{\mathrm{T}}$ and 0.0284 for $\upsilon_{21}^{\mathrm{T}}$ which is also the smallest value we can obtain in theory.

Table 1. EPRs and IPEs of SWCNTs with minimum diameter for a certain m (m = 1 to 6).

(n, m)	**(4, 1)**	**(4, 2)**	**(4, 3)**	**(5, 4)**	**(6, 5)**	**(7, 6)**
Chiral angle	10.89°	19.11°	25.28°	26.33°	27.00°	27.46°
Diameter (Å)	3.59	4.14	4.76	6.11	7.47	8.82
EPR $\upsilon_{12}^{\mathrm{T}}$	0.0385	0.0694	0.1436	0.2194	0.2591	0.2802
EPR $\upsilon_{21}^{\mathrm{T}}$	0.0284	0.0556	0.1206	0.1817	0.2247	0.2508
IPE (eV)	−7.048	−7.152	−7.223	−7.301	−7.341	−7.364

Next, the $\upsilon_{12}^{\mathrm{T}}$ of the SWCNTs with fixed 10.89° chiral angle but different diameters are compared in Table 2. The comparison between different potentials of AIREBO and REBO is also listed in Table 2. Owing to lack of items of E^{LJ} and $E^{TORSION}$ [62], the $\upsilon_{12}^{\mathrm{T}}$ for REBO is higher than those for AIREBO. The absolute value of $\upsilon_{12}^{\mathrm{T}}$ increases with the enlargement of diameter uniformly for 2 potentials. However, on the rising trend of EPR for a larger diameter, they are the same.

Table 2. In 10.89° chiral angle family, υ_{12}^{T} and IPEs acquired with AIREBO and REBO potentials.

(*n*, *m*)	(4, 1)	(8, 2)	(12, 3)	(16, 4)	(20, 5)	(24, 6)
Diameter (Å)	3.59°	7.18°	10.76°	14.35°	17.94°	21.53°
EPR(AIREBO)	0.0385	0.1483	0.2065	0.2332	0.2491	0.2588
IPE (AIREBO, eV)	−7.048	−7.333	−7.384	−7.402	−7.410	−7.415
EPR (REBO)	0.0518	0.1678	0.2211	0.2511	0.2693	0.2806
IPE (REBO, eV)	−6.899	−7.260	−7.333	−7.359	−7.372	−7.379

3.2. EPR and Ratio of Young's Moduli under Tension and Compression

As aforementioned, the increase of chiral index *m* or *n* would reduce the variation range of EPR, which is concluded from Figure 2. Since increasing *m* or *n* leads to the same result of larger diameter, what happens if the diameter of SWCNT is enough big? A conclusion form Figure 3 is that the υ_{12} and υ_{21} would infinitely approach the values of 0.29 and 0.33, respectively, when the SWCNT is large enough. In fact, the IPE has a similar trend shown in Figure 4b, which can explain the conclusion. Moreover, the sensibility of EPR to index m is further decreased and the dependence is mainly on the chiral angle. Moreover, very few researches have made the comparison between υ_{12} and υ_{21} with load type (tension and compression). Hence, we collect and plot EPRs of SWCNTs with diameter of 20 ± 1 Å subjected to tension or compression separately in Figure 5a, which encompasses 40 configurations. It can be seen that the sensibility of EPR to index *m* is further decreased and the dependence is mainly on chiral angle. There are two obvious trends exhibited in Figure 5a: (1) For tensile deformation, as the chiral angle increases from zero degree (zigzag) to thirty degrees (armchair), υ_{12}^{T} increases, while υ_{21}^{T} decreases. Armchair SWCNTs have bigger υ_{12}^{T} but smaller υ_{21}^{T} than zigzag ones. The former academic view is also reported by studies [18,20,35,61]; (2) In contrast to tension, SWCNTs with compressive deformation display the exact opposite behavior [18,19]. The υ_{12}^{C} decreases, while υ_{21}^{C} increases when the chiral angle is increased. Thus, the highest υ_{12}^{C} and lowest υ_{21}^{C} are observed in zigzag SWCNTs, which is in close agreement with Ref. [59]. In addition, EPRs subjected to compression is generally larger than tension.

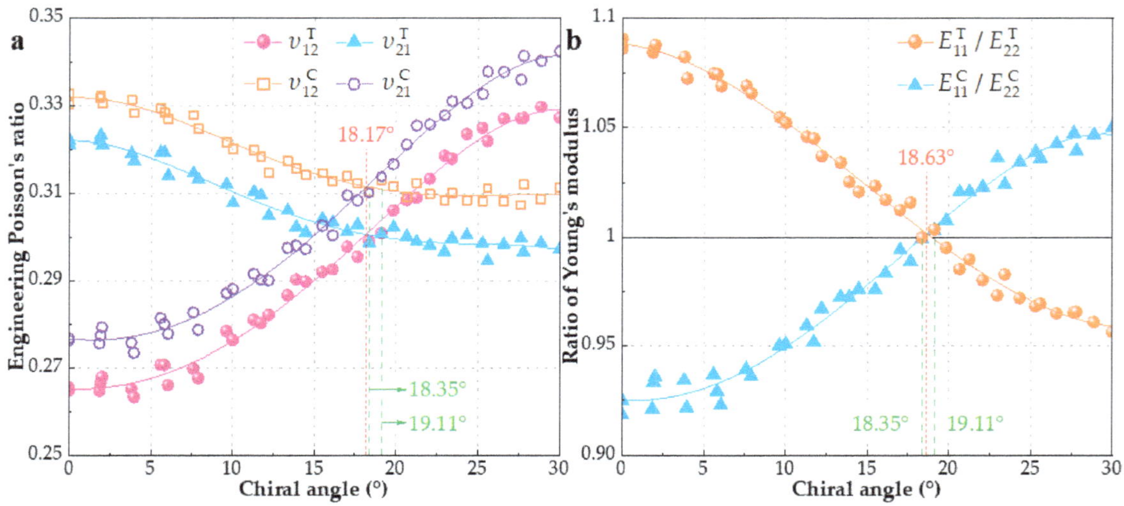

Figure 5. EPRs and ratio of Young's moduli of 40 SWCNTs with diameter of 20 ± 1 Å are compared according to chiral angle (**a**) υ_{12} vs. υ_{21} under tension and compression (**b**) ratio of Young's moduli under tension and compression.

Note that the ratio of Young's moduli E_{11} to E_{22}, equal to ratio of rigidities C_{11} to C_{22}, is derived from the Equation (6), namely $E_{11}/E_{22} = C_{11}/C_{22}$. The rigidity C_{11} and C_{22} are calculated by Equations (7) and (9), using the output parameter stress tensor S of LAMMPS, which is really a <stress*volume> formulation. Figure 5b depicts the change of ratio of Young's moduli E_{11}/E_{22} with chiral angle. It can be clearly observed that E_{11}/E_{22} shows the similar dependence on the chiral angle to EPRs and the opposite trend under different loading type (tension and compression). Ratio of Young's moduli reduces from larger than 1 to less than 1 as the chiral angle increases under tension, which means that only a few configurations match $E_{11}/E_{22} = 1$, i.e., $E_{11} = E_{22}$ (the deviation within −0.4%~0.4%). Particularly, these configurations labeled in green in Figure 5a,b whose chiral angle is 18.38° and 19.11° also satisfy $\upsilon_{12} = \upsilon_{21}$. EPRs and rigidity of these distinct SWCNTs under different loading type are presented in Tables 3 and 4 with other three groups G25, G15, G10 chosen in Figure 3. There are also two chiral angles marked in red representing the intersection of two fitting lines (υ_{12} and υ_{21}) or the intersection of fitting line and x-axis (Y = 1). Apparently, the above chiral angles are only in theory and do not correspond to integral chiral indices (n, m). In addition, the loading type has little influence on the intersection. It should be pointed out that when chiral angle is before the intersection, υ_{12}^{T} is less than υ_{21}^{T} and E_{11}^{T}/E_{22}^{T} is greater than 1, which means most SWCNTs do not meet Maxwell's theorem, i.e., $E_{11}\,\upsilon_{21} = E_{22}\,\upsilon_{12}$. Results of SWCNTs subjected to compression also indicate the above observation. Thus, it is inappropriate to treat all SWCNTs as transversely isotropic or simply orthogonal anisotropic material.

Table 3. EPRs and rigidities of distinctive SWCNTs satisfying $\upsilon_{12}^{T} = \upsilon_{21}^{T}$ and $E_{11}^{T} = E_{22}^{T}$ under tension.

Group	(n, m)	θ (°)	D (Å)	υ_{12}^{T}	υ_{21}^{T}	$\triangle\upsilon$ *	C_{11}^{T} (TPa·nm)	C_{22}^{T} (TPa·nm)	$\triangle C$ **
G25	(26, 11)	16.83°	25.76	0.3070	0.3075	−0.16%	0.2961	0.2954	0.24%
	(24, 11)	17.90°	24.27	0.3072	0.3061	0.36%	0.2957	0.2966	−0.30%
G20	(19, 9)	18.35°	19.38	0.2991	0.2987	0.13%	0.2983	0.2984	−0.03%
	(20, 10)	19.11°	20.71	0.3009	0.3007	0.07%	0.2965	0.2955	0.34%
G15	(14, 7)	19.11°	14.50	0.2877	0.2873	0.14%	0.3040	0.3049	−0.30%
G10	(10, 3)	12.73	9.23	0.2043	0.2046	−0.15%	0.3333	0.3343	−0.15%
	(11, 4)	14.92	10.53	0.2341	0.2343	−0.09%	0.3231	0.3242	−0.34%

* $\triangle\upsilon = [(\upsilon_{12}^{T} - \upsilon_{21}^{T})/\upsilon_{12}^{T}] \times 100\%$. ** $\triangle C = [(C_{11}^{T} - C_{22}^{T})/C_{11}^{T}] \times 100\%$.

Table 4. EPRs and rigidity of SWCNTs satisfying $\upsilon_{12}^{C} = \upsilon_{21}^{C}$ and $E_{11}^{C} = E_{22}^{C}$ under compression.

Group	(n, m)	θ (°)	D (Å)	υ_{12}^{C}	υ_{21}^{C}	$\triangle\upsilon$ *	C_{11}^{C} (TPa·nm)	C_{22}^{C} (TPa·nm)	$\triangle C$ **
G25	(26, 11)	16.83°	25.76	0.3174	0.3162	0.38%	0.3017	0.3026	−0.30%
	(24, 11)	17.90°	24.27	0.3168	0.3177	−0.28%	0.3029	0.3019	0.33%
G20	(19, 9)	18.35°	19.38	0.3103	0.3100	0.10%	0.3047	0.3049	−0.07%
	(20, 10)	19.11°	20.71	0.3130	0.3138	−0.26%	0.3042	0.3033	0.30%
G15	(14, 7)	19.11°	14.50	0.2943	0.2945	−0.07%	0.3119	0.3122	−0.10%
G10	(10, 3)	12.73	9.23	0.2134	0.2141	−0.33%	0.3418	0.3429	−0.32%
	(11, 4)	14.92	10.53	0.2404	0.2408	−0.17%	0.3340	0.3336	0.12%

* $\triangle\upsilon = [(\upsilon_{12}^{C} - \upsilon_{21}^{C})/\upsilon_{12}^{C}] \times 100\%$. ** $\triangle C = [(C_{11}^{C} - C_{22}^{C})/C_{11}^{C}] \times 100\%$.

3.3. Incremental Poisson's Ratio (IPR) under Tension

IPR, which reflects the variation of EPR in stretch, can be obtained from the historical EPR curve. We investigate the three IPRs υ_{12}^{in} of the chiral SWCNTs with chiral angle of 0° (zigzag), 10.89°, and 30° (armchair) in Figure 6a, and find that they would drop to be negative at the strain of 5.93%, 10.60%, and 17.33%, respectively. We notice that Jiang et al. [21] predicted the IPR of the zigzag graphene would turn to be negative at about 6% tensile strain, which is in perfect agreement with our simulation result of the

(25, 0) zigzag SWCNT. Our new finding is that the (24, 6) chiral and (14, 14) armchair SWCNTs have a different critical strain, which is the turning point of IPR from positive to negative. To further study the effect of chirality on critical strain, a group of 68 SWCNTs with various chiral angles from zero degree to thirty degrees are selected to be stretched by MD simulation and their critical strain ε_{11}^{cr} and ε_{22}^{cr} is shown in Figure 6b. If we take 15° chiral angle as a dividing line (green dashed line), the rise of ε_{11}^{cr} is remarkable at the left of the line, but the trend is extremely slowed at the right. Conversely, ε_{22}^{cr} increases slightly as the chiral angle increases until 13.37°, corresponding to configuration (22, 7), and then drops sharply to 6% at chiral angle 30° (armchair). Both minimum values of ε_{11}^{cr} and ε_{22}^{cr} are nearly 6%. The intersection of two critical strain fitting lines is about 15.9°. Furthermore, an interesting finding is that parts of ε_{11}^{cr} (0~10°) and ε_{22}^{cr} (20~30°) are approximately symmetric about the 15° axis.

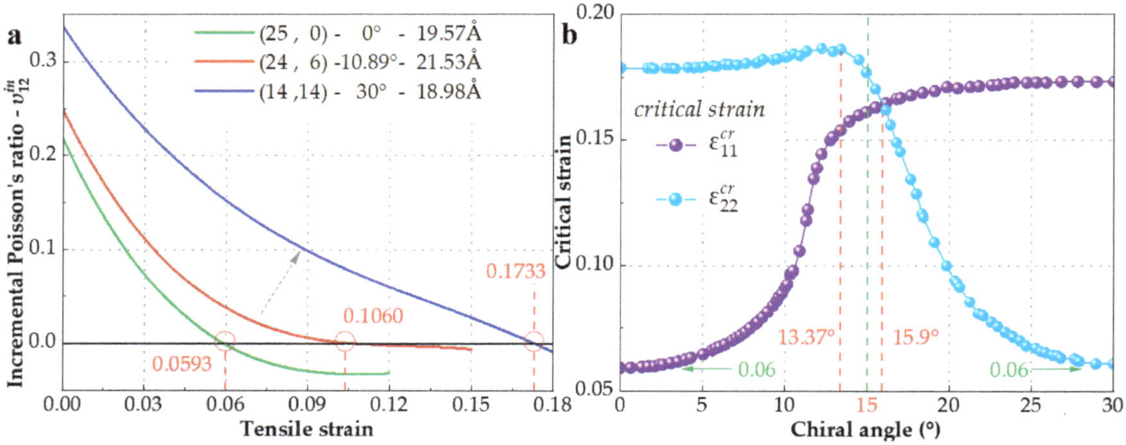

Figure 6. (**a**) The IPRs v_{12}^{in} of three SWCNTs with 0° (zigzag), 10.89°, and 30° (armchair) chiral angle. The values of their critical strains are given and highlighted in red. (**b**) The critical strain of the SWCNTs with various chiral angle between 0° and 30°.

Numerical previous studies [5,9,11,63–67] proved that bond stretching and angle variation are two main deformation modes affecting Poisson's ratio. By assessing the two deformation modes, Jiang et al. [21] well explained the auxetic phenomenon of the graphene and the corresponding critical strain. The situation, however, is more complex in SWCNT because of its geometric characteristic with curvature. It should be mentioned that bonds and angles are both varying in the stretch and their effect on IPR, therefore, is intercoupled. So, the auxetic mechanism can only be found by comparing their independent contributions to IPR. According to the definition of IPR in Equation (4), radial strain differential directly determines its plus or minus. Then, the problem is simplified to find the radial strains respectively induced by bond stretching and angle variation. To fully analyze the influence of the variation of bond length and angle on radial strain, a representative unit consisting of four carbon atoms for SWCNT is proposed in Figure 7a. In that unit, there are three bonds with length of a, b, and c and three angles formed by these bonds. It is worth mentioning that a, b, and c are different when the SWCNT is stretched. In a chiral SWCNT, an arbitrary selected section should include n bonds of length a, m bonds of length b, and $(n + m)$ bonds of length c. Since the projections of these bonds in x-z plane can construct a roughly circle, we can derive the following equation:

$$2n\left(\arcsin\frac{\bar{a}}{2r}\right)+2m\left(\arcsin\frac{\bar{b}}{2r}\right)+2(n+m)\left(\arcsin\frac{\bar{c}}{2r}\right)=2\pi, \tag{12}$$

in which $\bar{a}$, $\bar{b}$, and $\bar{c}$ are the projective lengths corresponding to the bonds with length a, b, and c, respectively, and expressed as:

$$\bar{a} = a\cos\Big(\frac{\alpha}{2} - \frac{\pi}{6} + \theta\Big), \tag{13}$$

$$\bar{b} = b\cos\Big(\frac{\alpha}{2} + \frac{\pi}{6} - \theta\Big), \tag{14}$$

$$\bar{c} = c\cos\Big(\frac{\pi}{6} - \theta\Big). \tag{15}$$

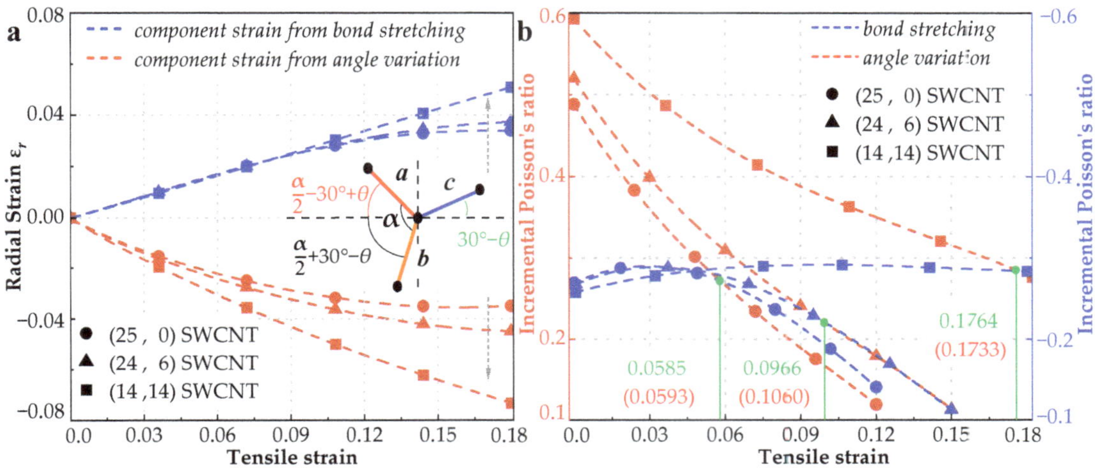

Figure 7. (**a**) Radial strain ε_r owing to bond stretching and angle variation corresponding to chiral 0°(zigzag), 10.89°, and 30°(armchair). (**b**) IPRs owing to bond stretching and angle variation and the tensile strain of the intersection points (green) are compared with the critical strains (red) shown in Figure 6a.

The contributions to radial strains can be obtained by solving Equations (12)–(15) using Newton's method.

Figure 7 illustrates the radial strain for three SWCNTs, respectively, induced by bond stretching and angle variation, as well as their IPRs. It is found from Figure 7 that bond stretching leads to positive radial strain resulting in negative Poisson's ratio, while angle variation contributes negative radial strain resulting in positive Poisson's ratio. Likewise, Jiang et al. [21], through very simplified algebra, obtained Poisson's ratio for bond stretching as $-1/3$ and for angle variation as 1. With the increase of chiral angle, the contributions to radial strain from bond stretching and angle variation are both increased. Their corresponding contributions to IPR, shown in Figure 7b, reveal the reason of critical strain variation for different SWCNTs. For example, the IPR contributed by angle variation for the (25, 0) SWCNT is dominant at the beginning, but its declining rate is higher than that contributed by bond stretching. The intersection point for the two curves is at the tensile strain of 5.85%, which is a good match with the critical strain (5.93%) for the same SWCNT depicted in Figure 4a. After that, the effect of bond stretching is dominant instead, the auxetic phenomenon, therefore, appears, i.e., IPR turns out to be negative. Following this way, the cases of (24, 6) and (14, 14) SWCNTs can also be successfully explained. In summary, the critical strain for negative IPR depends on which mode (bond stretching or angle variation) is dominant.

Tensile behavior curves of (24, 6) chiral SWCNT IPR under different temperatures (4.3 K, 300 K, and 500 K) are plotted in Figure 8a. The remarkable drop of IPR is observed with temperature increase when the tensile strain is lower by about 5%. The corresponding change with temperature of critical strain ε_{11}^{cr} of (24, 6) SWCNT as well as (14, 14) and (25, 0) SWCNTs are illustrated in Figure 8b. From the results, the effect of variation of temperature from 4.3 K to 500 K on critical strain is very limited, in other words, IPRs seem to be insensitive to temperature changes.

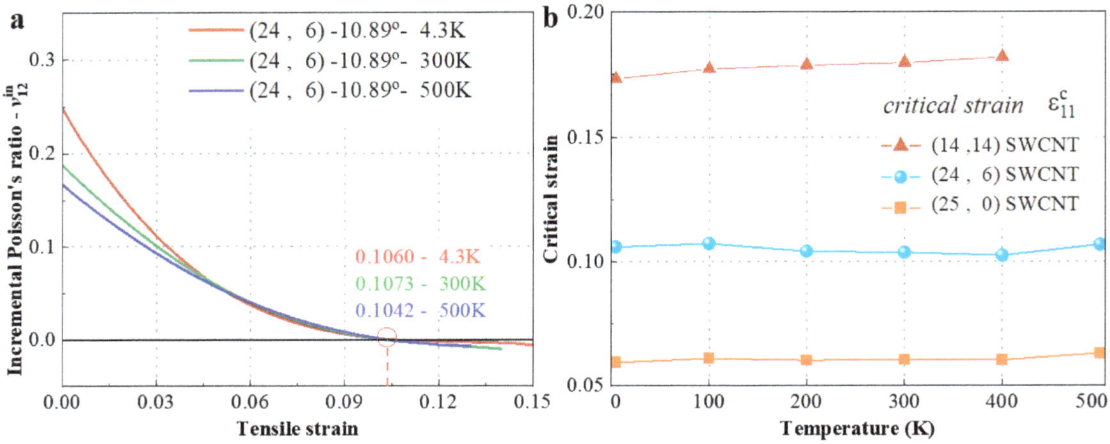

Figure 8. (**a**) The IPRs υ_{12}^{in} of (24, 6) SWCNT under the temperatures of 4.3 K, 300 K, and 500 K. (**b**) Variation of the critical strain with temperature for three SWCNTs (0°, 10.89°, and 30°).

4. Conclusions

In this work, we discussed in detail the effect of chirality on the EPR, rigidity and IPR of SWCNT under different loading and revealed the auxetic mechanism of SWCNT above critical strain based on an MD simulation. First of all, we obtained the minimum EPR υ_{12}^{T} of 0.0385 and υ_{21}^{T} of 0.0284 for (4, 1) SWCNTs in theory. Then, we observed that the distribution and trend of EPRs υ_{12}^{T} and υ_{21}^{T} are sensitive to chiral index m when the diameter is small. However, this sensibility would be reduced with the increase of diameter. The insight of the regularity of EPR on m was partly revealed by IPE representing resistance to deformation. EPR is mainly affected by chiral angle instead when the diameter is greater than 20 Å. Secondly, we noticed that a SWCNT under different load mode could have an exactly opposite EPR trend, so that armchair SWCNTs under tension own the greatest υ_{12}^{T} but smallest υ_{21}^{T}, while under compression, zigzag structure has the greatest υ_{12}^{C} but smallest υ_{21}^{C}. Importantly, after obtaining the ratios of Young's moduli and EPR for various chirality, we verified Maxwell's theorem ($E_{11}\ \upsilon_{21}= E_{22}\ \upsilon_{12}$) and found that most SWCNTs appear anisotropic characteristics in mechanics. Finally, IPRs for both υ_{12}^{in} and υ_{21}^{in} of SWCNT would be negative once tensile strain exceeds critical strain. The minimum of critical strain ε_{11}^{cr} and ε_{22}^{cr} of zigzag or armchair SWCNT are both 6%, close to those of zigzag or armchair graphene. The ε_{11}^{cr} and ε_{22}^{cr} are also influenced by chiral angle and the greatest value can approach 0.18. The comparison of the contributions to IPR from two different deformation modes, bond stretch and angle variation, can give a good explanation of different critical strains for SWCNTs.

Author Contributions: Conceptualization, Y.F. and H.-S.S.; methodology, H.-N.Z. and Y.F.; software, H.-N.Z.; validation, H.-N.Z., Y.F. and H.-S.S.; formal analysis, H.-N.Z., Y.F. and H.-S.S.; investigation, H.-N.Z., Y.F. and H.-S.S.; resources, Y.F.; data curation, H.-N.Z.; writing—original draft, H.-N.Z. and Y.F.; writing—review and editing, H.-N.Z., Y.F. and H.-S.S.; supervision, Y.F.; project administration, Y.F.; funding acquisition, Y.F. All authors have read and agreed to the published version of the manuscript.

Funding: The work is sponsored by the National Natural Science Foundation of China (NSFC) under Grant No. 12102255 and Natural Science Foundation of Shanghai under Grant No. 22ZR1429500.

Institutional Review Board Statement: Not applicable.

Informed Consent Statement: Not applicable.

Data Availability Statement: Not applicable.

Acknowledgments: The computations in this work were run on the π 2.0 cluster supported by the Center for High Performance Computing at Shanghai Jiao Tong University.

Conflicts of Interest: The authors declare that they have no known competing financial interests or personal relationships that could have appeared to influence the work reported in this paper.

References

1. Iijima, S. Helical microtubules of graphitic carbon. *Nature* **1991**, *354*, 56–58. [CrossRef]
2. Fan, Y.; Xiang, Y.; Shen, H.-S. Temperature-Dependent Mechanical Properties of Graphene/Cu Nanocomposites with In-Plane Negative Poisson's Ratios. *Research* **2020**, *2020*, 5618021. [CrossRef]
3. Bian, L.C.; Li, H.; Cheng, Y. Temperature and size-dependent modeling for predicting mechanical properties of carbon nanotubes. *Appl. Math. Model.* **2021**, *98*, 518–536. [CrossRef]
4. Zhu, S.Q.; Wang, X. Effect of environmental temperatures on elastic properties of single-walled carbon nanotube. *J. Therm. Stresses* **2007**, *30*, 1195–1210. [CrossRef]
5. Natsuki, T.; Tantrakarn, K.; Endo, M. Effects of carbon nanotube structures on mechanical properties. *Appl. Phys. A-Mater.* **2004**, *79*, 117–124. [CrossRef]
6. Lei, X.W.; Natsuki, T.; Shi, J.X.; Ni, Q.Q. Analysis of Carbon Nanotubes on the Mechanical Properties at Atomic Scale. *J. Nanomater.* **2011**, *2011*, 805313. [CrossRef]
7. Eberhardt, O.; Wallmersperger, T. Energy consistent modified molecular structural mechanics model for the determination of the elastic properties of single wall carbon nanotubes. *Carbon* **2015**, *95*, 166–180. [CrossRef]
8. Bian, L.C.; Gao, M. Nanomechanics model for properties of carbon nanotubes under a thermal environment. *Acta Mech.* **2018**, *229*, 4521–4538. [CrossRef]
9. Xiao, J.R.; Gama, B.A.; Gillespie, J.W. An analytical molecular structural mechanics model for the mechanical properties of carbon nanotubes. *Int. J. Solids Struct.* **2005**, *42*, 3075–3092. [CrossRef]
10. Popov, V.N.; Van Doren, V.E.; Balkanski, M. Elastic properties of single-walled carbon nanotubes. *Phys. Rev. B* **2000**, *61*, 3078–3084. [CrossRef]
11. Shen, L.X.; Li, J. Transversely isotropic elastic properties of single-walled carbon nanotubes. *Phys. Rev. B* **2004**, *69*, 045414. [CrossRef]
12. Zhang, H.W.; Wang, J.B.; Ye, H.F. Influence of inversion energy on elastic properties of single-walled carbon nanotubes. *Mater. Sci. Eng. A-Struct.* **2007**, *467*, 78–88. [CrossRef]
13. Hwu, C.B.; Yeh, Y.K. Explicit expressions of mechanical properties for graphene sheets and carbon nanotubes via a molecular-continuum model. *Appl. Phys. A-Mater.* **2014**, *116*, 125–140. [CrossRef]
14. Sanchez-Portal, D.; Artacho, E.; Soler, J.M.; Rubio, A.; Ordejon, P. Ab initio structural, elastic, and vibrational properties of carbon nanotubes. *Phys. Rev. B* **1999**, *59*, 12678–12688. [CrossRef]
15. Zhang, C.L.; Shen, H.-S. Predicting the elastic properties of double-walled carbon nanotubes by molecular dynamics simulation. *J. Phys. D-Appl. Phys.* **2008**, *41*, 055404. [CrossRef]
16. Fan, Y.; Xiang, Y.; Shen, H.-S. Temperature-dependent negative Poisson's ratio of monolayer graphene: Prediction from molecular dynamics simulations. *Nanotechnol. Rev.* **2019**, *8*, 415–421. [CrossRef]
17. Brenner, D.W.; Shenderova, O.A.; Harrison, J.A.; Stuart, S.J.; Ni, B.; Sinnott, S.B. A second-generation reactive empirical bond order (REBO) potential energy expression for hydrocarbons. *J. Phys.-Condens. Mater.* **2002**, *14*, 783–802. [CrossRef]
18. Ni, X.G.; Zhang, Z.; Wang, Y.; Wang, X.X. Atomistic study of the strain- and size-dependence of Poisson's ratio of single-walled carbon nanotubes. *Solid State Phenom.* **2007**, *121–123*, 1021–1024. [CrossRef]
19. Shintani, K.; Narita, T. Atomistic study of strain dependence of Poisson's ratio of single-walled carbon nanotubes. *Surf. Sci.* **2003**, *532*, 862–868. [CrossRef]
20. Canadija, M. Deep learning framework for carbon nanotubes: Mechanical properties and modeling strategies. *Carbon* **2021**, *184*, 891–901. [CrossRef]
21. Jiang, J.-W.; Chang, T.; Guo, X.; Park, H.S. Intrinsic Negative Poisson's Ratio for Single-Layer Graphene. *Nano Lett.* **2016**, *16*, 5286–5290. [CrossRef] [PubMed]
22. Chang, T.C.; Geng, J.Y.; Guo, X.M. Prediction of chirality- and size-dependent elastic properties of single-walled carbon nanotubes via a molecular mechanics model. *Proc. R. Soc. A Math. Phys. Eng. Sci.* **2006**, *462*, 2523–2540. [CrossRef]
23. Li, H.J.; Guo, W.L. Transversely isotropic elastic properties of single-walled carbon nanotubes by a rectangular beam model for the C-C bonds. *J. Appl. Phys.* **2008**, *103*, 103501. [CrossRef]
24. Wang, L.F.; Zheng, Q.S.; Liu, J.Z.; Jiang, Q. Size dependence of the thin-shell model for carbon nanotubes. *Phys. Rev. Lett.* **2005**, *95*, 105501. [CrossRef]

25. Li, C.Y.; Chou, T.W. Elastic properties of single-walled carbon nanotubes in transverse directions. *Phys. Rev. B* **2004**, *69*, 073401. [CrossRef]
26. Pereira, A.F.G.; Fernandes, J.V.; Antunes, J.M.; Sakharova, N.A. Shear modulus and Poisson's ratio of single-walled carbon nanotubes: Numerical evaluation. *Phys. Status Solidi B* **2016**, *253*, 366–376. [CrossRef]
27. Jalan, S.K.; NageswaraRao, B.; Gopalakrishnan, S. Development of Empirical Relations for the Transversely Isotropic Properties of Zigzag, Armchair and Chiral Single-Walled Carbon Nanotubes. *Adv. Compos. Lett.* **2012**, *21*, 93–99. [CrossRef]
28. Ghavamian, A.; Rahmandoust, M.; Ochsner, A. On the determination of the shear modulus of carbon nanotubes. *Compos. Part B-Eng.* **2013**, *44*, 52–59. [CrossRef]
29. Shokrieh, M.M.; Rafiee, R. A review of the mechanical properties of isolated carbon nanotubes and carbon nanotube composites. *Mech. Compos. Mater.* **2010**, *46*, 155–172. [CrossRef]
30. Zhang, L.C. On the mechanics of single-walled carbon nanotubes. *J. Mater. Process. Technol.* **2009**, *209*, 4223–4228. [CrossRef]
31. Vodenitcharova, T.; Zhang, L.C. Effective wall thickness of a single-walled carbon nanotube. *Phys. Rev. B* **2003**, *68*, 165401. [CrossRef]
32. Odegard, G.M.; Gates, T.S.; Nicholson, L.M.; Wise, K.E. Equivalent-continuum modeling of nano-structured materials. *Compos. Sci. Technol.* **2002**, *62*, 1869–1880. [CrossRef]
33. Bao, W.X.; Zhu, C.C.; Cui, W.Z. Simulation of Young's modulus of single-walled carbon nanotubes by molecular dynamics. *Phys. B* **2004**, *352*, 156–163. [CrossRef]
34. Tserpes, K.I.; Papanikos, P. Finite element modeling of single-walled carbon nanotubes. *Compos. Part B-Eng.* **2005**, *36*, 468–477. [CrossRef]
35. Avila, A.F.; Lacerda, G.S.R. Molecular Mechanics Applied to Single-Walled Carbon Nanotubes. *Mater. Res.* **2008**, *11*, 325–333. [CrossRef]
36. Ababtin, S.; Adibi, S.; Mun, S.; Carino, R.L.; Dickel, D.E.; Gwaltney, S.R.; Novotny, M.A.; Baskes, M.I.; Horstemeyer, M.F. Single-wall carbon nanotube mechanical behavior using the modified embedded atom method with bond order (MEAM-BO). *Model. Simul. Mater. Sci. Eng.* **2022**, *30*, 035004. [CrossRef]
37. Thamaraikannan, S.; Sunny, M.R.; Pradhan, S.C. Chirality dependent mechanical properties of carbon nano-structures. *Mater. Res. Express* **2019**, *6*, 095018. [CrossRef]
38. Yakobson, B.I.; Brabec, C.J.; Berhnolc, J. Nanomechanics of carbon tubes: Instabilities beyond linear response. *Phys. Rev. Lett.* **1996**, *76*, 2511–2514. [CrossRef]
39. Zhou, X.; Zhou, J.J.; Ou-Yang, Z.C. Strain energy and Young's modulus of single-wall carbon nanotubes calculated from electronic energy-band theory. *Phys. Rev. B* **2000**, *62*, 13692–13696. [CrossRef]
40. Pantano, A.; Parks, D.M.; Boyce, M.C. Mechanics of deformation of single- and multi-wall carbon nanotubes. *J. Mech. Phys. Solids* **2004**, *52*, 789–821. [CrossRef]
41. Huang, Y.; Wu, J.; Hwang, K.C. Thickness of graphene and single-wall carbon nanotubes. *Phys. Rev. B* **2006**, *74*, 245413. [CrossRef]
42. Mylvaganam, K.; Zhang, L.C. Important issues in a molecular dynamics simulation for characterising the mechanical properties of carbon nanotubes. *Carbon* **2004**, *42*, 2025–2032. [CrossRef]
43. Fan, Y.; Shen, H.-S. Non-symmetric stiffness of origami-graphene metamaterial plates. *Compos. Struct.* **2022**, *297*, 115974. [CrossRef]
44. Stuart, S.J.; Tutein, A.B.; Harrison, J.A. A reactive potential for hydrocarbons with intermolecular interactions. *J. Chem. Phys.* **2000**, *112*, 6472–6486. [CrossRef]
45. Galiakhmetova, L.K.; Korznikova, E.A.; Kudreyko, A.A.; Dmitriev, S.V. Negative Thermal Expansion of Carbon Nanotube Bundles. *Phys. Status Solidi RRL Rapid Res. Lett.* **2022**, *16*, 2100415. [CrossRef]
46. Korznikova, E.A.; Zhou, K.; Galiakhmetova, L.K.; Soboleva, E.G.; Kudreyko, A.A.; Dmitriev, S.V. Partial Auxeticity of Laterally Compressed Carbon Nanotube Bundles. *Phys. Status Solidi RRL Rapid Res. Lett.* **2022**, *16*, 2100189. [CrossRef]
47. Thompson, A.P.; Plimpton, S.J.; Mattson, W. General formulation of pressure and stress tensor for arbitrary many-body interaction potentials under periodic boundary conditions. *J. Chem. Phys.* **2009**, *131*, 154107. [CrossRef] [PubMed]
48. Shi, X.; He, X.Q.; Wang, L.F.; Sun, L.G. Hierarchical-structure induced adjustable deformation of super carbon nanotubes with radial shrinkage up to 66%. *Carbon* **2017**, *125*, 289–298. [CrossRef]
49. Goel, M.; Harsha, S.P.; Singh, S.; Sahani, A.K. Analysis of temperature, helicity and size effect on the mechanical properties of carbon nanotubes using molecular dynamics simulation. *Mater. Today Proc.* **2020**, *26*, 897–904. [CrossRef]
50. Lin, Y.W.; Xu, K.; Cao, X.Z.; Zhang, Z.S.; Wu, J.Y. Role of nanotube chirality on the mechanical characteristics of pillared graphene. *Mech. Mater.* **2021**, *162*, 104035. [CrossRef]
51. Chen, Y.; Qin, H.S.; Liu, Y.L.; Pei, Q.X.; Zhang, Y.W. Modeling and Analysis of the Geometry-Dependent Mechanical and Thermal Properties of Coiled Carbon Nanotubes. *Phys. Status Solidi RRL Rapid Res. Lett.* **2022**, *16*, 2100360. [CrossRef]
52. Qian, C.; McLean, B.; Hedman, D.; Ding, F. A comprehensive assessment of empirical potentials for carbon materials. *APL Mater.* **2021**, *9*, 061102. [CrossRef]
53. O'Connor, T.C.; Andzelm, J.; Robbins, M.O. AIREBO-M: A reactive model for hydrocarbons at extreme pressures. *J. Chem. Phys.* **2015**, *142*, 024903. [CrossRef] [PubMed]
54. Dilrukshi, K.G.S.; Dewapriya, M.A.N.; Puswewala, U.G.A. Size dependency and potential field influence on deriving mechanical properties of carbon nanotubes using molecular dynamics. *Theor. Appl. Mech. Lett.* **2015**, *5*, 167–172. [CrossRef]

55. Thompson, A.P.; Aktulga, H.M.; Berger, R.; Bolintineanu, D.S.; Brown, W.M.; Crozier, P.S.; in't Veld, P.J.; Kohlmeyer, A.; Moore, S.G.; Nguyen, T.D.; et al. LAMMPS-a flexible simulation tool for particle-based materials modeling at the atomic, meso, and continuum scales. *Comput. Phys. Commun.* **2022**, *271*, 108171. [CrossRef]
56. Ansari, R.; Ajori, S.; Ameri, A. On the vibrational characteristics of single- and double-walled carbon nanotubes containing ice nanotube in aqueous environment. *Appl. Phys. A-Mater.* **2015**, *121*, 223–232. [CrossRef]
57. Jin, Y.; Yuan, F.G. Simulation of elastic properties of single-walled carbon nanotubes. *Compos. Sci. Technol.* **2003**, *63*, 1507–1515. [CrossRef]
58. Peng, L.M.; Zhang, Z.L.; Xue, Z.Q.; Wu, Q.D.; Gu, Z.N.; Pettifor, D.G. Stability of carbon nanotubes: How small can they be? *Phys. Rev. Lett.* **2000**, *85*, 3249–3252. [CrossRef]
59. Zhang, C.-L.; Shen, H.-S. Temperature-dependent elastic properties of single-walled carbon nanotubes: Prediction from molecular dynamics simulation. *Appl. Phys. Lett.* **2006**, *89*, 081904. [CrossRef]
60. Gamboa, A.; Vignoles, G.L.; Leyssale, J.-M. On the prediction of graphene's elastic properties with reactive empirical bond order potentials. *Carbon* **2015**, *89*, 176–187. [CrossRef]
61. Suzuki, K.; Nomura, S. On elastic properties of single-walled carbon nanotubes as composite reinforcing fillers. *J. Compos. Mater.* **2007**, *41*, 1123–1135. [CrossRef]
62. Chen, W.H.; Cheng, H.C.; Hsu, Y.C. Mechanical properties of carbon nanotubes using molecular dynamics simulations with the inlayer van der Waals interactions. *CMES Comp. Model. Eng.* **2007**, *20*, 123–145. [CrossRef]
63. Chang, T.C.; Gao, H.J. Size-dependent elastic properties of a single-walled carbon nanotube via a molecular mechanics model. *J. Mech. Phys. Solids* **2003**, *51*, 1059–1074. [CrossRef]
64. Yao, Y.T.; Alderson, A.; Alderson, K.L. Can nanotubes display auxetic behaviour? *Phys. Status Solidi B* **2008**, *245*, 2373–2382. [CrossRef]
65. Natsuki, T.; Tantrakarn, K.; Endo, M. Prediction of elastic properties for single-walled carbon nanotubes. *Carbon* **2004**, *42*, 39–45. [CrossRef]
66. Chang, T.C.; Geng, J.Y.; Guo, X.M. Chirality- and size-dependent elastic properties of single-walled carbon nanotubes. *Appl. Phys. Lett.* **2005**, *87*, 251929. [CrossRef]
67. Ajori, S.; Eftekharfar, A.R. Mechanical properties and fracture analysis of defective penta-graphene under temperature variation: Insight from molecular dynamics. *Diamond Relat. Mater.* **2022**, *124*, 108956. [CrossRef]

materials

MDPI

Article

Auxetic Composite Laminates with Through-Thickness Negative Poisson's Ratio for Mitigating Low Velocity Impact Damage: A Numerical Study

Yeqing Wang

Department of Mechanical & Aerospace Engineering, Syracuse University, Syracuse, NY 13244, USA; ywang261@syr.edu; Tel.: +1-(315)-443-2341

Abstract: Auxetic materials are those that exhibit negative Poisson's ratios. Such a unique property was shown to improve the indentation and impact resistances. Angle-ply composite laminates can be designed to produce negative Poisson's ratio at the laminate level due to the large anisotropicity of the individual layer and the strain mismatch between adjacent layers. This paper investigates the effect of through-thickness negative Poisson's ratio on the low velocity impact behaviors of carbon fiber reinforced polymer matrix composite laminates, including the global impact behaviors, as well as the delamination, and the fiber and matrix damage. Results from numerical investigations show consistently reduced fiber and matrix tensile damage in the auxetic laminate in all plies, in comparison to the non-auxetic counterpart laminates (up to 40% on average). However, the auxetic laminate does not present a clear advantage on mitigating the delamination damage or the matrix compressive damage.

Keywords: auxetic composite laminates; carbon fiber composite; negative Poisson's ratio; low velocity impact; finite element analysis

1. Introduction

Carbon fiber reinforced polymer (CFRP) matrix composites are increasingly used in a variety of industries, such as aerospace, marine, automotive, energy, civil infrastructure, and high-end sports. This is due to their significant weight-saving capability and extraordinary properties, including the high specific stiffness and specific strength, excellent fatigue and corrosion resistance, and low coefficient of thermal expansion. However, despite having extraordinary properties, these composites are susceptible to the low velocity impact of foreign objects in service life (e.g., tool drop impact, and the impact of debris from the runway) [1–5]. The impact will cause delamination, matrix cracking, and fiber breakage, which results in significant degradations in mechanical properties [6–9]. For example, the compression strength after impact experienced a reduction of 34.5% and 60.2% for a 16-layer CFRP composite plate, when subjected to an impact energy of 17 J and 29.5 J, respectively [10].

One potential approach to improve the low velocity impact damage tolerance of CFRP composites is to exploit the laminate-level negative Poisson's ratios. Materials or structures that exhibit negative Poisson's ratios are known as auxetic materials or structures [11]. Typically, materials contract transversely under uniaxial tension and expand transversely under uniaxial compression. Auxetic materials/structures exhibit counterintuitive behavior as they expand transversely after tension and contract transversely after compression. Such materials/structures are rare in nature, and thus, are often artificially engineered. There are several ways of engineering auxetic structures. The most common way is to use porous structures, such as re-entrant or chiral structures [12–14]. Another way is to use non-porous multidirectional layered composite structures [15–18]. Additionally, auxetic materials and structures have been developed and studied on different levels and various length scales [19–32]. Examples include auxetic structures that are achieved at the molecular level [23] and through microscopic structure modification [24]. Moreover, researchers have also studied cubic metals that exhibit auxetic behaviors [27] and methods to tailor graphene to achieve negative Poisson's ratio [31]. The current study focuses on the non-porous multidirectional layered composite structures, which are designed to produce negative Poisson's ratio at the macroscopic continuum level.

Previous studies have shown that auxetic materials/structures demonstrate performance enhancements in many properties, such as indentation resistance [13,33], impact resistance [13,15,16,34–36], energy absorption [37], shock wave absorption [38], and the sensitivity of strain sensing [39,40], when compared to their non-auxetic counterparts. For example, the experimental test data reported by Ref. [13] showed that the indentation stiffness of an auxetic lattice reinforced composite structure with a Poisson's ratio of −0.4 is three times that of the composite with a positive Poisson's ratio of 1. Figure 1 shows a comparison of the indentation behaviors between non-auxetic and auxetic materials. Specifically, the through-thickness compression that occurs during mechanical impact will cause auxetic materials to contract in their in-plane directions. Material, therefore, flows into the vicinity of the impact site and the density locally increases. The result is that an auxetic material will have an improved capability to withstand the localized deformation, in comparison to an otherwise identical non-auxetic one.

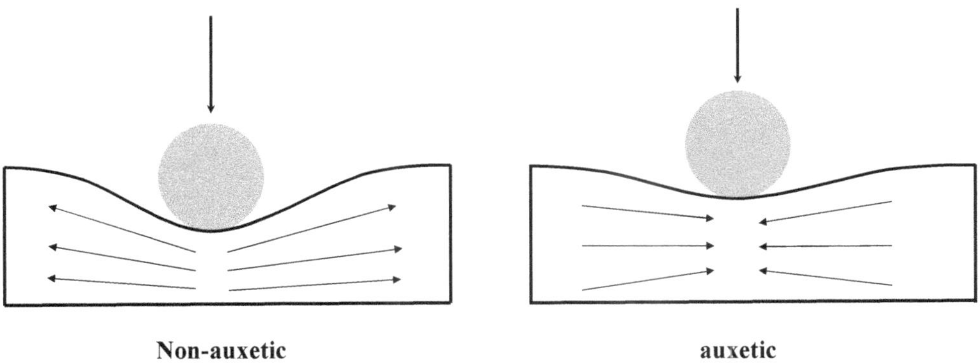

Figure 1. A comparison of indentation behavior between a non-auxetic material (**left**) and an auxetic material (**right**) [41].

Although experimental evidence exists, the number of corresponding modeling studies is scarce, especially for those focusing on the non-porous layered auxetic composite structures. The effect of laminate-level negative Poisson's ratio on the global impact response and damage behavior of the composite structures remains largely unknown. Without such understanding, it is challenging to exploit the negative Poisson's ratio as a design constraint to achieve controllable performance enhancements in layered composite structures. To fill this knowledge gap, numerical simulations are performed in this study to understand the effect of laminate-level negative Poisson's ratio on the low velocity impact behavior of CFRP composite laminates at elevated impact energies.

2. Layups of CFRP Composite Laminates That Allow to Produce Negative Poisson's Ratios

By leveraging the anisotropicity of the individual CFRP composite ply (i.e., a large ratio between E_{11} and E_{22}) and the strain mismatch between adjacent plies, laminate-level (or effective) negative Poisson's ratios can be produced by tuning the layup of the composite laminates [18]. According to the Classical Lamination Theory, the laminate-level through-thickness Poisson's ratio is [35,36],

$$\nu_{13}^{e} = -\frac{J_{31}}{J_{11}} \quad (1)$$

where J_{11} and J_{31} are elements of the **J** matrix,

$$\mathbf{J} = \mathbf{A}^{-1} + \mathbf{A}^{-1}\mathbf{B}\left(\mathbf{D} - \mathbf{B}\mathbf{A}^{-1}\mathbf{B}\right)^{-1}\mathbf{B}\mathbf{A}^{-1} \quad (2)$$

where **A**, **B**, and **D** are the extensional stiffness, extensional-bending coupling stiffness, and bending stiffness matrices [42,43].

As shown in Equations (1) and (2), the negative Poisson's ratio reflects the combined effect of the **A**, **B**, and **D** matrices. Using these equations, the layups for laminated composites to produce negative Poisson's ratios in the through-thickness direction can be identified. Note that fundamental coupon-level tensile tests, conducted by the author's group [44] and other researchers [17,18,45,46], have proved that the laminate-level negative Poisson's ratios predicted analytically, based on the **A**, **B**, and **D** matrices, agree favorably with the experimental data. Table 1 below shows the engineering constants of the IM7/977-3 CFRP composite laminate [47–49] that were used in the calculation of the **A**, **B**, and **D** matrices. Figure 2 shows the calculated through-thickness Poisson's ratio, ν_{13}^{e}, for the laminate with a layup of $[\theta_2/-\theta_2/\theta_2/-\theta_2/\theta_2]$ as the ply angle, θ, changes between 0 and 90 degrees. As we can see, the negative through-thickness Poisson's ratios are produced when the ply angle is between 15 and 40 degrees. The largest through-thickness negative Poisson's ratio is found at the ply angle of 25 degrees. It is worth noting that the CFRP composite laminate is anisotropic. The layups marked in the red dashed box in Figure 2 only allow negative Poisson's ratios in the through-thickness direction. In the in-plane directions, the Poisson's ratios remain positive. Since the auxeticity is produced in only one direction, when adopting the terminology proposed by Branka et al. [50], these composite structures can be considered partially auxetic.

Table 1. Material properties of IM7/977-3 CFRP composites [47–49].

Composite lamina properties	Density	ρ = 1600 kg/m^3
	Elastic moduli	E_{11} = 159 GPa, E_{22} = E_{33} = 9.2 GPa
		G_{12} = G_{13} = 4.37 GPa, G_{23} = 2.57 GPa
	Poisson's ratio	ν_{12} = ν_{13} = 0.253, ν_{23} = 0.456
	Strength	X_T = 2275 MPa, X_C = 1680 MPa, Y_T = 64 MPa, Y_C = 168 MPa. S_{xy} = 121 MPa, S_{yz} = S_{zx} = 127 MPa
	Fracture energy	G_{ft} = 133 N/mm, G_{fc} = 40 N/mm, G_{mt} = 0.6 N/mm, G_{mc} = 2.1 N/mm
Interface properties	Modulus	E = 5 GPa
	Strength	N = S = 30 MPa
	Fracture energy	G_n^C = 0.6 N/mm (normal), G_s^C = 2.1 N/mm (shear)

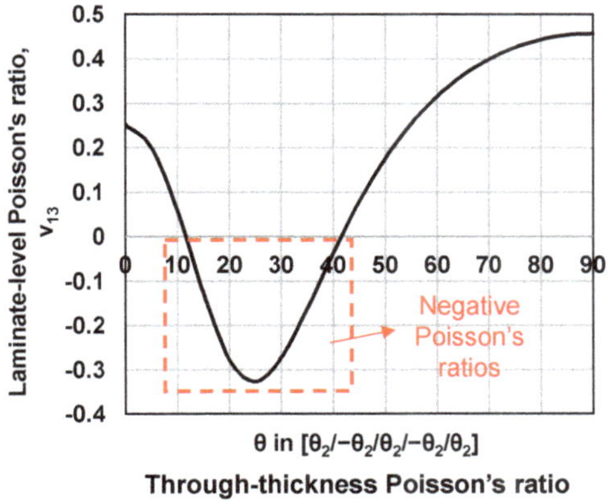

Figure 2. Predicted layups to produce laminate-level through-thickness negative Poisson's ratios in IM7/977-3 CFRP composite laminates with a layup of $[\theta_2/-\theta_2/\theta_2/-\theta_2/\theta_2]$ as the ply angle, θ, changes from 0 to 90 degrees.

3. Layups of Non-Auxetic CFRP Counterpart Laminates with Positive Poisson's Ratios

To ensure a meaningful comparison, the layups of the counterpart CFRP laminates are identified such that they allow them to produce positive Poisson's ratios and, at the same time, produce identical effective moduli to those of the auxetic laminates. The effective moduli of the laminate are calculated using the well-validated analytical equations proposed by Sun and Li [42]. It is worth noting that no layups exist that would allow the non-auxetic counterpart laminate to 100% match all effective moduli in three directions of the auxetic composite laminate (i.e., E_1^{eff}, E_2^{eff}, and E_3^{eff}). Therefore, two matching configurations are used for best approximations: Configuration 1 strictly matches both the longitudinal and the through-thickness effective moduli (E_1^{eff} and E_3^{eff}) of those of the auxetic laminate with very low tolerances (≤0.7%), and without constraining the transverse effective modulus (E_2^{eff}), while Configuration 2 is the best available configuration that matches all effective moduli of the auxetic laminate in three directions with relatively higher tolerances. Table 2 shows our identified layups of the non-auxetic counterpart composite laminates in both. The reason for choosing the layups of $[25_2/-25_2/25_2/-25_2/25_2]$ for the auxetic composite is that this layup will allow the laminate to produce the largest through-thickness negative Poisson's ratio, as shown in Figure 2, which is expected to provide the most significant enhancement in the low velocity impact resistance. Note that, this layup is an unbalanced layup and does not necessarily represent an optimum layup for practical engineering applications. It is only used here to study the effect of the laminate-level negative Poisson's ratio on the impact resistance of CFRP composites.

Table 2. Layups of the auxetic laminate and the corresponding non-auxetic counterpart composite laminates in two configurations.

	Through-Thickness Auxetic CFRP Laminate	Configuration 1: Non-Auxetic Counterpart CFRP Laminate (with Strictly Matched E_1^{eff} and E_3^{eff})	Configuration 2: Non-Auxetic Counterpart CFRP Laminate (with Weakly Matched E_1^{eff}, E_2^{eff}, and E_3^{eff})
Layup	$[25_2/-25_2/25_2/-25_2/25_2]$	$[50_2/0_2/50_2/0_2/50_2]$	$[20_2/10_2/5_2/10_2/20_2]$
ν_{13}^e	−0.327	0.264	0.260
E_1^{eff} (GPa)	70.83	70.33 (−0.7%)	71.88 (+1.5%)
E_2^{eff} (GPa)	9.45	14.23 (+50.6%)	9.46 (+0.1%)
E_3^{eff} (GPa)	9.95	9.95 (0)	9.24 (−7.1%)

4. Low Velocity Impact Model for CFRP Composite Laminates

To study the effect of the laminate-level negative Poisson's ratio on the low velocity impact behavior of the composite laminates, a well-validated progressive damage modeling approach [7,10,47,49,51–53] is used. The primary components of the low velocity impact model adopted throughout this study include: (1) the Hashin damage criteria, which are used to predict the initiation of the fiber tensile and compressive failure, and the matrix tensile and compressive failure [54]; (2) the linear stiffness degradation function, based on the equivalent strain method [49], which is used to track the damage evolution in each failure mode [49]; and (3) the Benzeggagh and Kenane (B—K) delamination criterion along with mixed-mode fracture energy laws, which are used to model the initiation and evolution of the delamination damage [55]. The low velocity impact model is implemented using finite element analysis (FEA) with the general-purpose FEA software, ABAQUS. Specifically, the above-mentioned stiffness degradation law, damage initiation, and damage evolution are implemented using a VUMAT subroutine while the delamination damage is modeled by defining cohesive surface contacts between adjacent laminate plies. Note that finite element methods have been widely adopted for analyses of mechanics of composite laminates [56,57].

4.1. Model Verification Using a Benchmark Low Velocity Impact Problem

To verify the model, the low velocity impact problem reported by Ref. [52] is used as a benchmark. The reason for choosing this problem is due to the completeness of the experimental test results, including the load vs. time curve, the load vs. displacement curve, and the delamination patterns per layer. This problem has been widely used as a benchmark in many studies for the verification of low velocity impact models [10,49,53,58]. Below is a brief description of this benchmark problem, followed by the verification results.

The CFRP composite laminate considered in this benchmark is a T700CG/M21 carbon fiber epoxy resin composite laminate with a layup of $[0_2/45_2/90_2/-45_2]_S$ and a dimension of 150 × 100 × 4.16 mm. The impactor is made of steel, has a mass of 2 kg, and has a semi-spherical head with a diameter of 16 mm. The velocity of the impactor is 5 m/s, representing an impact energy of 25 J. The composite laminate is placed on top of a supporting plate with an inner open-cut window of 125 × 75 mm. The impactor and the supporting plate are modeled as discrete rigid bodies using R3D4 elements while the CFRP composite laminate is modeled using the C3D8R elements (i.e., eight-node linear brick, reduced integration elements). Figure 3 shows the schematic of the problem setup and the mesh used for the impactor, the CFRP composite, and the supporting plate. The global seed sizes of the impactor and the supporting plate defined in ABAQUS are 0.5 and 3 mm, respectively. For the CFRP composite, the mesh size at the center region (72 × 36 mm) directly under the impactor is refined to 0.9 × 0.9 mm, whereas the mesh at the regions far away from the impact site is created using a global seed size of 3.5 mm to reduce the computational time. To model the delamination, the interfaces between each adjacent ply pairs are assigned using cohesive surface contacts. The material parameters, including the density, Young's modulus, Poisson's ratios, strength, fracture energy, as well as the interface properties, are taken from Ref. [49]. The verification results are discussed in the section below.

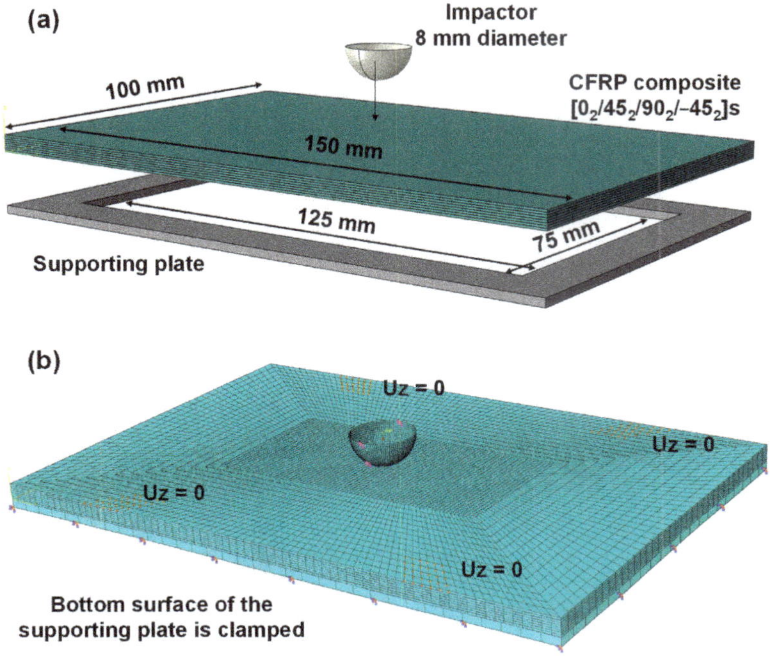

Figure 3. Verification of the low velocity impact model using a benchmark simulation problem for a CFRP composite laminate under an impact energy of 25 J: (**a**) problem setup and (**b**) mesh and boundary conditions used in the finite element analysis. Note: the x direction represents the longitudinal direction or direction 1, y direction represents the transverse direction or direction 2, and z direction represents the through-thickness direction or direction 3 in the laminate coordinate system.

4.2. Model Verification Results

The model verification results are shown in Figures 4 and 5. Specifically, Figure 4a shows the comparison between the simulation result of the force history during the impact and the experimental test result provided by Ref. [52]. As one can see, the predicted force history, including the peak load, the time to peak load, and the duration of the impact agree well with the experimental test data. Additionally, Figure 4b,c show comparisons of the force vs. displacement and the overlapped delamination patterns between the simulation results and the reported experimental test results, in which good agreements can be observed. Specifically, the experimental data of the delamination patterns are shown on the left side of Figure 4c. The data are taken from Ref. [52], which were obtained using an ultrasonic C-scan. The delamination patterns in interfaces of the composite laminate between plies with different angles are overlapped and illustrated in the C-scan contour. The color legend of the contour indicates the location of the delamination in the through-thickness direction, where the greenish color indicates a location near the top surface (i.e., impact side), while the reddish color indicates a location near the bottom surface. On the other hand, the predicted delamination patterns are shown on the right-hand side of Figure 4c, in which the color legend indicates the degree of delamination, where the red color indicates complete delamination while the blue color indicates no delamination.

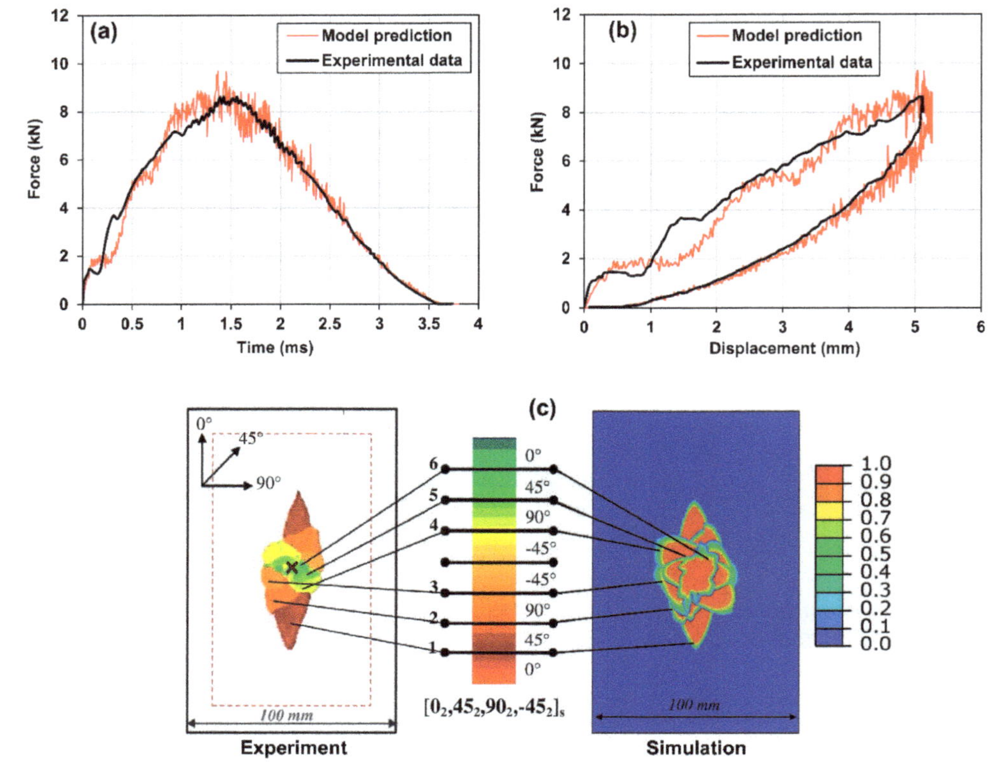

Figure 4. Verification of the low velocity impact model: comparison between the experimental and simulation results for a 25 J impact: (**a**) force history, (**b**) force vs. displacement, and (**c**) overlapped delamination areas in interfaces of the CFRP laminate between plies with different angles, where the experimental data were obtained using ultrasonic C-scan taken from Ref. [52]. The color legend of the C-scan contour (left) indicates the location of the delamination in the through-thickness direction, where greenish indicates the location near the top surface (impact side) while the reddish indicates the location near the bottom surface. The color legend in the simulation contour (right) indicates the degree of the predicted delamination, where the red color indicates complete delamination while the blue color indicates no delamination.

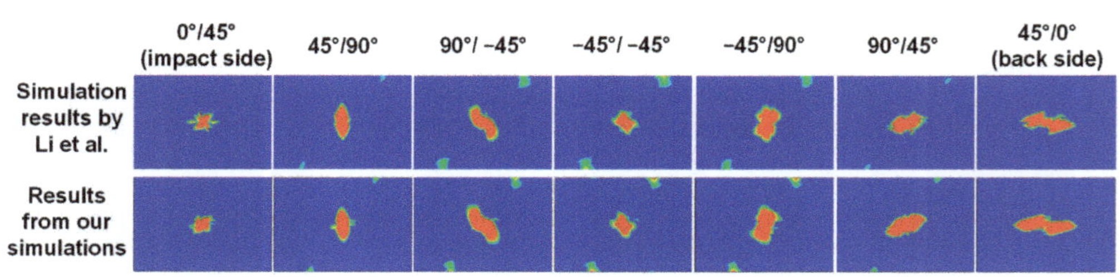

Figure 5. Verification of the low velocity impact model: comparison between simulation results reported in Ref. [49] (first row) and our simulation results (second row) for the same benchmark problem (i.e., low velocity impact at an impact energy of 25 J for a CFRP composite laminate of 150×100 mm and a layup of $[0_2/45_2/90_2/-45_2]_S$). The contour plots from left to right show the delamination patterns at each interface of the composite laminate. The red color indicates complete delamination while the blue color indicates no delamination.

To further illustrate the predictive accuracy of the low velocity impact model, the predicted ply-by-ply delamination patterns are compared with the simulation results reported by Ref. [52], as shown in Figure 5. It can be seen that the predicted delamination patterns are identical.

5. Results and Discussion

After the impact model was verified, it was employed to study the effect of laminate-level Poisson's ratios, i.e., the combined effect of the **A**, **B**, and **D** matrices (see Equations (1) and (2)), on the low velocity impact behavior of the auxetic and non-auxetic IM7/977-3 layered CFRP composites. To achieve this, the same computational setup shown in Figure 3 was used, except that the number of layups was reduced from 16 layers to 10 layers, and the layups of the laminates follow those identified in Table 2. The ply-level engineering constants of the composite lamina used in the simulation studies are shown in Table 1. The simulations were conducted at three elevated impact energy levels, i.e., 3, 5, and 8 J. The choice of these three energy levels is because they were found, from preliminary trial-and-error simulation studies, to produce minimum, intermediate, and maximum damage without causing the laminates to penetrate.

5.1. Effect on the Global Response during Low Velocity Impact

As shown in Figure 6a, the auxetic laminate shows consistently higher impact forces at all three energy levels, when compared to the non-auxetic counterparts. Meanwhile, the difference in the impact force between the auxetic laminate and the non-auxetic counterparts becomes more significant as the impact energy increases. Specifically, at 3 J, the impact force of the auxetic laminate is 8.7% and 33.3% higher than those of the non-auxetic Configuration 1 and Configuration 2 laminates, respectively. These differences increase to 14.0% and 81.0% as the impact energy increases from 3 J to 8 J. Moreover, it can be observed in Figure 6a that the predicted impact forces of the non-auxetic Configuration 1 laminate are much higher than those of non-auxetic Configuration 2 laminate and are relatively closer to those of the auxetic laminate. This is due to the higher effective moduli in the transverse and through-thickness directions of the non-auxetic Configuration 1 laminate than those of the Configuration 2 laminate (see Table 2). Note that mechanical impact is a contact problem where the effective contact modulus is a function of the through-thickness modulus [36]. Moreover, impact also involves biaxial bending where transverse stiffness also plays an important role. Therefore, the higher effective moduli in the transverse and through-thickness directions can cause a higher impact force.

In addition, the auxetic laminate shows consistently shorter impact times as shown in Figure 6b. Similar to the result of the impact force, the difference in the impact time between the auxetic laminate and the non-auxetic counterparts becomes more significant as the impact energy increases. Furthermore, the impact times of the non-auxetic Configuration 1 laminate are consistently much shorter than those of the non-auxetic Configuration 2 laminate and are closer to those of the auxetic laminate. This, together with the result of the impact force, implies that, although the through-thickness negative Poisson's ratio increases the impact force and decreases the impact time, such an effect can be mitigated if the transverse or through-thickness effective moduli of the laminate decrease after the laminate layup is tuned to produce the through-thickness negative Poisson's ratio.

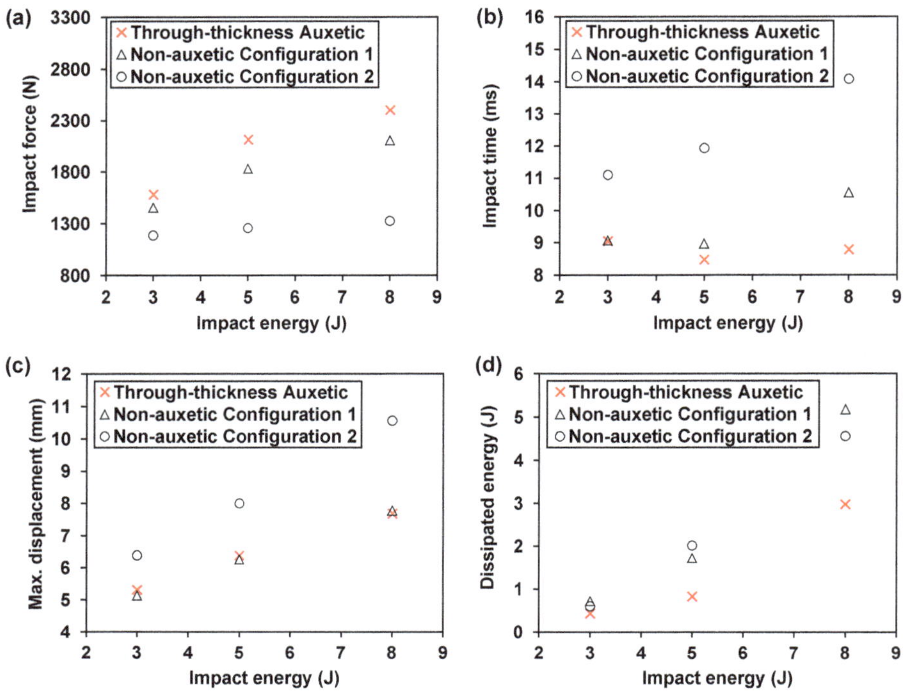

Figure 6. Effect of the through-thickness negative Poisson's ratio on the global impact response: comparison between the auxetic and corresponding non-auxetic CFRP composite laminates: (**a**) impact load, (**b**) impact time, (**c**) maximum displacement, and (**d**) dissipated energy.

Furthermore, the auxetic laminate shows consistently much lower maximum displacements than those of the non-auxetic Configuration 2 laminate, as depicted in Figure 6c. The reductions are 16.8%, 20.3%, and 27.4% at 3, 5, and 8 J, respectively. This indicates that producing the through-thickness negative Poisson's ratio is beneficial for reducing the maximum displacement during low velocity impact. Moreover, this positive effect is more significant as the impact energy increases. For the Configuration 1 laminate, the maximum displacements are slightly lower (i.e., 3.2% and 1.7%) than those of the auxetic laminate at 3 and 5 J. However, when the impact energy rises to 8 J, the maximum displacement of the Configuration 1 non-auxetic laminate exceeds that of the auxetic laminate by 1.4%. The result suggests that both the negative Poisson's ratio and the increased effective modulus (i.e., the transverse modulus of Configuration 1 laminate is higher than that of the auxetic laminate, see Table 2) are beneficial for reducing the maximum displacement. Specifically, at lower energy levels (i.e., 3 and 5 J), the increased transverse effective modulus is a more dominating factor, whereas at a higher energy level (i.e., 8 J), the effect of the through-thickness negative Poisson's ratio becomes more significant than the effect of the increased modulus, which contributes to a more significant reduction of the maximum displacement.

Moreover, Figure 6d illustrates that the auxetic laminate consistently exhibits lower dissipated energies than those of the non-auxetic laminates. Since the energy dissipation is closely related to the damage behavior of the composite laminates, the effect of the through-thickness negative Poisson's ratio on the damage behaviors, including delamination and fiber and matrix damage, is examined in the following sections.

5.2. Effect on the Delamination Damage

The through-thickness negative Poisson's ratio, ν_{13}^e, have restrained the delamination growth in the transverse direction but encouraged the growth in the longitudinal direction. Figure 7 illustrates the comparison of the predicted delamination patterns in each interface of the auxetic laminate and the corresponding non-auxetic laminates at an impact energy of 8 J. Specifically, the delamination patterns in the auxetic laminate follow a diamond shape where the transverse diagonal length is much shorter than the longitudinal diagonal length. The delamination patterns in the non-auxetic Configuration 1 laminate also exhibit a diamond shape, but the lengths of the two diagonals are almost identical. As for the non-auxetic Configuration 2 laminate, the shapes of delamination at the top and bottom interfaces are like those of the non-auxetic Configuration 1 laminate, while the shapes at the two middle interfaces are similar to those of the auxetic laminate.

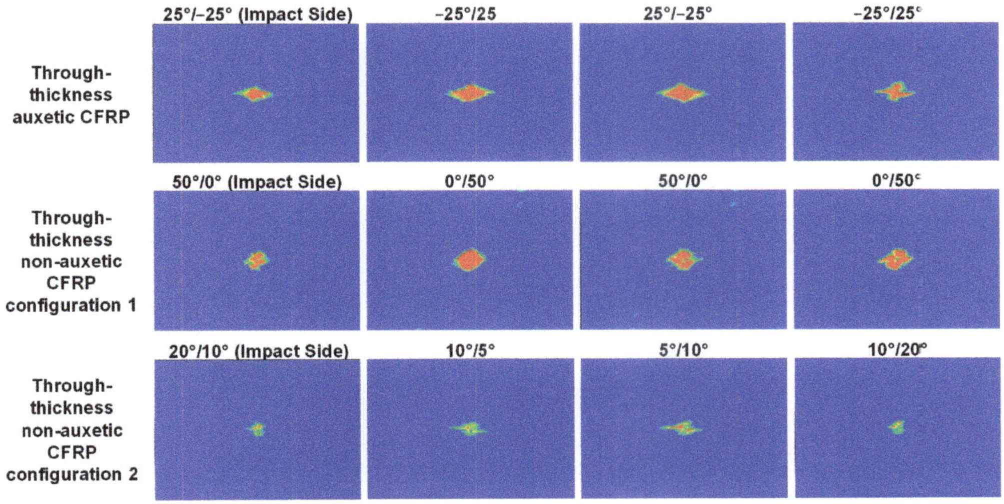

Figure 7. Effect of the through-thickness negative Poisson's ratio on the delamination: comparison of predicted delamination pattern in each interface of auxetic CFRP composite (layup: $[25_2/-25_2/25_2/-25_2/25_2]$, results shown in the first row) and the corresponding non-auxetic laminates in two configurations (layups: $[50_2/0_2/50_2/0_2/50_2]$ and $[20_2/10_2/5_2/10_2/20_2]$, results shown in the second and third rows), at an 8 J impact. The red color indicates complete delamination while the blue color indicates no delamination.

Figure 8 provides a quantitative comparison of the delamination area between the auxetic and non-auxetic laminates. At an impact energy of 3 J, the delamination areas of the auxetic laminate at all interfaces are close to those of the non-auxetic Configuration 2 laminate, as shown in Figure 8a. At 5 and 8 J, the auxetic laminate exhibits larger delamination areas than the non-auxetic Configuration 2 laminate, as shown in Figure 8b,c. The difference increases as the impact energy increases. At 8 J, the delamination areas of the auxetic laminate are 2.1, 1.9, 1.8, and 2.5 times those of the non-auxetic Configuration 2 laminate at the four interfaces, respectively. Results indicate that producing the through-thickness negative Poisson's ratio does not present a clear advantage in mitigating the delamination damage.

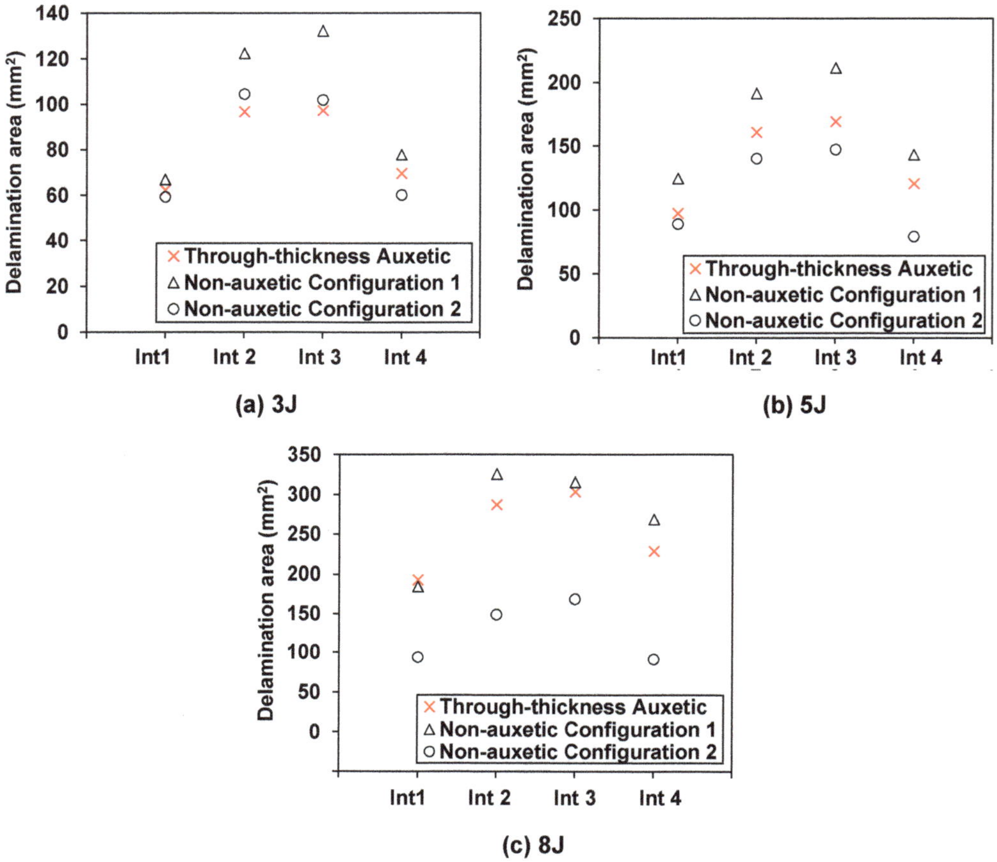

Figure 8. Effect of the through-thickness negative Poisson's ratio on the delamination: comparison of the predicted delamination area in each interface of the auxetic and corresponding non-auxetic CFRP composite laminates at: (**a**) 3 J, (**b**) 5 J, and (**c**) 8 J, where "int" in the horizontal axis denotes the interface of the composite laminate.

As for the non-auxetic Configuration 1 laminate, the delamination areas are much larger than those of the non-auxetic Configuration 2 laminate and are generally larger than those of the auxetic laminate, as shown in Figure 8a–c. This is due to the high transverse effective modulus of the non-auxetic Configuration 1 laminate. Additionally, it can be observed that, as the impact energy increases, the difference in the delamination areas between the non-auxetic Configuration 1 laminate and the auxetic laminate becomes smaller and smaller. At 8 J, the delamination areas of the two laminates are identical, as shown in Figures 7 and 8c. This implies that both producing through-thickness negative Poisson's ratio and increasing the transverse effective modulus could lead to an adverse effect on the delamination propagation. As the impact energy increases, the adverse effect due to the through-thickness negative Poisson's ratio becomes more pronounced.

Note that although the delamination damage becomes more pronounced as the impact energy increases, the dissipated energy, as shown in Figure 6d, is still consistently lower for the auxetic laminate. This indicates that the other damage modes (e.g., fiber and matrix damage) in the auxetic laminate are more confined as the impact energy increases, when compared to the non-auxetic laminates, which will be discussed as follows.

5.3. Effect on the Matrix and Fiber Damage

The matrix tensile, the fiber tensile, and the matrix compressive damage are commonly observed damage modes in CFRP composites under low velocity impact (fiber compressive damage is often negligible [49]). The effect of the through-thickness negative Poisson's ratio on the matrix tensile damage is illustrated in Figures 9 and 10. Specifically, Figure 9 shows the comparison of the predicted patterns of the matrix tensile damage in each ply of the auxetic laminate and non-auxetic laminates at an impact energy of 8 J. It can be clearly observed that the auxetic laminate exhibits consistently much smaller matrix tensile damage in all plies, when compared to the two non-auxetic laminates. This is also true for cases with impact energies of 3 and 5 J, as shown in Figure 10.

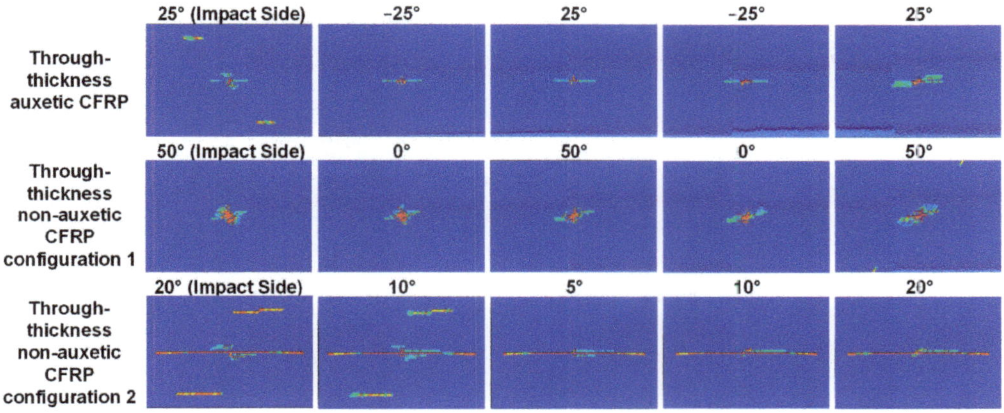

Figure 9. Effect of the through-thickness negative Poisson's ratio on the matrix tensile damage: comparison of predicted matrix tensile damage pattern in each ply of the auxetic CFRP composite (layup: $[25_2/-25_2/25_2/-25_2/25_2]$, results shown in the first row) and the corresponding non-auxetic laminates in two configurations (layups: $[50_2/0_2/50_2/0_2/50_2]$ and $[20_2/10_2/5_2/10_2/20_2]$, results shown in the second and third rows), at an 8 J impact. The red color indicates complete damage while the blue color indicates no damage.

Figure 10a–c show that the auxetic laminate exhibits consistently much smaller matrix tensile damage areas in all plies, when compared to the non-auxetic laminates at all impact energy levels. For instance, at 8 J, the predicted matrix tensile damage areas in the auxetic laminate are shown to reduce by 49.3%, 86.5%, 68.5%, 63.6%, and 7.0%, in the five plies, respectively, in comparison to the non-auxetic Configuration 1 laminate, and are shown to reduce by 5.2%, 5.3%, 54.0%, 51.8%, and 20.0%, in the five plies, respectively, in comparison to the non-auxetic Configuration 2 laminate. This suggests that producing the through-thickness negative Poisson's ratio is beneficial in suppressing the propagation of the matrix tensile damage, as the contraction in the through-thickness direction during the impact leads to the contraction in the in-plane direction caused by the negative Poisson's ratio (see Figure 1), thereby mitigating the tensile damage. Furthermore, although the non-auxetic Configuration 1 laminate has a higher transverse effective modulus than the auxetic laminate, the predicted matrix tensile damage areas are still consistently much larger than those of the auxetic laminate. This implies that the matrix tensile damage is much more sensitive to the negative Poisson's ratio than the increase of the transverse effective modulus.

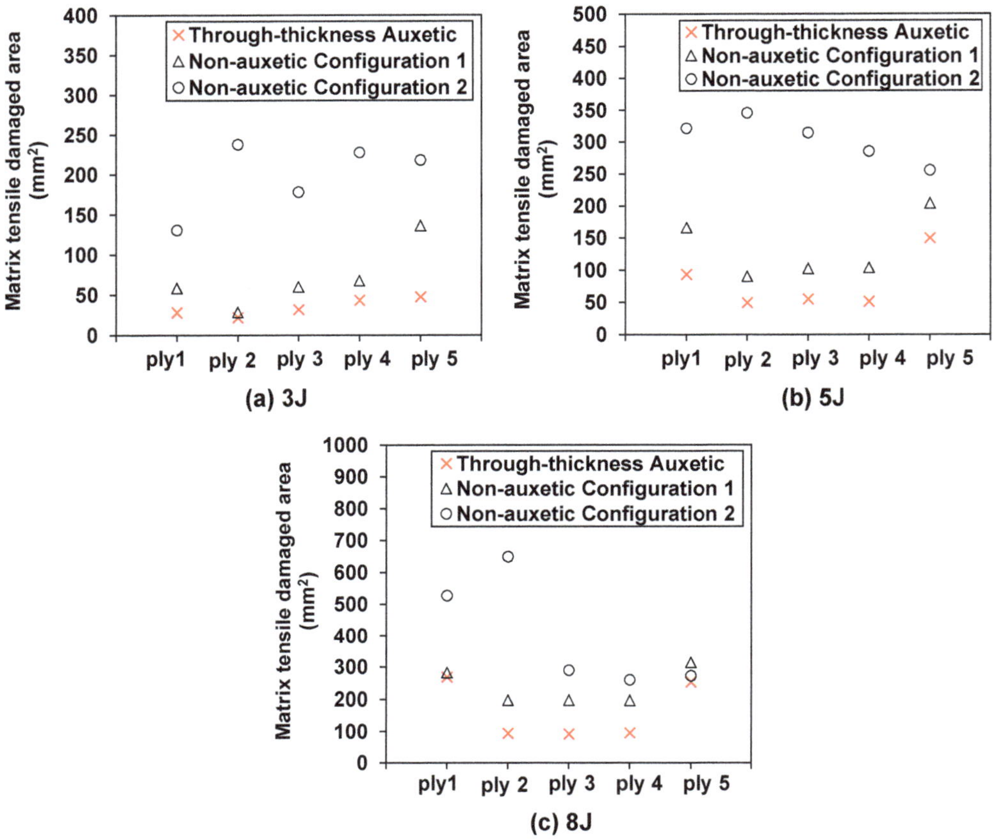

Figure 10. Effect of the through-thickness negative Poisson's ratio on the matrix tensile damage: comparison of the predicted damage area in each ply of the auxetic and non-auxetic CFRP composite laminates at: (**a**) 3 J, (**b**) 5 J, and (**c**) 8 J.

The influence of the through-thickness negative Poisson's ratio on the fiber tensile damage is shown in Figures 11 and 12a. As illustrated in Figure 11, the auxetic laminate shows consistently much reduced fiber tensile damage than that of the two non-auxetic laminates. Note that the fiber tensile damage is negligible at lower impact energies (i.e., 3 and 5 J) and therefore only the predictions at 8 J are presented. It can be seen from Figure 12a, the reductions in the fiber tensile damaged areas of the auxetic laminate are 46.2%, 34.0%, 54.7%, 41.7%, and 39.5% for the five plies, respectively, in comparison to those of the non-auxetic Configuration 1 laminate, and are 58.4%, 35.1%, 24.8%, 44.6%, and 45.7% for the five plies, respectively, in comparison to those of the non-auxetic Configuration 2 laminate. The results suggest that producing the through-thickness negative Poisson's ratio significantly reduces the fiber tensile damage. This could also be due to the contraction in the in-plane direction caused by the negative Poisson's ratio during the impact event (see Figure 1).

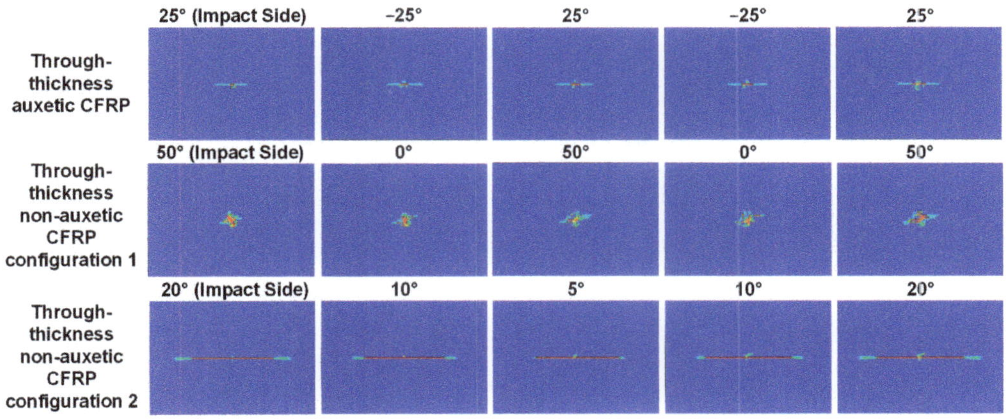

Figure 11. Effect of the through-thickness negative Poisson's ratio on the fiber tensile damage: comparison of the predicted fiber tensile damage pattern in each ply of the auxetic CFRP composite laminate (layup: $[25_2/-25_2/25_2/-25_2/25_2]$, results shown in the first row) and the corresponding non-auxetic CFRP composite laminates (layups: $[50_2/0_2/50_2/0_2/50_2]$ and $[20_2/10_2/5_2/10_2/20_2]$, results shown in the second and third rows), at an 8 J impact. The red color indicates complete damage while the blue color indicates no damage.

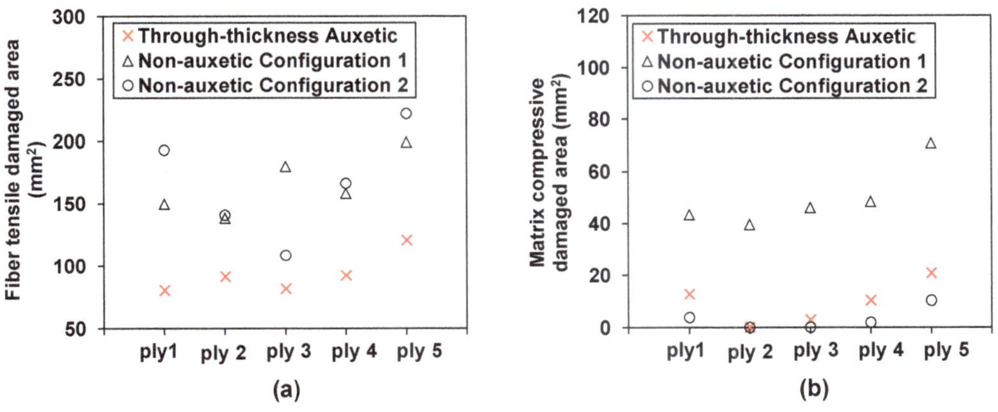

Figure 12. Effect of the through-thickness negative Poisson's ratio on the fiber tensile damage and matrix compressive damage: comparison of the predicted damage areas in each ply of the auxetic laminate and the corresponding non-auxetic CFRP composite laminates at an impact energy of 8 J: (**a**) fiber tensile damaged area and (**b**) matrix compressive damaged area.

The auxetic laminate generally shows a larger matrix compressive damaged area than the non-auxetic Configuration 2 laminate, as shown in Figure 12b. Specifically, at ply 1, ply 3, and ply 5, the matrix compressive damaged areas of the auxetic laminate are 3.4, 5.3, and 2.0 times of those of the non-auxetic Configuration 2 laminate (note that ply 2 and ply 3 have no matrix compressive damage in the non-auxetic Configuration 2 laminate). This implies that the through-thickness negative Poisson's ratio could unfavorably extend the matrix compressive damaged areas. This could be because the contraction in the in-plane direction during the impact event caused by the negative Poisson's ratio exacerbates the compressive damage (see Figure 1). As for the non-auxetic Configuration 1 laminate, interestingly, the matrix compressive damaged areas are significantly larger than those of the auxetic laminate and those of the non-auxetic Configuration 2 laminate. This implies that the matrix compressive damaged area is sensitive to the transverse effective modulus. As it increases, the laminate is prone to experience more extensive matrix compressive damaged areas.

To briefly summarize, the through-thickness negative Poisson's ratio shows large influences on the global impact response and damage behavior due to the unique triaxial state of stresses produced in the auxetic laminate. For the global response, the simulation results suggest that producing the negative Poisson's ratio in the laminate could result in an increase of the impact force and reductions in the impact time, maximum displacement, and the dissipated energy. For the damage behavior, the negative Poisson's ratio could lead to increases in the delamination areas and matrix compressive damaged areas and significant reductions in the matrix tensile and fiber tensile damaged areas.

6. Conclusions

The effect of laminate-level through-thickness negative Poisson's ratio on the low velocity impact behavior of the CFRP composite laminate is investigated using numerical simulations. The layups of the auxetic laminates are identified based on the Classical Lamination Theory. The auxetic laminate has a laminate-level through-thickness Poisson's ratio of -0.327. The layups of the non-auxetic counterpart laminates are determined by matching the effective moduli in three directions while allowing them to produce positive Poisson's ratios. The main conclusions are summarized below.

The through-thickness negative Poisson's ratio reflects the combined effect of the **A**, **B**, and **D** stiffness matrices. It largely influences the global impact response and the damage behavior. The auxetic laminate with a through-thickness negative Poisson's ratio shows consistently higher impact forces, shorter impact times, reduced maximum displacements, and lower dissipated energies at elevated impact energy levels. For the damage behavior, the through-thickness negative Poisson's ratio does not present a clear advantage in mitigating the delamination damage, but it results in consistent and significant reductions in the matrix tensile damage and fiber tensile damage (i.e., 40% reduction in matrix tensile damage on average and 42% reduction in fiber tensile damage on average). At the same time, it could unfavorably yield an increase in the matrix compressive damage (i.e., 1.5 times increase).

Future studies are recommended to investigate the effect of the elevated auxeticity on the impact behavior and if the findings presented in this study are still valid for thick auxetic laminates, as well as the interlaminar cohesive behaviors of auxetic laminates.

Funding: This research was funded by National Science Foundation under Award No. CMMI-2202737. The APC was funded by Syracuse University.

Data Availability Statement: The VUMAT subroutine and datasets generated during and/or analyzed during the current study are available from the corresponding author on reasonable request.

Acknowledgments: The author would like to thank many helpful discussions with Barry Davidson (Emeritus Professor at Syracuse University).

Conflicts of Interest: The author declares no conflict of interest.

References

1. Shah, S.Z.H.; Karuppanan, S.; Megat-Yusoff, P.S.M.; Sajid, Z. Impact resistance and damage tolerance of fiber reinforced composites: A review. *Compos. Struct.* **2019**, *217*, 100–121. [CrossRef]
2. Liao, B.; Zhou, J.; Li, Y.; Wang, P.; Xi, L.; Gao, R.; Bo, K.; Fang, D. Damage accumulation mechanism of composite laminates subjected to repeated low velocity impacts. *Int. J. Mech. Sci.* **2020**, *182*, 105783. [CrossRef]
3. Zhang, L.; Chen, Z.; Habibi, M.; Ghabussi, A.; Alyousef, R. Low-velocity impact, resonance, and frequency responses of FG-GPLRC viscoelastic doubly curved panel. *Compos. Struct.* **2021**, *269*, 114000. [CrossRef]
4. Hill, C.B.; Wang, Y.; Zhupanska, O.I. Impact response of CFRP laminates with CNT buckypaper layers. In Proceedings of the 54th AIAA/ASME/ASCE/AHS/ASC Structures, Structural Dynamics, and Materials Conference, Boston, MA, USA, 8–11 April 2013; p. 1617. [CrossRef]

5. Hill, C.B.; Wang, Y.; Zhupanska, O.I. Effects of carbon nanotube buckypaper layers on the electrical and impact response of IM7/977-3 composite laminates. In Proceedings of the 27th Annual Technical Conference of the American Society for Composites, Arlington, TX, USA, 1–3 October 2012; pp. 1205–1221.
6. Abrate, S. *Impact Engineering of Composite Structures*; Springer Science & Business Media: Berlin/Heidelberg, Germany, 2011.
7. González, E.V.; Maimí, P.; Camanho, P.P.; Turon, A.; Mayugo, J.A. Simulation of drop-weight impact and compression after impact tests on composite laminates. *Compos. Struct.* **2012**, *94*, 3364–3378. [CrossRef]
8. Wagih, A.; Maimí, P.; Blanco, N.; Costa, J. A quasi-static indentation test to elucidate the sequence of damage events in low velocity impacts on composite laminates. *Compos. Part A Appl. Sci. Manuf.* **2016**, *82*, 180–189. [CrossRef]
9. Lin, W.; Wang, Y.; Lampkin, S.; Philips, W.; Prabhakar, S.; Smith, R.; Whittington, L.; Fan, Y.; Wolz, R.; Whittington, W.; et al. Hail impact testing of stitched carbon fiber epoxy composite laminates. In Proceedings of the 35th Annual American Society for Composites Technical Conference, ASC 2020, Online, 14–17 September 2020; DEStech Publications: Lancaster, PA, USA, 2020; pp. 731–745. [CrossRef]
10. Tan, W.; Falzon, B.G.; Chiu, L.N.S.; Price, M. Predicting low velocity impact damage and Compression-After-Impact (CAI) behaviour of composite laminates. *Compos. Part A Appl. Sci. Manuf.* **2015**, *71*, 212–226. [CrossRef]
11. Evans, K.E. Auxetic polymers: A new range of materials. *Endeavour* **1991**, *15*, 170–174. [CrossRef]
12. Hou, X.; Silberschmidt, V.V. Metamaterials with negative poisson's ratio: A review of mechanical properties and deformation mechanisms. In *Mechanics of Advanced Materials*; Springer: Berlin/Heidelberg, Germany, 2015; pp. 155–179.
13. Li, T.; Liu, F.; Wang, L. Enhancing indentation and impact resistance in auxetic composite materials. *Compos Part B Eng.* **2020**, *198*, 108229. [CrossRef]
14. Mir, M.; Ali, M.N.; Sami, J.; Ansari, U. Review of mechanics and applications of auxetic structures. *Adv. Mater. Sci. Eng.* **2014**, *2014*, 753496. [CrossRef]
15. Allen, T.; Shepherd, J.; Hewage, T.A.M.; Senior, T.; Foster, L.; Alderson, A. Low-kinetic energy impact response of auxetic and conventional open-cell polyurethane foams. *Phys. Status Solidi (B)* **2015**, *252*, 1631–1639. [CrossRef]
16. Zhou, L.; Zeng, J.; Jiang, L.; Hu, H. Low-velocity impact properties of 3D auxetic textile composite. *J. Mater. Sci.* **2018**, *53*, 3899–3914. [CrossRef]
17. Clarke, J.F.; Duckett, R.A.; Hine, P.J.; Hutchinson, I.J.; Ward, I.M. Negative Poisson's ratios in angle-ply laminates: Theory and experiment. *Composites* **1994**, *25*, 863–868. [CrossRef]
18. Alderson, K.L.; Simkins, V.R.; Coenen, V.L.; Davies, P.J.; Alderson, A.; Evans, K.E. How to make auxetic fibre reinforced composites. *Phys. Status Solidi (B)* **2005**, *242*, 509–518. [CrossRef]
19. Almgren, R. An isotropic three-dimensional structure with Poisson's ratio = −1. *J. Elast.* **1985**, *15*, 427–430.
20. Lakes, R. Foam structures with a negative Poisson's ratio. *Science* **1987**, *235*, 1038–1040. [CrossRef]
21. Wojciechowski, K.W. Constant thermodynamic tension Monte Carlo studies of elastic properties of a two-dimensional system of hard cyclic hexamers. *Mol. Phys.* **1987**, *61*, 1247–1258. [CrossRef]
22. Wojciechowski, K.W. Two-dimensional isotropic system with a negative Poisson ratio. *Phys. Lett. A* **1989**, *137*, 60–64. [CrossRef]
23. Pigłowski, P.M.; Wojciechowski, K.W.; Tretiakov, K.V. Partial auxeticity induced by nanoslits in the Yukawa crystal. *Phys. Status Solidi (RRL)–Rapid Res. Lett.* **2016**, *10*, 566–569. [CrossRef]
24. Narojczyk, J.W.; Wojciechowski, K.W.; Tretiakov, K.V.; Smardzewski, J.; Scarpa, F.; Pigłowski, P.M.; Kowalik, M.; Imre, A.R.; Bilski, M. Auxetic Properties of a fcc Crystal of Hard Spheres with an Array of -Nanochannels Filled by Hard Spheres of Another Diameter. *Phys. Status Solidi (B)* **2019**, *256*, 1800611. [CrossRef]
25. Poźniak, A.A.; Wojciechowski, K.W.; Grima, J.N.; Mizzi, L. Planar auxeticity from elliptic inclusions. *Compos. Part B Eng.* **2016**, *94*, 379–388. [CrossRef]
26. Milton, G.W. Composite materials with Poisson's ratios close to—1. *J. Mech. Phys. Solids* **1992**, *40*, 1105–1137. [CrossRef]
27. Baughman, R.H.; Shacklette, J.M.; Zakhidov, A.A.; Stafström, S. Negative Poisson's ratios as a common feature of cubic metals. *Nature* **1998**, *392*, 362–365. [CrossRef]
28. Kimizuka, H.; Kaburaki, H.; Kogure, Y. Mechanism for negative Poisson ratios over the α-β transition of cristobalite, SiO_2: A molecular-dynamics study. *Phys. Rev. Lett.* **2000**, *84*, 5548. [CrossRef]
29. Peel, L.D. Exploration of high and negative Poisson's ratio elastomer-matrix laminates. *Phys. Status Solidi (B)* **2007**, *244*, 988–1003. [CrossRef]
30. Hoover, W.G.; Hoover, C.G. Searching for auxetics with DYNA3D and ParaDyn. *Phys. Status Solidi (B)* **2005**, *242*, 585–594. [CrossRef]
31. Grima, J.N.; Winczewski, S.; Mizzi, L.; Grech, M.C.; Cauchi, R.; Gatt, R.; Attard, D.; Wojciechowski, K.W.; Rybicki, J. Tailoring graphene to achieve negative Poisson's ratio properties. *Adv. Mater.* **2015**, *27*, 1455–1459. [CrossRef]
32. Bilski, M.; Wojciechowski, K.W.; Stręk, T.; Kędziora, P.; Grima-Cornish, J.N.; Dudek, M.R. Extremely non-auxetic behavior of a typical auxetic microstructure due to its material properties. *Materials* **2021**, *14*, 7837. [CrossRef] [PubMed]
33. Coenen, V.L.; Alderson, K.L. Mechanisms of failure in the static indentation resistance of auxetic carbon fibre laminates. *Phys. Status Solidi (B)* **2011**, *248*, 66–72. [CrossRef]
34. Hou, S.; Li, T.; Jia, Z.; Wang, L. Mechanical properties of sandwich composites with 3d-printed auxetic and non-auxetic lattice cores under low velocity impact. *Mater. Des.* **2018**, *160*, 1305–1321. [CrossRef]

35. Fan, Y.; Wang, Y. A Study on Effect of Auxeticity on Impact Resistance of Carbon Nanotube Reinforced Composite Laminates. In Proceedings of the 35th Annual American Society for Composites Technical Conference, ASC 2020, Online, 14–17 September 2020; DEStech Publications: Lancaster, PA, USA, 2020. [CrossRef]
36. Fan, Y.; Wang, Y. The effect of negative Poisson's ratio on the low-velocity impact response of an auxetic nanocomposite laminate beam. *Int. J. Mech. Mater. Des.* **2021**, *17*, 153–169. [CrossRef]
37. Han, S.C.; Kang, D.S.; Kang, K. Two nature-mimicking auxetic materials with potential for high energy absorption. *Mater. Today* **2019**, *26*, 30–39. [CrossRef]
38. Oh, J.-H.; Kim, J.-S.; Oh, I.-K. Auxetic graphene oxide-porous foam for acoustic wave and shock energy dissipation. *Compos. Part B Eng.* **2020**, *186*, 107817. [CrossRef]
39. Ferguson, W.J.G.; Kuang, Y.; Evans, K.E.; Smith, C.W.; Zhu, M. Auxetic structure for increased power output of strain vibration energy harvester. *Sens. Actuators A Phys.* **2018**, *282*, 90–96. [CrossRef]
40. Jiang, Y.; Liu, Z.; Matsuhisa, N.; Qi, D.; Leow, W.R.; Yang, H.; Yu, J.; Chen, G.; Liu, Y.; Wan, C.; et al. Auxetic mechanical metamaterials to enhance sensitivity of stretchable strain sensors. *Adv. Mater.* **2018**, *30*, 1706589. [CrossRef] [PubMed]
41. Evans, K.E.; Alderson, A. Auxetic materials: Functional materials and structures from lateral thinking! *Adv. Mater.* **2000**, *12*, 617–628. [CrossRef]
42. Sun, C.T.; Li, S. Three-dimensional effective elastic constants for thick laminates. *J. Compos. Mater.* **1988**, *22*, 629–639. [CrossRef]
43. Agarwal, B.D.; Broutman, L.J.; Chandrashekhara, K. *Analysis and Performance of Fiber Composites*; John Wiley & Sons: Hoboken, NJ, USA, 2017.
44. Lin, W.; Wang, Y. Effect of Negative Poisson's Ratio on the Tensile Properties of Auxetic CFRP Composites. In Proceedings of the 37th American Society for Composites Conference and ASTM D30 Meeting, Tucson, AZ, USA, 19–21 September 2022.
45. Alderson, K.L.; Coenen, V.L. The low velocity impact response of auxetic carbon fibre laminates. *Phys. Status Solidi (B)* **2008**, *245*, 489–496. [CrossRef]
46. Aziz, S. An Investigation into the Post Impact and Post Indentation Behavior of Auxetic Composites. Ph.D. Thesis, University of Bolton, Bolton, UK, 2016.
47. Zhang, J.; Zhang, X. An efficient approach for predicting low-velocity impact force and damage in composite laminates. *Compos. Struct.* **2015**, *130*, 85–94. [CrossRef]
48. Mohammadi, B.; Rohanifar, M.; Salimi-Majd, D.; Farrokhabadi, A. Micromechanical prediction of damage due to transverse ply cracking under fatigue loading in composite laminates. *J. Reinf. Plast. Compos.* **2017**, *36*, 377–395. [CrossRef]
49. Li, X.; Ma, D.; Liu, H.; Tan, W.; Gong, X.; Zhang, C.; Li, Y. Assessment of failure criteria and damage evolution methods for composite laminates under low-velocity impact. *Compos. Struct.* **2019**, *207*, 727–739. [CrossRef]
50. Brańka, A.C.; Heyes, D.M.; Wojciechowski, K.W. Auxeticity of cubic materials. *Phys. Status Solidi (B)* **2009**, *246*, 2063–2071. [CrossRef]
51. Wang, F.; Wang, B.; Kong, F.; Ouyang, J.; Ma, T.; Chen, Y. Assessment of degraded stiffness matrices for composite laminates under low-velocity impact based on modified characteristic length model. *Compos. Struct.* **2021**, *272*, 114145. [CrossRef]
52. Hongkarnjanakul, N.; Bouvet, C.; Rivallant, S. Validation of low velocity impact modelling on different stacking sequences of CFRP laminates and influence of fibre failure. *Compos. Struct.* **2013**, *106*, 549–559. [CrossRef]
53. Rivallant, S.; Bouvet, C.; Hongkarnjanakul, N. Failure analysis of CFRP laminates subjected to compression after impact: FE simulation using discrete interface elements. *Compos. Part A Appl. Sci. Manuf.* **2013**, *55*, 83–93. [CrossRef]
54. Hashin, Z. Failure criteria for unidirectional fiber composites. *J. Appl. Mech.* **1980**, *47*, 329–334. [CrossRef]
55. Benzeggagh, M.L.; Kenane, M. Measurement of mixed-mode delamination fracture toughness of unidirectional glass/epoxy composites with mixed-mode bending apparatus. *Compos. Sci. Technol.* **1996**, *56*, 439–449. [CrossRef]
56. Özütok, A.; Madenci, E. Static analysis of laminated composite beams based on higher-order shear deformation theory by using mixed-type finite element method. *Int. J. Mech. Sci.* **2017**, *130*, 234–243. [CrossRef]
57. Madenci, E.; Özütok, A. Variational approximate for high order bending analysis of laminated composite plates. *Struct. Eng. Mech. Int'l J.* **2020**, *73*, 97–108.
58. Zhou, J.; Wen, P.; Wang, S. Numerical investigation on the repeated low-velocity impact behavior of composite laminates. *Compos. Part B Eng.* **2020**, *185*, 107771. [CrossRef]

materials

MDPI

Article

Novel Alumina Matrix Composites Reinforced with MAX Phases—Microstructure Analysis and Mechanical Properties

Mateusz Petrus [1], Jaroslaw Wozniak [1,*], Tomasz Cygan [1], Wojciech Pawlak [2] and Andrzej Olszyna [1]

[1] Faculty of Material Science and Engineering, Warsaw University of Technology, ul. Wołoska 141, 02-507 Warsaw, Poland

[2] Faculty of Mechanical Engineering, Lodz University of Technology, Stefanowskiego 1/15 St, 90-924 Lodz, Poland

* Correspondence: jaroslaw.wozniak@pw.edu.pl

Abstract: This article describes the manufacturing of alumina composites with the addition of titanium aluminum carbide Ti_3AlC_2, known as MAX phases. The composites were obtained by the powder metallurgy technique with three types of mill (horizontal mill, attritor mill, and planetary mill), and were consolidated with the use of the Spark Plasma Sintering method at 1400 °C, with dwelling time 10 min. The influence of the Ti_3AlC_2 MAX phase addition on the microstructure and mechanical properties of the obtained composites was analyzed. The structure of the MAX phase after the sintering process was also investigated. The chemical composition and phase composition analysis showed that the Ti_3AlC_2 addition preserved its structure after the sintering process. The increase in fracture toughness for all series of composites has been noted (over 20% compared to reference samples). Detailed stereological analysis of the obtained microstructures also could determine the influence of the applied mill on the homogeneity of the final microstructure and the properties of obtained composites.

Keywords: sintering; composites; mechanical properties; Al_2O_3; MAX phases

1. Introduction

Alumina is one of the most commonly used ceramics. This is due to its unique properties, such as high hardness, good thermal and chemical stability, high strength, high young modulus and low price [1,2]. These advantages allow monolithic ceramics to be used in electronics, power generation, cutting tools, aerospace, military or biomedical [3,4]. However, one of the main disadvantages, which limits the potential use, is low fracture toughness. The method to improve the fracture toughness of ceramics is the production and development of new multiphase composites [5,6]. One of the most interesting groups of materials, characterized by a combination of the properties of ceramics and metals that can potentially be used as a phase reinforcing the ceramic matrix, is the MAX phases [7–11]. The MAX phases constitute a large group of anisotropic crystalline materials. Their name reflects their composition $M_{n+1}AX_n$ where M—light transition metal, e.g., Ti, Nb, V; A—metals from group 13 or 14, whereas X is carbon or nitrogen [12,13]. The MAX phases exhibit a layered structure with a metallic layer between layers M–X. Since the discovery of the first MAX phases in 1967 by Nowotny's team, more than 60 different, thermodynamically stable MAX phases have been developed [10,14]. Due to the presence of covalent–metallic–ionic bonds, the MAX phases are characterized, like ceramics, by high hardness, high strength properties, good thermal properties and like metals, they display good electrical and thermal conductivity, good thermal shock, high fracture toughness and are relatively easily mechanically machined [15–17].

In recent years, there has been a growing interest in using MAX phases as an addition to the ceramic matrix to improve ceramic composites' sinterability and mechanical properties. Titanium aluminum carbide (Ti_3AlC_2) is a typical member of the MAX phases with a density of 4.2 g/cm^3, fracture toughness of 9.1 $MPa \cdot m^{0.5}$ and flexural strength over 550 MPa [18]. However, Ti_3AlC_2 is unstable and decomposes above 1300 °C. Nayebi et al. [19] used the addition of the Ti_3AlC_2 phase to lower the sintering temperature and produce TiB_2-based composites using the SPS sintering technology. During the sintering, the authors confirmed the decomposition of the Ti_3AlC_2 phase with the production of Al, Ti, and $TiC_{0.67}$. Similar studies have been conducted for ZrB_2 [20]. The decomposition of Ti_3AlC_2 during sintering promoted the reactive sintering mechanism by forming several in-situ synthesized secondary phases. The aforementioned phases can be categorized into three main groups: Al-, Ti- and Zr-rich reinforcement, such as Al_2O_3, AlOC, Al_2OC, TiB_2, and ZrC. The presence of TiC, which is a decomposition product, was also confirmed in the microstructure. The reactive sintering mechanism in the B_4C-Ti_3AlC_2 systems was also noted [21,22]. These studies also indicated the problem of TiC agglomerates formation, which has not been solved yet. Thus far, the research on the Ti_3AlC_2 phase has focused mainly on non-oxide ceramics, which require the use of temperatures higher than the MAX phase thermal stability [23]. They used the MAX phase decomposition to form aluminum and TiC, which reacted with the sintered material. These studies also suggest a problem with the agglomeration of the MAX phases during the production of composites, but thus far, it has not been thoroughly tested and described. However, there are no studies on the use of the Ti_3AlC_2 phase as an additive to aluminum oxide, the sintering temperature of which is much lower than that of non-oxide ceramics and amounts to 1300–1400 °C [24]. It should prevent the decomposition of the MAX phases, thus enabling the development of a new family of ceramic composites.

This article aims to produce alumina matrix composites reinforced with Ti_3AlC_2. The MAX phases were obtained through the Self-Propagating High-Temperature Synthesis (SHS) using the Spark Plasma Sintering Method (SPS) from pure elements. The effect of the addition of Ti_3AlC_2 MAX phase and the type of applied mill type on the microstructure and mechanical properties of the composites were investigated.

2. Materials and Methods

The commercial powders of titanium aluminum, synthetic graphite and alumina were used in the process of producing the Al_2O_3—Ti_3AlC_2 composites (Table 1).

Table 1. The parameters of used powders. APS—average particle size.

Powder	Purity	APS *	Manufacturer
Titanium	99.6%	<20 μm	GoodFellow, Cambridge, UK
Aluminum	99.7%	6.74 μm	Bend-Lutz Co. Skawina, Poland
Synthetic graphite	99.9%	<20 μm	Sigma-Aldrich, St. Louis, MO, USA
Alumina	99.99%	0.125 nm	Taimei Chemicals Co., Ltd., Tokyo, Japan

* average particles size.

To prepare MAX phases, the titanium, aluminum, and carbon powders were wet blended in propan-2-ol (no. 1759, Stanlab, Lublin, Poland) with the use of a ball-type mill (Fritsch Pulverisette, Fritsch, Idar-Oberstein, Germany). After that, the obtained powder mixture was dried and sieved (# 300 μm). The molar ratio of Ti:Al:C = 3:1:1.9 was applied. The composition of the mixture was developed on the basis of optimization tests, taking into account the phase purity of the obtained products. The reactive synthesis process of synthesis the Ti_3AlC_2 phase was performed using the Spark Plasma Sintering technique (FCT Systeme GMBH, Effelder Rauenstein, Germany). The following parameters were applied: temperature: 1300 °C, heating rate: 250 °C/min, vacuum atmosphere ($p = 5 \times 10^{-2}$ mbar). The MAX phase was ground with an automatic mortar grinder (Retsch KM100, Retsch GmbH, Haan, Germany) below 45 μm (grinding bowl speed = 70 rpm, applied force 50–125 N).

The Al_2O_3—Ti_3AlC_2 composites were prepared with the powder metallurgy technique and sintered with the SPS method. First, the Al_2O_3-xTi_3AlC_2 powder mixtures (where x = 5, 10, 15, 20 wt%) were wet blended in three different ball-type mills: horizontal mill, attritor mill and planetary mill in propan-2-ol and alumina grinding balls (Nikkato, Osaka, Japan). The obtained powder was dried and manually sieved (# 300 μm). Composites were sintered with the use of SPS method. The process parameters were as follows: sintering temperature: 1400 °C, heating, and cooling rate: 250 °C/min, 4 min dwell time, 50 MPa applied pressure and vacuum ($p = 5 \times 10^{-2}$ mbar). As a reference sample, pure alumina sinters were prepared.

The density of the specimens was examined using the helium pycnometer (Ultrapycnometer 1000, Quantachrome Instruments, Boynton Beach, FL, USA). Hardness of the obtained composites was measured under the load of 49.05 N with the Vickers Hardness Tester (FV-700e, Future-Tech, Kawasaki, Japan). The Vickers Indentation Fracture (VIF) of the produced composites was determined based on the crack length generated at the corners of the Vickers indenterunder the load of 49.05 N. The Niihara, Morena, Hasselman equation was used. For each sample, 20 hardness measurements and 12 crack length measurements were performed. The phase composition of composites was analyzed with X-ray diffraction (XRD, PANalytical Empyrean, Malvern Panalytical, Malvern, UK), using CuKα radiation at a wavelength of 0.15406 nm. The parameters of this test were as follows: voltage: 45 kV, current: 40 mA, angular range: 10–155 deg with step 0.03. The microstructure observations were performed on a scanning electron microscope (SEM Hitachi 5500, Hitachi, Tokyo, Japan). The observations were carried out at a 20 kV accelerating voltage. The quantitative description of the microstructure was performed on binarized images using the following parameters: average equivalent diameter of matrix grains size, average equivalent diameter of agglomerates $\overline{d^2}$, a circularity of agglomerates ($\eta = 4\pi S/O^2$), elongation of agglomerates defined as the ratio of the maximum to the minimum Feret diameter. The homogeneity of the microstructure in terms of the agglomerate distribution of the Ti_3AlC_2 phase was estimated using the skeletization by the function of SKIZ (skeleton by influence zone). In this method, the fields of influence of structure elements are determined. By determining the mean equivalent diameter of these fields and subjecting it to statistical analysis, it is possible to compare the homogeneity of the structures of the tested materials. Stereological analysis was performed using the NIS ELEMENTS BR 5.30-03 software (Nikon, Tokio, Japan).

3. Results

3.1. Characterization of MAX Phases

The morphology of the obtained Ti_3AlC_2 MAX phase is presented in Figure 1a. The layered structure, characteristic of this group of materials, can be observed, which verifies the correct occurrence of the MAX phase synthesis process. This is also confirmed by the results of the phase composition of powders (Figure 1b). The XRD analysis exhibited the presence of Ti_3AlC_2 as the main phase and slight amounts of TiC, Ti_2AlC and graphite.

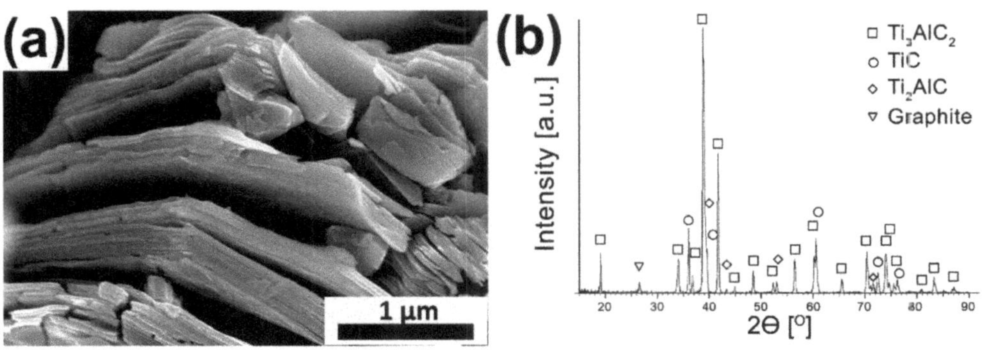

Figure 1. The (**a**) morphology and (**b**) XRD analysis of synthesized Ti_3AlC_2 MAX phase powders.

3.2. Microstructure Analysis

Figure 2 presents the relative density of obtained Al_2O_3-Ti_3AlC_2 composites. The reference samples have a relative density of 99.3%. It can be seen that the composites obtained with the use of the planetary mill and the attritor type mill show a similar course of density changes as a function of the Ti_3AlC_2 MAX phase content: a decrease in the value for the 5 wt% additive, then an increase until reaching the maximum for the 15% additive and a decrease for the 20 wt% additive. For composites obtained with a horizontal mill, the decrease for 5 wt% is smaller and the subsequent increase in density with increasing Ti_3AlC_2 phase addition becomes linear until reaching the maximum for 20 wt%. The composite with the addition of 15 wt% of Ti_3AlC_2 obtained using a planetary mill shows the highest relative density.

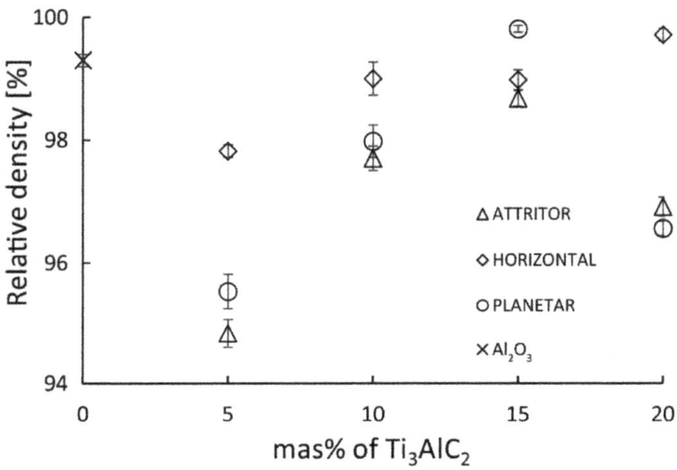

Figure 2. The relative density of Al_2O_3/Ti_3AlC_2 composites.

The fracture surfaces of obtained sinters are shown in Figure 3. A presence of numerous fine pores characterizes the microstructure of the reference sample. In the case of composites with the addition of 10 wt% Ti_3AlC_2 phase, irrespective of the mill used, a finer grain was observed, as well as the presence of pores and voids. Comparing the fracture surfaces in terms of the mill used, we can see that the finest grains and, simultaneously, the highest amount of pores and voids occur in the composite obtained using a planetary mill. In the case of composites, there were also areas where the fracture mechanism changed from transcrystalline to intracrystalline (white arrows).

Observations of the composites microstructure also showed the presence of an additional phase with a layered structure. Its morphology corresponds to that used as the starting powder of the MAX Ti_3AlC_2 phase (Figure 4a). Depending on the mill used, a different degree of bonding of the observed phase with the matrix can be seen. More pores and discontinuities at the interface were observed in the composites obtained with the use of the attritor mill (Figure 4b—white arrows). In the case of composites obtained with the use of a horizontal mill, the presence of a significant amount of second phase agglomerates with sizes exceeding 100 μm was also observed (Figure 4c).

The analysis of the chemical composition of the observed layered structures (Figure 5) showed that they consist of titanium, aluminum and carbon, which corresponds to the chemical composition of the used Ti_3AlC_2 phases. It suggests that the MAX phases did not decompose during sintering and retained their original structure. The study of phase composition confirmed the above finding (Figure 6). XRD analysis showed that the obtained composites are mainly composed of Al_2O_3, Ti_3AlC_2, and a small amount of TiC. The presence of TiC may be related to the fouling of the starting powder of the Ti_3AlC_2

(Figure 1b) phase or the possibility of partial decomposition of the Ti_3AlC_2 phase during the sintering process.

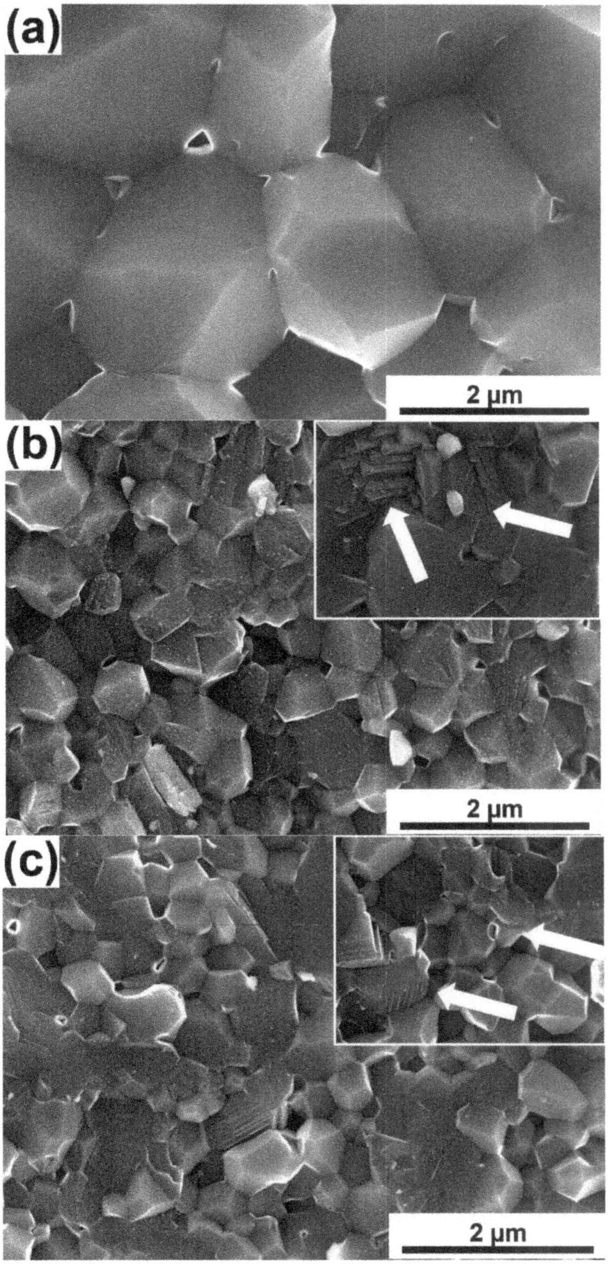

Figure 3. *Cont.*

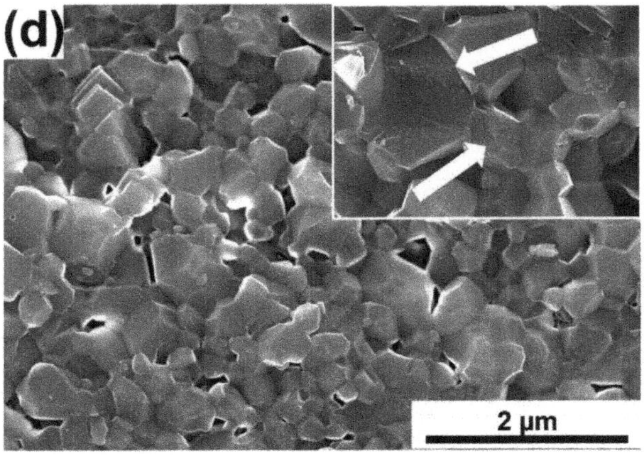

Figure 3. The fracture surface of (**a**) Al_2O_3, (**b**) Al_2O_3—10 wt% of Ti_3AlC_2 from horizontal mill, (**c**) Al_2O_3—10 wt% of Ti_3AlC_2 from Atrittor mill, (**d**) Al_2O_3—10 wt% of Ti_3AlC_2 from planetary mill (the arrows mark areas with a variable fracture mechanism).

3.3. Stereological Analysis

A complete stereological analysis of the microstructures of the obtained composites was carried out. Figure 7 shows the mean equivalent grain size of the produced composites. As can be seen, the addition of 5 wt% of the Ti_3AlC_2 phase causes a decrease in the average grain size by over 80% (2.2 μm and 0.42 μm for reference Al_2O_3 and Al_2O_3—5 wt% Ti_3AlC_2 sinter, respectively). A further increase in the MAX phase content does not cause a further decrease in d_2, all produced composites, regardless of the composition and type of the mill used, show an average grain size oscillating between 0.3–0.5 μm.

Figure 8 shows the average equivalent diameter of the Ti_3AlC_2 phase agglomerates. In the case of composites made with a horizontal mill and an attritor mill, an increase in the average size of agglomerates was observed with the increase in Ti_3AlC_2 phase addition up to 15 wt% additive by nearly 80%, compared to 5 wt% additive. For the 20 wt% addition of the Ti_3AlC_2 phase, an insignificant decrease in the average size of the agglomerates was noted. A slightly different trend was observed for composites obtained using a planetary mill. In the analyzed range, much smaller average sizes of the formed agglomerates were noted. The increase in size with the increase in the Ti_3AlC_2 phase was insignificant and amounted to a maximum of 20% for the composite with the addition of 20 wt%.

Figure 4. *Cont.*

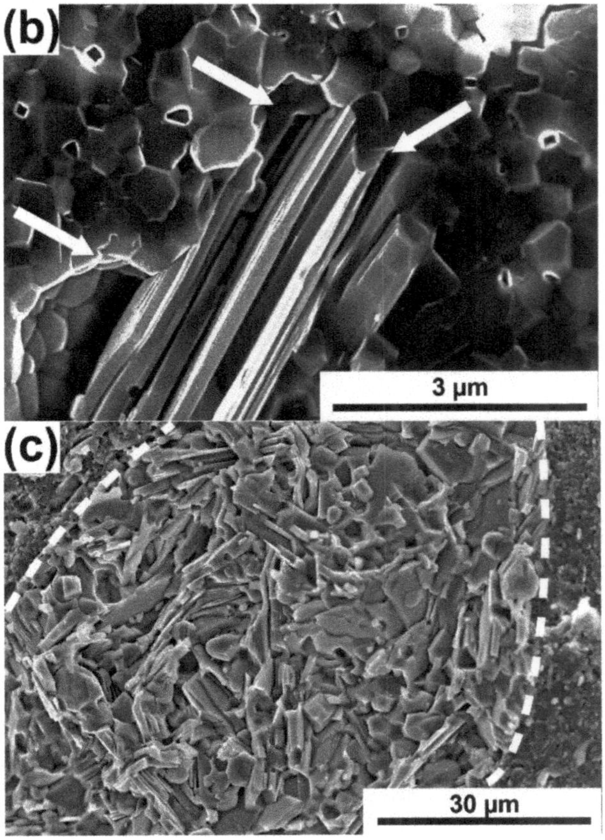

Figure 4. The microstructure of (**a**) Al_2O_3—5 wt% of Ti_3AlC_2 from planetary mill, (**b**) Al_2O_3—5 wt% of Ti_3AlC_2 from Atrittor mill (**c**) Al_2O_3—5 wt% of Ti_3AlC_2 from horizontal mill (the arrows mark the porosity on the interface).

The aspect ratio of the formed agglomerates was also analyzed (Figure 8). The circularity of the Ti_3AlC_2 phase agglomerates in all the obtained composites oscillates around 0.8 and shows no changes depending on the amount of addition of the MAX phase (Figure 9a). The elongation behaves similarly, which for all composites in the entire analyzed range fluctuates around 1.5, which indicates a slight elongation of the agglomerates (Figure 9b).

Figure 10 shows the coefficients of variability of the surface area of cells resulting from SKIZ tessellation (images of macro-regions analyzed by SKIZ tessellation were presented in Supplementary Material Scheme S1a–l). In the case of composites obtained with the use of the horizontal and the attritor mill, an increase in the coefficient of variation is observed with the addition of the Ti_3AlC_2 phase. In contrast, in the case of composites from the attritor, this increase is more significant. This proves the decrease in the homogeneity of the Ti_3AlC_2 phase distribution in the discussed composites. In the case of composites obtained with a planetary mill, no decrease in the homogeneity of the microstructure as a function of the composition content was observed. In the entire analyzed range, these composites were characterized by the lowest CV coefficient and, thus, the highest degree of homogeneity of the microstructure.

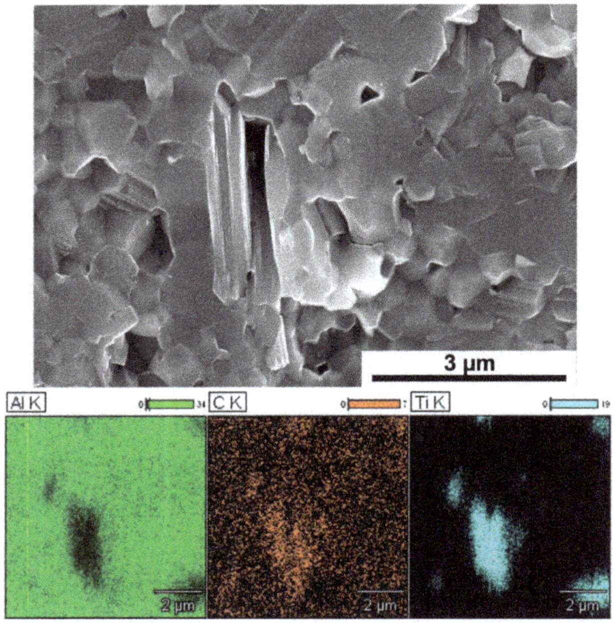

Figure 5. The EDS (Energy-dispersive X-ray spectroscopy) elemental map of Al_2O_3—5wt% of Ti_3AlC_2 composites.

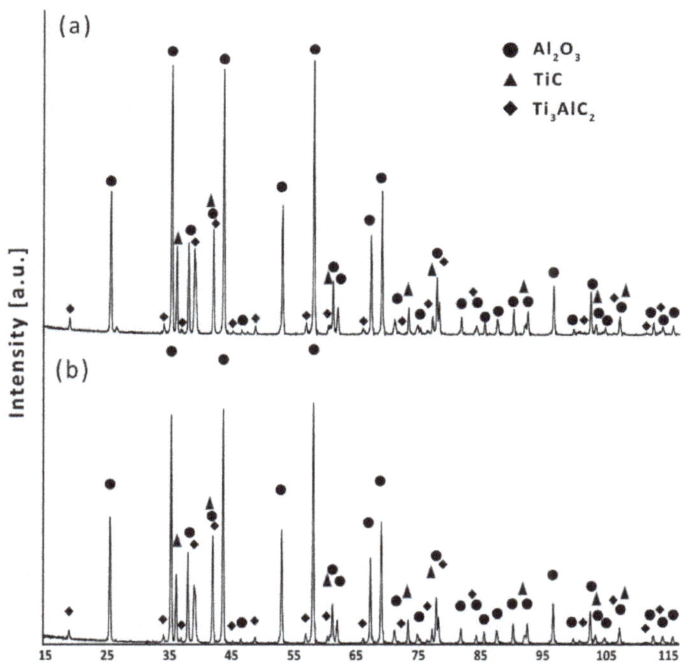

Figure 6. XRD analysis of (**a**) Al_2O_3 + 20 wt% of Ti_3AlC_2 addition from horizontal mill, (**b**) Al_2O_3 + 20 wt% of Ti_3AlC_2 addition from planetary mill.

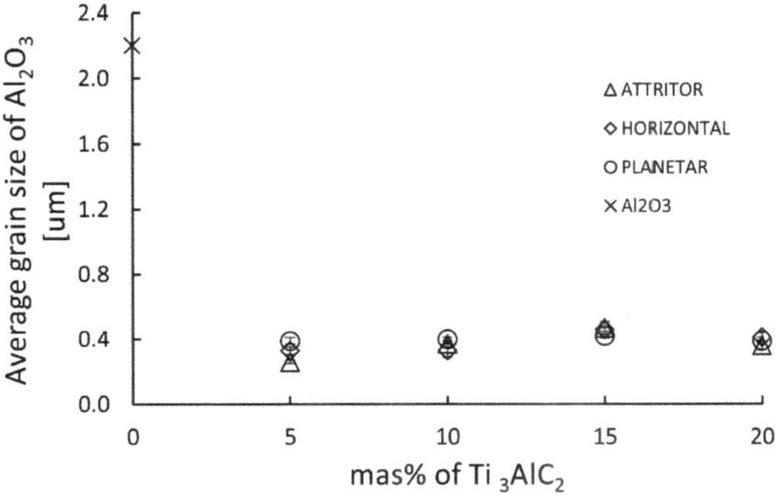

Figure 7. Average grain size of alumina in Al_2O_3-Ti_3AlC_2 composites.

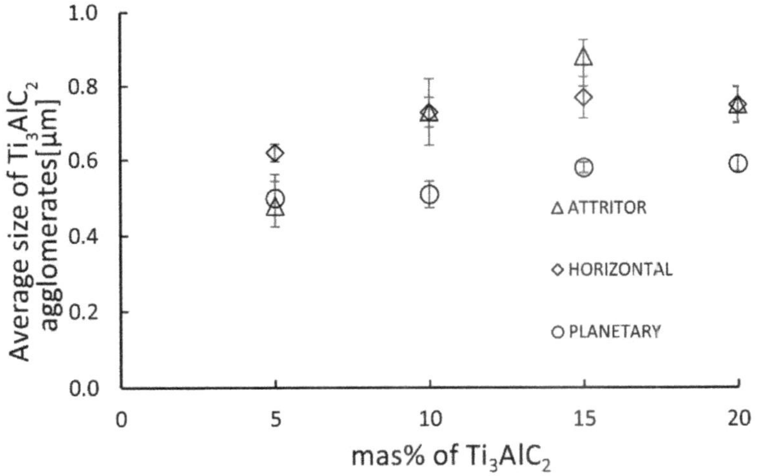

Figure 8. Average grain size of Ti_3AlC_2 agglomerates in Al_2O_3-Ti_3AlC_2 composites.

3.4. Mechanical Properties

Figure 11 shows the mechanical properties of the obtained composites. In the case of hardness, we can see that the reference sample is characterized by a high hardness value of 1843 HV. The hardness of composites, in most cases, achieved a lower value than the reference samples. Only samples containing 5, 10 and 15 wt% MAX phases mixed in a horizontal mill and a 15 wt% MAX phase sample mixed in a planetary mill achieved higher hardness values than the unreinforced sinter. A similar trend can be seen for all series of composites. The hardness value increases with an increase in the content of the reinforcing phase to about 15 wt%, and then, a slight decrease is observed. The fracture toughness of obtained composites is shown in Figure 11b. All produced composites show higher fracture toughness compared to the reference sample. The highest value measured was 5.7 MPa·$m^{0.5}$ for the composite with the addition of 15 wt% Ti_3AlC_2 obtained using an attritor mill.

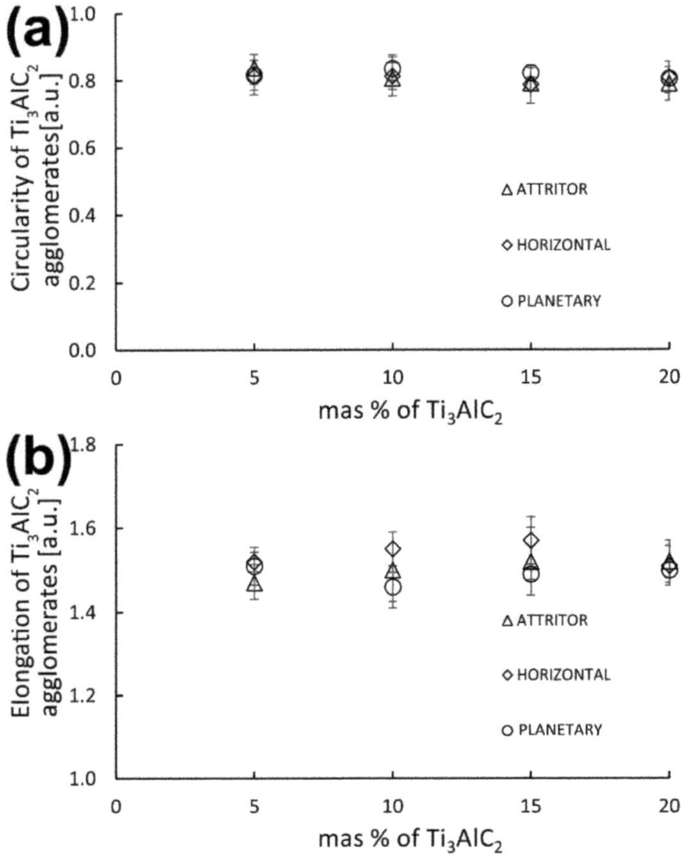

Figure 9. The (**a**) circularity and (**b**) elongation of Ti_3AlC_2 agglomerates.

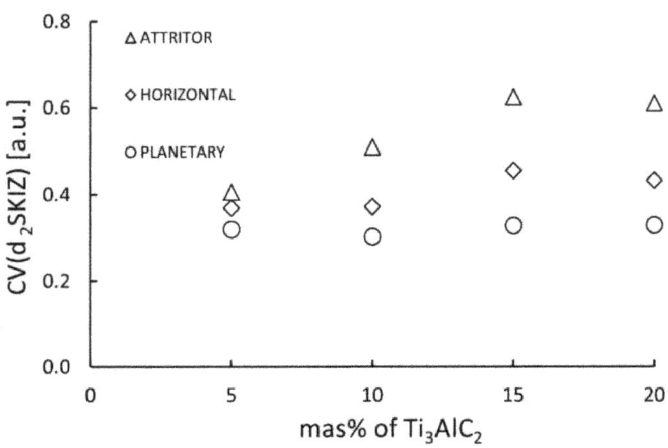

Figure 10. The CV_{SKIZ} (coefficient of SKIZ variation) of homogeneity of the microstructure of obtained composites.

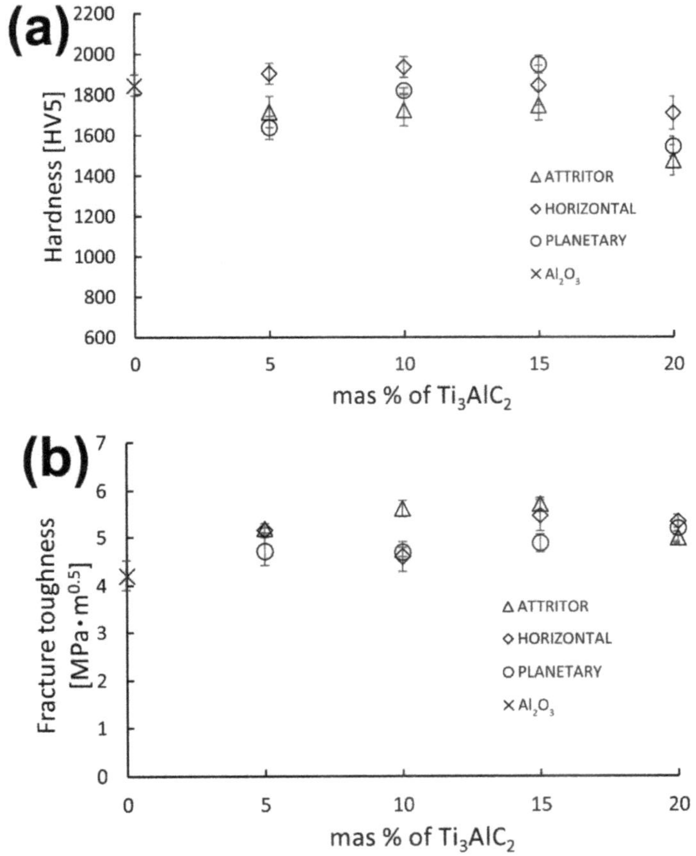

Figure 11. (**a**) Hardness of Al_2O_3-Ti_3AlC_2 composites; (**b**) fracture toughness of Al_2O_3-Ti_3AlC_2 composites.

4. Discussion

The SEM observations of the microstructure, the chemical composition analysis and the phase composition analysis showed that it is possible to obtain composites based on aluminum oxide reinforced with the Ti_3AlC_2 phase. Despite the use of the sintering temperature exceeding the temperature of the Ti_3AlC_2 phase decomposition, their original structure can be preserved. The small amount of TiC found during the phase composition analysis may come from both contamination of the starting powder of the MAX phase and may indicate the decomposition of a small amount of the MAX phase during the sintering process. However, no characteristic clusters of TiC grains signaled in other works were found [25], which indicates that the decomposition process has been successfully limited. It is probably related to the applied SPS sintering technology. Shortening the sintering time to a few minutes limits the decomposition process [26]. All the produced composites were characterized by a relative density above 95%, while for the 5 wt% addition of the Ti_3AlC_2 phase, a clear decrease in the mean value of the relative density was observed. This decrease is particularly noticeable for the series of composites obtained with the use of a horizontal and planetary mill. Similar changes were observed in the case of composites with the addition of 20 wt% Ti_3AlC_2 phase. Microscopic observations indicate that porosity is generated mainly within the flakes of the Ti_3AlC_2 phase, at the interface and between the matrix grains. As already mentioned, for composites with a small content of the Ti_3AlC_2 phases, the observed changes in relative density can be caused by several factors. A small amount of MAX phases may undergo thermal decomposition during sintering. The amount of phase that decomposes and the amount of products formed is small, but it can affect the calculated theoretical density and thus distort it. In addition, three different mills with different mixing energies were used during the composites manufacturing process. This may affect the different degree of contamination of composites with the dope coming from the used grinding balls (Al_2O_3) and the blades in the attritor (ZrO_2). This contamination may also cause changes in theoretical density, which will affect the theoretical density of composites, and thus affect the density of the measured relative density [27,28]. The microstructure analysis also shows that, regardless of the mill used in the technological process, the Ti_3AlC_2 phase distributed among the matrix grains effectively reduces the growth of the Al_2O_3 matrix grain in relation to the reference sample. The phenomenon of limiting the grain growth in composites on a ceramic matrix, even for a small amount of additives, has been signaled in the literature [29]. The addition of the Ti_3AlC_2 phase also changes the nature of the cracking of the composites from intergranular to transgranular (mark by arrow at Figure 3a–d). The reference sample breaks almost completely intergranular, while with the addition of the reinforcing phase increase, the proportion of transgranular cracking increases. This proves that the cohesive forces are increased in the case of composites compared to the unreinforced sinter. Significant differences were also found between individual series of composites depending on the mill used. The highest content of transgranular cracking was observed for composites mixed in the attritor. These differences are manifested both on the macro and micro scale. In the case of composites obtained with the attritor mill, pores and voids were found at the Al_2O_3-Ti_3AlC_2 interface, which suggests poor bonding of the reinforcement phase flakes with the matrix (Figure 4b). Moreover, the type of mill used has a significant influence on the formation of agglomerates. This aspect was signaled in the literature as one of the main technological problems in the case of this type of material [30]. In the case of composites obtained with a horizontal mill and an attritor, an increase in the average size of Ti_3AlC_2 phase agglomerates is observed with the increase in Ti_3AlC_2 phase addition. At the same time, the coefficient CV_{SKIZ} increases, which indicates a decrease in the homogeneity of the obtained microstructure. The increase in the size of agglomerates in composites is not reflected in the results of density measurements. Increasing the dimensions and content of agglomerates is usually associated with a decrease in density caused by the occurrence of voids between the particles inside the agglomerates [31,32]. In the case of the produced composites, due to the similar sintering temperature of Al_2O_3 and Ti_3AlC_2, the particles of the MAX phase were consolidated inside the agglomerates (Figure 4c). Thus, they did not lower the density of the composites. A much more significant impact on the density change in individual composites has the content of impurities remaining from the starting powders and a slight decomposition of the MAX phases during the sintering process.

The analysis of the shape of the formed agglomerates shows that it does not change with the increase in the average size of the agglomerates. Composites made using a planetary mill have a completely different character. The observed increase in the average size of agglomerates is slight, while the obtained microstructures are characterized by a high degree of homogeneity, regardless of the amount of Ti_3AlC_2 phase addition. High mixing energy contributes to the more effective breaking of the agglomerates at this stage, which reduces their average size and effects on, increasing the homogeneity of the mixture and, consequently, increasing the homogeneity of the obtained microstructure [33].

Despite significant differences in the microstructure of the obtained composites, the observed differences in mechanical properties are lesser. In the case of hardness, the results for composites will be related to several factors. Namely, the average size of the matrix grain, the average hardness of the individual components and the degree of homogeneity of the matrix. In the case of Al_2O_3 and Ti_3AlC_2, these phases' hardness differences are significant; the crystal hardness is 22 GPa and 2.7 GPa, respectively [34]. This suggests that the hardness of the produced composites should be much lower than indicated by the research. The high hardness of the tested composites, close to the hardness of the reference sample sintered at 1400 °C, is probably associated with a clear decrease in the average grain size of the matrix resulting from the addition of the Ti_3AlC_2 phase (Figure 7) and the strengthening effect caused by the presence of an additional phase at the grain boundaries. The homogeneity of the Ti_3AlC_2 phase distribution is also important in this case. This is confirmed by the lowest standard deviations of the averaged hardness of composites obtained with a planetary mill, which are also characterized by the highest degree of homogeneity. Fluctuations in the hardness obtained in the case of the remaining series probably result from the overlapping of the effects mentioned above, the decrease in the degree of microstructure homogeneity and an increase in the average size of agglomerates. The situation is similar in the case of fracture toughness. All produced composites are characterized by a higher VIF than the reference samples, which confirms that adding the Ti_3AlC_2 phase effectively increases the fracture toughness. Improving the fracture toughness of composites reinforced with MAX phases is related to their structure. Due to the alternating occurrence of metallic and ceramic layers, the propagated fracture can be bridged, can deflect causing delamination of individual layers or lower the fracture energy by plastic deformation of the metallic layers [35]. However, the fracture toughness of the tested composites, despite a significant increase, does not show significant changes depending on the used mill or the amount of reinforcement.

5. Conclusions

Based on the obtained results, it can be concluded that it is possible to obtain dense alumina composites reinforced with Ti_3AlC_2 MAX phases. SPS sintering technology effectively reduces the decomposition of the used MAX phase. In the case of the production of this type of composite, it is important to select the appropriate type of mill. The most homogeneous microstructure in terms of the size and distribution of Ti_3AlC_2 agglomerates was obtained using the planetary mill with the highest mixing energy. The application of the MAX phases as a reinforcing phase in the alumina allows one to obtain sinters with high hardness, similar to the hardness of pure sinter while improving the fracture toughness. The composites showed almost 20% higher fracture toughness while maintaining high hardness. Considering the obtained mechanical properties of the composites and the possibility of maintaining the MAX phase structure after the sintering process, new opportunities for producing a new family of ceramic materials appear.

Supplementary Materials: The following supporting information can be downloaded at: https://www.mdpi.com/article/10.3390/ma15196909/s1, Scheme S1: The images present the distribution of Ti_3AlC_2 phase for obtained composites. (a) Al_2O_3—5wt% Ti_3AlC_2 from attritor mill, (b) Al_2O_3—5wt% Ti_3AlC_2 from horizontal mill, (c) Al_2O_3—5wt% Ti_3AlC_2 from planetary mill, (d) Al_2O_3—10wt% Ti_3AlC_2 from attritor mill, (e) Al_2O_3—10wt% Ti_3AlC_2 from horizontal mill, (f) Al_2O_3—10wt% Ti_3AlC_2 from planetary mill, (g) Al_2O_3—15wt% Ti_3AlC_2 from attritor mill, (h) Al_2O_3—15wt% Ti_3AlC_2 from horizontal mill, (i) Al_2O_3—15wt% Ti_3AlC_2 from planetary mill, (j) Al_2O_3—20wt% Ti_3AlC_2 from attritor mill, (k) Al_2O_3—20wt% Ti_3AlC_2 from horizontal mill, (l) Al_2O_3—20wt% Ti_3AlC_2 from planetary mill.

Author Contributions: M.P. manuscript preparation, results analysis, MAX preparation; J.W. article concept, manuscript preparation, results analysis, image analysis, SEM observations; T.C. composites sintering, MAX preparation; W.P. phase analysis of composites, results analysis; A.O. results analysis, review and editing. All authors have read and agreed to the published version of the manuscript.

Funding: The study was accomplished thanks to the funds allotted be the Ministry of Science and Higher Education Republic of Poland.

Institutional Review Board Statement: Not applicable.

Informed Consent Statement: Not applicable.

Data Availability Statement: All the data is available within the manuscript.

Conflicts of Interest: The authors declare no conflict of interest.

References

1. Koli, D.K.; Agnihotri, G.; Purohit, R. A Review on Properties, Behaviour and Processing Methods for Al-Nano Al_2O_3 Composites. *Procedia Mater. Sci.* **2014**, *6*, 567–589. [CrossRef]
2. Munro, M. Evaluated Material Properties for a Sintered alpha-Alumina. *J. Am. Ceram. Soc.* **1997**, *80*, 1919–1928. [CrossRef]
3. Sternitzke, M. Structural ceramic nanocomposites. *J. Eur. Ceram. Soc.* **1997**, *17*, 1061–1082. [CrossRef]
4. Niihara, K. New Design Concept of Structural Ceramics. *J. Ceram. Soc. Jpn.* **1991**, *99*, 974–982. [CrossRef]
5. Petrus, M.; Wozniak, J.; Cygan, T.; Kostecki, M.; Cygan, S.; Jaworska, L.; Teklińska, D.; Olszyna, A. Comprehensive study on graphene-based reinforcements in Al_2O_3–ZrO_2 and Al_2O_3–Ti(C,N) systems and their effect on mechanical and tribological properties. *Ceram. Int.* **2019**, *45*, 21742–21750. [CrossRef]
6. Roy, R. Purposive Design of Nanocomposites: Entire Class of New Materials. *Ceram. Microstruct.* **1987**, *86*, 25–32. [CrossRef]
7. Porwal, H.; Grasso, S.; Reece, M.J. Review of graphene–ceramic matrix composites. *Adv. Appl. Ceram.* **2013**, *112*, 443–454. [CrossRef]
8. Li, L. A Micromechanical Fatigue Limit Stress Model of Fiber-Reinforced Ceramic-Matrix Composites under Stochastic Overloading Stress. *Materials* **2020**, *13*, 3304. [CrossRef]
9. Ling, Z.; Ren, C.E.; Zhao, M.-Q.; Yang, J.; Giammarco, J.M.; Qiu, J.; Barsoum, M.W.; Gogotsi, Y. Flexible and conductive MXene films and nanocomposites with high capacitance. *Proc. Natl. Acad. Sci. USA* **2014**, *111*, 16676–16681. [CrossRef]
10. Barsoum, M.W.; Radovic, M. Elastic and Mechanical Properties of the MAX Phases. *Annu. Rev. Mater. Sci.* **2011**, *41*, 195–227. [CrossRef]
11. Naguib, M.; Kurtoglu, M.; Presser, V.; Lu, J.; Niu, J.; Heon, M.; Hultman, L.; Gogotsi, Y.; Barsoum, M.W. Two-Dimensional Nanocrystals Produced by Exfoliation of Ti3AlC2. *Adv. Mater.* **2011**, *23*, 4248–4253. [CrossRef]
12. Lei, J.-C.; Zhang, X.; Zhou, Z. Recent advances in MXene: Preparation, properties, and applications. *Front. Phys.* **2015**, *10*, 276–286. [CrossRef]
13. Barsoum, M.; El-Raghy, T. The MAX Phases: Unique New Carbide and Nitride Materials: Tertiary ceramics are soft and machinable, yet heat-tolerant, strong and lightweight. *Am. Sci.* **2001**, *89*, 334–343. [CrossRef]
14. Song, G. *Self-Healing of MAX Phase Ceramics for High Temperature Applications: Evidence from Ti_3AlC_2*; Elsevier: Amsterdam, The Netherlands, 2012; pp. 271–288. [CrossRef]
15. Low, I.M.; Pang, W.K. Thermal Stability of MAX Phases. *Key Eng. Mater.* **2014**, *617*, 153–158. [CrossRef]
16. Lange, C.; Barsoum, M.W.; Schaaf, P. Towards the synthesis of MAX-phase functional coatings by pulsed laser deposition. *Appl. Surf. Sci.* **2007**, *254*, 1232–1235. [CrossRef]
17. Guo, C.; Wang, E.; Wang, S.; Hou, X.; He, Z.; Liang, T.; Chou, K.-C. Oxidation mechanism of MAX phases (Ti_3AlC_2 powders) with and without Sn doping. *Corros. Sci.* **2020**, *180*, 109197. [CrossRef]
18. Zhou, W.; Mei, B.; Zhu, J.; Hong, X. Synthesis of high-purity Ti_3SiC_2 and Ti_3AlC_2 by spark plasma sintering (SPS) technique. *J. Mater. Sci.* **2005**, *8*, 2099–2100. [CrossRef]
19. Nayebi, B.; Asl, M.S.; Akhlaghi, M.; Ahmadi, Z.; Tayebifard, S.A.; Salahi, E.; Shokouhimehr, M.; Mohammadi, M. Spark plasma sintering of TiB2-based ceramics with Ti_3AlC_2. *Ceram. Int.* **2021**, *47*, 11929–11934. [CrossRef]
20. Asl, M.S.; Nayebi, B.; Akhlaghi, M.; Ahmadi, Z.; Tayebifard, S.A.; Salahi, E.; Shokouhimehr, M.; Mohammadi, M. A novel ZrB2-based composite manufactured with Ti3AlC2 additive. *Ceram. Int.* **2020**, *47*, 817–827. [CrossRef]
21. Tan, Y.; Luo, H.; Zhang, H.; Zhou, X.; Peng, S. Fabrication of toughened B4C composites with high electrical conductivity using MAX phase as a novel sintering aid. *Ceram. Int.* **2016**, *42*, 7347–7352. [CrossRef]
22. Tan, Y.Q.; Chen, C.; Li, F.Z.; Zhang, H.B.; Zhang, G.J.; Peng, S.M. Enhancement of sinterability and mechanical properties of B_4C ceramics using Ti_3AlC_2 as a sintering aid. *RSC Adv.* **2015**, *5*, 76309–76314. [CrossRef]
23. Stobierski, L.; Gubernat, A. Sintering of silicon carbideI. Effect of carbon. *Ceram. Int.* **2003**, *29*, 287–292. [CrossRef]
24. Yang, H.; Shang, F.; Gao, L. Microstructure and mechanical properties of gas pressure sintered Al_2O_3/TiCN composite. *Ceram. Int.* **2007**, *33*, 1521–1524. [CrossRef]
25. Zheng, L.; Li, F.; Zhou, Y. Preparation, Microstructure, and Mechanical Properties of TiB2 Using Ti_3AlC2 as a Sintering Aid. *J. Am. Ceram. Soc.* **2012**, *95*, 2028–2034. [CrossRef]
26. Guillard, F.; Allemand, A.; Lulewicz, J.-D.; Galy, J. Densification of SiC by SPS-effects of time, temperature and pressure. *J. Eur. Ceram. Soc.* **2007**, *27*, 2725–2728. [CrossRef]
27. Suryanarayana, C. Mechanical alloying and milling. *Prog. Mater. Sci.* **2001**, *46*, 1–184. [CrossRef]
28. Sarwat, S.G. Contamination in wet-ball milling. *Powder Metall.* **2017**, *60*, 267–272. [CrossRef]
29. Petrus, M.; Woźniak, J.; Cygan, T.; Lachowski, A.; Rozmysłowska-Wojciechowska, A.; Wojciechowski, T.; Ziemkowska, W.; Chlubny, L.; Jastrzębska, A.; Adamczyk-Cieślak, B.; et al. Silicon carbide nanocomposites reinforced with disordered graphitic carbon formed in situ through oxidation of Ti_3C_2 MXene during sintering. *Arch. Civ. Mech. Eng.* **2021**, *21*, 87. [CrossRef]
30. Agne, M.T.; Anasori, B.; Barsoum, M.W. Reactions Between Ti_2AlC, B_4C, and Al and Phase Equilibria at 1000 °C in the Al-Ti-B-C Quaternary System. *J. Phase Equilibria Diffus.* **2015**, *36*, 169–182. [CrossRef]
31. Woźniak, J.; Kostecki, M.; Broniszewski, K.; Bochniak, W.; Olszyna, A. Influence of mixing parameters on homogeneity of Al/SiC composites. *Arch. Metall. Mater.* **2014**, *59*, 1493–14981. [CrossRef]
32. Wozniak, J.; Kurtycz, P.; Broniszewski, K.; Kostecki, M.; Morgiel, J.; Olszyna, A. Properties of Alumina Matrix Composites Reinforced with Nickel-coated Graphene. *Mater. Today: Proc.* **2015**, *2*, 376–382. [CrossRef]

33. Cygan, T.; Wozniak, J.; Petrus, M.; Adamczyk-Cieslak, B.; Kostecki, M.; Olszyna, A. The effect of microstructure evolution on mechanical properties in novel alumina-montmorillonite composites. *Int. J. Refract. Hard Met.* **2019**, *80*, 195–203. [CrossRef]
34. Wang, X.; Zhou, Y. Layered Machinable and Electrically Conductive Ti2AlC and Ti3AlC2 Ceramics: A Review. *J. Mater. Sci. Technol.* **2010**, *26*, 385–416. [CrossRef]
35. Chen, X.; Bei, G. Toughening Mechanisms in Nanolayered MAX Phase Ceramics—A Review. *Materials* **2017**, *10*, 366–372. [CrossRef]

materials

MDPI

Article

Tests and Numerical Study of Single-Lap Thermoplastic Composite Joints Bolted by Countersunk

Jian Zhang [1], Xiuhua Chen [2,*], Aiqin Tian [3,4] and Yin Fan [2,*]

1 School of Mechanical Engineering, Shanghai Jiao Tong University, Shanghai 200240, China; zhangjian970603@sjtu.edu.cn
2 School of Aeronautics and Astronautics, Shanghai Jiao Tong University, Shanghai 200240, China
3 School of Materials Science and Engineering, Shanghai Jiao Tong University, Shanghai 200240, China; tianaiqin@cqsf.com
4 CRRC Qingdao Sifang Co., Ltd., Qingdao 266111, China
* Correspondence: chenxiuhua@sjtu.edu.cn (X.C.); yfan1987@sjtu.edu.cn (Y.F.)

Abstract: Tensile tests were carried out to investigate the effect of stacking sequences on the bearing strength of single-lap thermoplastic composite countersunk bolted joints. A 3D elastoplastic model was built based on a plastic theory for numerical analysis. The damage initiation was judged based on LaRC05 criteria, and the damage propagation was described by using a nonlinear, gradual unloading method based on crack band theory. The accuracy of the present model was validated by comparing the numerical results to those from the tests. The test results showed that the effects of stacking sequences on the ultimate bearing strength and the 2% offset bearing strength are limited. Moreover, the numerical results depicted that the ultimate bearing strength and the 2% offset bearing strength reduce when the bolt-tightening torque or the bolt–hole clearance is increased.

Keywords: thermoplastic composite; single-lap countersunk bolted joint; 3D elastoplastic damage model; bolt-tightening torque; clearance

1. Introduction

Owing to their superior features, such as high specific strength and stiffness, high specific moduli, damage tolerance, and environmental resilience, carbon-fiber-reinforced thermoplastic (CFRTP) composites have been widely used in the aerospace industry [1–5]. Composite components are divided into numerous parts as per the requirements of structural design, manufacture, and maintenance of composites, as well as the limitations of the integrity molding technology. As a result, the joining technique plays a key role in the application of composites [6]. Usually, fasteners are used to connect joint parts and bolted joints are frequently used in aircraft structures because they are easy to produce, disassemble, repair, and replace [7]. The bolted joints, however, are usually the weakest part in a composite laminate structure since a composite material is anisotropic and relatively brittle [8]. Therefore, the failure mechanism of composite joints is of most concern in the design of composite structures.

In the past decades, many scholars have conducted in-depth research into thermoset composite bolted joints [9–25]. For example, Park [11] investigated the influence of stacking sequences on the delamination bearing strength of mechanically fastened joints. Results revealed that the delamination bearing strength of the lay-up with the 90° layers on the surface was stronger than the one with the 90° layers located at the center of the laminate. Using experiments and 3D finite element models, McCarthy et al. [12,13] studied the influence of bolt–hole clearance on composite protruding bolted joints. They found that the increase of the clearance might lead the protruding bolts to rotate and the contact area and stiffness were both degraded as a result. Most scholars [11–17] have used 3D models to simulate composite bolted joints. To improve computational efficiency, Liu et al. [25]

modified the bolt stiffness model based on the concept of control volume proposed by Gray and McCarthy [24]; Belardi et al. [23] proposed a novel composite bolted joint element. Both numerical methods had been adopted for accurately predicting the mechanical behaviors of composite bolted joints.

With the development of technology, thermoplastic composite materials have been increasingly used in modern industry and thermoplastic joints, such as pinned joints [7], riveted joints [26–28], and protruding bolted joints [29–32], have already been focused on. Ýýlmaz and Sýnmazçelik [7] carried out experimental methods to investigate the bearing performance of the pinned joints of carbon fiber/polyphenylene sulfide (CF/PPS). The findings revealed that the bearing strength of the pinned joints improved if enough end distance and width were provided. Gay et al. [26,27] investigated the performance of a PA6.6-GF (glass fiber/polyamide)/aluminum assembly joined by self-piercing rivets under static and fatigue loads. They found that the domed head rivet achieved higher fatigue limit strength at 2×10^6 cycles and higher bearing strength at static loading than that of the countersunk rivet and the failure mode under the fatigue load was determined by the load amplitude and the joint geometry. The influence of interference fit on the bearing strength of a carbon fiber/polyetheretherketone (CF/PEEK) riveted joint was investigated by Yao et al. [28] using both experimental and numerical approaches. According to their research, the extrusion between the rivet and the composite laminate created a plastic deformation on the hole wall, which would enhance the bearing capacity of the composite laminate; the acceptable interference-fit size was in the range of 2.0–2.6% for the CF/PEEK riveted joint. Vieille et al. [29–31] investigated the effect of environmental conditions on the mechanical behaviors of carbon-fiber-fabric-reinforced PPS laminates. It was found that the failure mode could be changed from brittle to progressive in a severe environment (120 °C after hygrothermal aging). The enhanced localized plasticization of the PPS and the degraded fiber/matrix interface dissipated a portion of the mechanical energy transmitted to the joint and contributed to easing the stress concentration, which led to an increases in the strength of the double-lap joints.

Although reports on mechanical behaviors of thermoplastic joints can be found, there is a shortage of studies on the tensile performance of thermoplastic composites with single-lap countersunk bolted joints. The thermoplastic composites made for single-lap countersunk bolted joints are first studied by tests in this paper, and then a 3D elastoplastic damage model that can properly replicate its mechanical behavior is proposed for analysis of their mechanical behaviors. The purpose of this paper is to reveal the failure mechanism and investigate the influence of bolt torque and bolt–hole clearance on the bearing strength of single-lap countersunk bolted joints of AS4/PEEK (carbon fiber/polyetheretherketone). Firstly, tensile tests of the single-lap countersunk bolted joints of laminates with three stacking sequences are performed to obtain load–displacement curves and failure modes. After that, a 3D elastoplastic model is developed based on a plastic theory proposed by Sun and Chen [33]. The LaRC05 criteria [34] are used to predict damage initiation, and a nonlinear gradual unloading method based on crack band theory [35] is used to simulate the damage extension. Then, the above 3D elastoplastic model, failure criteria, and damage propagation are all converted into a numerical simulation of single-lap countersunk bolted joints and this method is validated by comparing with the test results. Finally, the impacts of bolt-tightening torque and bolt–hole clearance on the ultimate bearing strength and the 2% offset bearing strength are investigated, respectively. The influence on the bearing strength of the single-lap countersunk joints is discussed in detail.

2. Test

2.1. Test Setup

The dimensions of a test specimen are shown in Figure 1a, which are recommended by ASTM D5961/D 5961M-17 [36]. The specimens operated in the tests were manufactured by unidirectional AS4/PEEK prepregs, where the average fiber volume fraction was 59%. Each specimen had 25 layers, and the nominal thickness of each layer was 0.14 mm. There were three different stacking sequences, $[-45/90/45/0/-45/0/45/0/-45/0/45/-45/\overline{90}]_s$, $[-45/90/45/0_2/-45/0/45/0/-45/0/45/\overline{90}]_s$, and $[-45/90/45/0_2/-45/0/45/0_2/-45/0/\overline{90}]_s$, and the overbar represents the center layer in symmetric lay-up with an odd number of piles. The names of these three stacking sequences were DJ-A, DJ-B, and DJ-C. The width-to-diameter ratio was 6, and the end-distance-to-diameter ratio was 3. The 100° countersunk bolt used in the test specimen is made of Ti-6Al-4V and fixed with steel nuts. As shown in Figure 1b, the cylindrical depth of the countersunk bolt in the upper plate was 1.3 mm; the bolt-tightening torque was 1.8 N·m, and there was no clearance between the countersunk bolt and the laminate. To ensure neutrality during the test, pads with the same material as the test specimen were bonded at both ends of the test specimen, and the size of the pads was 75 mm × 36 mm × 3.5 mm. For each layup configuration, we tested six specimens.

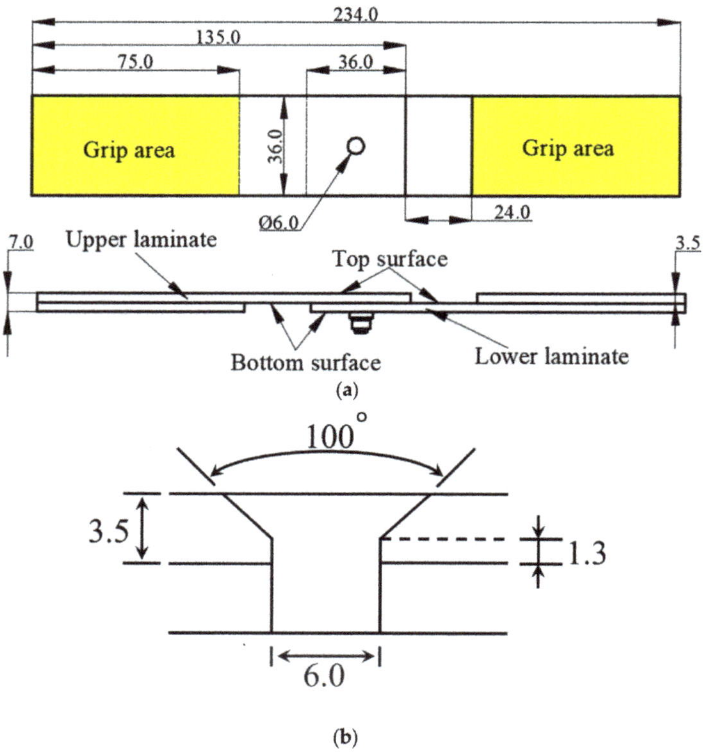

Figure 1. Geometry and dimensions (unit: mm) of the test specimen and the countersunk bolt: (**a**) test specimen and (**b**) countersunk bolt.

The test was carried out on the UTM5105X microcomputer—controlled electronic universal mechanical testing machine with 100 kN load capability. The laminates were clamped by the grip holder of the testing machine. The gripping area length was about 55 mm. An extensometer, was mounted on the specimen to measure the longitudinal hole deformation. The type of extensometer was SUNS-YSJ50/10 and the gauge distance was 50 mm. All above equipment was manufactured by Shenzhen SUNS Technology Stock CO., LTD. in Shenzhen, China. The test equipment is shown in Figure 2. The applied load was of a constant displacement rate of 2 mm/min, according to ASTM D5961/D 5961M-17 [36], up to the catastrophic failure. The data related to the applied load, grip holder displacement, and hole deformation were recorded automatically by the computer. After the tests, disassembling processes were executed to examine the damage in the laminates.

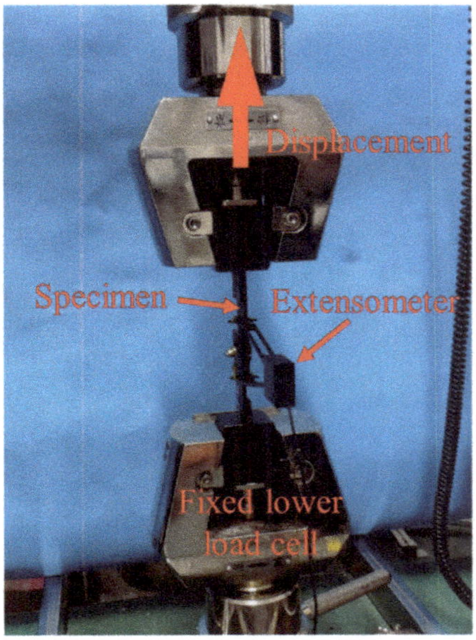

Figure 2. Test equipment of the AS4/PEEK countersunk single-lap bolted joint.

2.2. Test Results

Figure 3a depicts one of the load–displacement curves obtained from the tests to demonstrate the mechanical responses of the countersunk single-lap bolted joints. The bearing strength and strain were calculated from the load–displacement data by using the equations $\sigma_{br} = F/(k \times D \times h)$ and $\varepsilon_{br} = \delta/(K \times D)$, where σ_{br} and ε_{br} are the bearing stress and strain, respectively; F and δ represent the load and the displacement measured during the bearing tests, respectively; D and h are the bolt hole diameter and the specimen thickness, respectively; k is the force per hole factor (1.0 for single-fastener or pin tests and 2.0 for double-fastener tests); and K is 2 in this test [36]. The 2% offset bearing strength was determined by the intersection point of the stress–strain curve and the translated chord stiffness line, as depicted in Figure 3b. The ultimate bearing strength and the 2% offset bearing strength corresponding to the test specimens are summarized in Table 1. The differences between the ultimate bearing strength of DJ-B and DJ-C and that of DJ-A were −0.33% and 0.70%, respectively, and the differences between the 2% offset bearing strength were 1.45% and −3.54%, respectively. It is apparent that the stacking sequences had a limited effect on the ultimate bearing strength and the 2% offset bearing strength of the single-lap countersunk joints.

Table 1. The bearing strength of AS4/PEEK single-lap countersunk bolted joints with three stacking sequences.

	Ultimate Bearing Strength (MPa)	2% Offset Bearing Strength (MPa)
DJ-A	560.50	542.20
DJ-B	562.36	534.33
DJ-C	564.40	561.40

The damage mainly occurred in the laminates, while no obvious damage occurred on the countersunk bolt. The photographs from different perspectives and the micrograph of the test specimen are shown in Figure 4. Around the bolt hole, we could see fiber-matrix splitting on both sides of the upper laminate and the bottom surface of the lower laminate. This damage is caused by the tensile bending stress arising from transverse loading [37].

The micrograph shows an evident compressive area in the top half of the lower laminate, whereas the lower section of the laminate was mainly the matrix cracking.

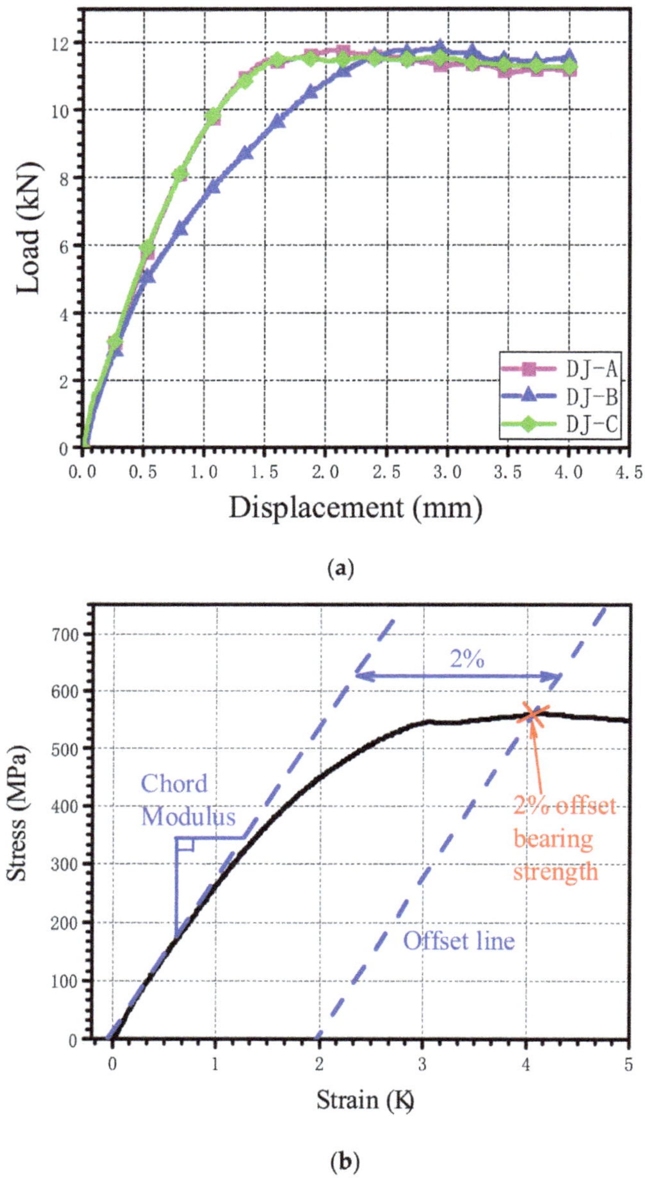

Figure 3. (**a**) Load–displacement curves of AS4/PEEK single-lap countersunk bolted joints with three stacking sequences. (**b**) Determination of the 2% offset bearing strength.

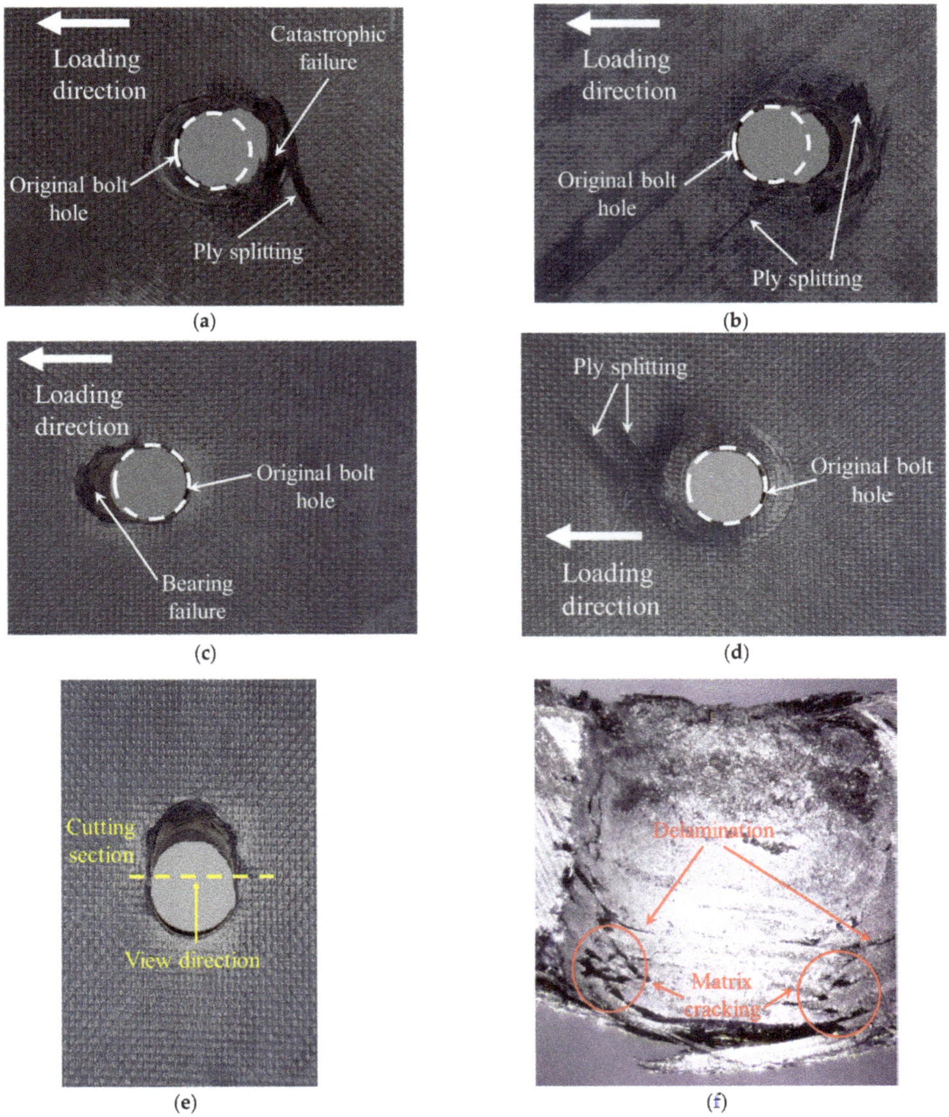

Figure 4. Damage to the specimens after tensile loading: (**a**) Upper laminate, top surface; (**b**) upper laminate, bottom surface; (**c**) lower laminate, top surface; (**d**) lower laminate, bottom surface; (**e**) cutting section and view direction; and (**f**) microscopic photograph.

3. Numerical Model

3.1. Constitutive Relationship

For the unidirectional composite materials that exhibit plasticity and progressive degradation of material properties, this paper built a 3D elastoplastic damage model. It was developed for an elementary orthotropic ply and consisted of a plastic part that describes the plastic behavior of composites, failure criteria for predicting damage initiation and damage evolution laws that account for the development of the failure process. The typical stress–strain curve of elastoplastic materials is shown in Figure 5a, which has three regions: an elastic region, an elastoplastic region, and a damage evolution region. Note that

1-direction is the longitudinal direction (fiber direction) and the 2-direction and 3-direction are the transverse directions in this paper.

3.1.1. Elastic Model

Assuming that the laminate is homogeneous and orthotropic, the relationship between elastic stress and strain is formulated as:

$$\sigma = C : \varepsilon^e = C : (\varepsilon - \varepsilon^p), \tag{1}$$

where σ is the stress tensor; C is the stiffness tensor; ε, ε^e, and ε^p are the total strain tensor, the elastic strain tensor, and the plastic strain tensor, respectively; and the symbol (:) denotes the inner product of two tensors with double contraction.

3.1.2. Plastic Model

In this model, it is assumed that the plastic deformation occurs in the undamaged area of the laminates and the plastic flow rule and the hardening law are expressed in terms of effective stresses, equivalent plastic strain, and equivalent stress, which are based on the effective stress space concept [38].

The yield function used in this model is expressed as follows:

$$F(\sigma, \overline{\varepsilon^p}) = F^p(\sigma) - \kappa(\overline{\varepsilon^p}), \tag{2}$$

where $\overline{\varepsilon^p}$ is the equivalent plastic strain; $F^p(\sigma)$ is the plastic potential function; and κ is the hardening parameter, which depends on the plastic deformations and is expressed in terms of equivalent plastic strain.

Due to its simplicity and accuracy, the 3D plastic potential function proposed by Sun and Chen [33] was adopted in this study. Weeks and Sun [39] assumed that the composite material is transversely isotropic in the 2–3 plane and linearly elastic in the fiber direction. Therefore, the plastic potential function is reduced to:

$$F^p(\sigma) = \sqrt{\frac{3}{2}(\sigma_{22}^2 + \sigma_{33}^2) - 3\sigma_{22}\sigma_{33} + 6\sigma_{23}^2 + 3a_{66}\sigma_{12}^2 + 3a_{66}\sigma_{13}^2}, \tag{3}$$

where a_{66} is a coefficient describing the amount of anisotropy in the plasticity. Its value can be determined using an approach based on the linear regression analyses of the off-axis tensile tests performed on the unidirectional composite laminates [33,40,41]. It should be noted that Figure 5a is only appropriate for describing the mechanical behaviors of non-fiber directions. The plastic strain increment can be calculated by the associated plastic flow rule:

$$d\varepsilon^p = d\lambda \frac{\partial F^p}{\partial \sigma}, \tag{4}$$

where $d\lambda$ is a non-negative plastic consistency parameter. Let the equivalent stress be defined as:

$$\overline{\sigma} = F^p. \tag{5}$$

The effective plastic strain increment $d\overline{\varepsilon^p}$ can be defined through the consideration of plastic work increment:

$$dW^p = \sigma \cdot d\varepsilon^p = \overline{\sigma} d\overline{\varepsilon^p}. \tag{6}$$

Substitution of Equations (3)–(5) into (6) yields:

$$d\overline{\varepsilon^p} = d\lambda. \tag{7}$$

Experimental results have proved that thermoplastic composites do not have a well-defined yield point [39]. For this reason, the yield stress σ_s is always assumed to be 0 in many works [28,42–45]. An isotropic hardening function expressed by the equivalent plastic strain was adopted in this study:

$$\kappa\left(\overline{\varepsilon^p}\right) = \beta\left(\overline{\varepsilon^p}\right)^n, \tag{8}$$

where β and n are determined by fitting the experimental hardening curve. These values are shown in Table 2.

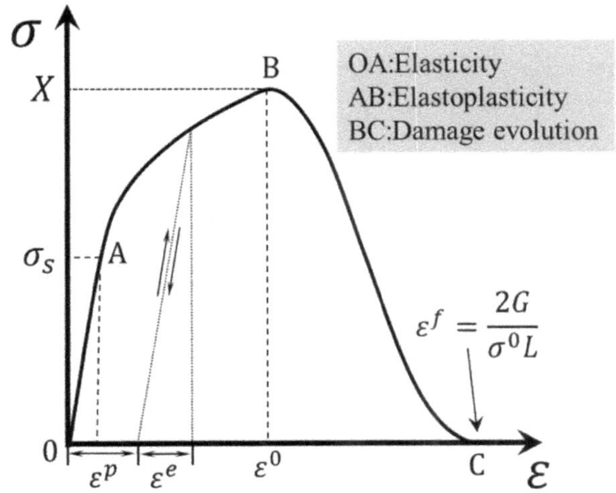

(a)

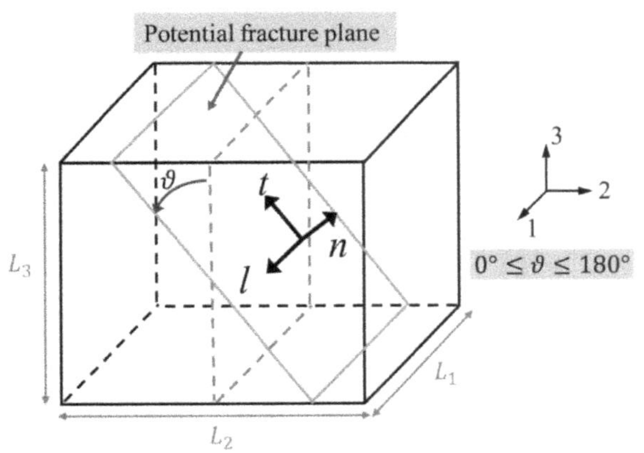

(b)

Figure 5. An illustration of (**a**) a typical stress–strain curve of elastoplastic materials, where X denotes the strength, and (**b**) calculation of tractions on any fracture plane.

3.1.3. Damage Initiation and Evolution Model

The maximum stress criteria were applied to predict the failure in the fiber direction:

$$f_{ft} = \frac{\sigma_{11}}{X_t}(\sigma_{11} > 0), \tag{9}$$

$$f_{fc} = -\frac{\sigma_{11}}{X_c}(\sigma_{11} < 0), \tag{10}$$

where X_t and X_c are the tensile and compressive failure strength in the fiber direction, respectively, and f_{ft} and f_{fc} are the fiber tensile damage index and the fiber compressive damage index, respectively. Fiber tensile failure will occur if $f_{ft} \geq 1$, and fiber compressive failure will occur when $f_{fc} \geq 1$.

To predict the failure initiation of a matrix-dominated failure, the LaRC05 criteria [34] were adopted in this study. The failure criteria assume that failure may appear on an arbitrary fracture plane parallel to the fiber direction depending on the combination of shear tractions (σ_{nT} and σ_{nL}) and normal traction (σ_n) on the corresponding plane [46]. When the failure index on a certain fracture plane reaches 1, the composites fail on this plane. As shown in Figure 5b, the tractions on the potential fracture plane can be transformed from the material coordinate system using the following equations:

$$\begin{cases} \sigma_n(\vartheta) = \frac{\sigma_{22}+\sigma_{33}}{2} + \frac{\sigma_{22}-\sigma_{33}}{2}\cos(2\vartheta) + \sigma_{23}\sin(2\vartheta) \\ \sigma_{nT}(\vartheta) = \frac{\sigma_{22}-\sigma_{33}}{2}\sin(2\vartheta) + \sigma_{23}\cos(2\vartheta) \\ \sigma_{nL}(\vartheta) = \sigma_{12}\cos(\vartheta) + \sigma_{13}\sin(\vartheta) \end{cases}, \tag{11}$$

where ϑ is the angle of the fracture plane, ranging from 0° to 180°. In these criteria, two typical failure modes are considered, depending on the value of the normal traction, and the failure indexes are defined as:

$$f_{mt} = \left(\frac{\sigma_{nT}(\vartheta)}{S_T - \mu_T\sigma_n(\vartheta)}\right)^2 + \left(\frac{\sigma_{nL}(\vartheta)}{S_L - \mu_L\sigma_n(\vartheta)}\right)^2 + \left(\frac{\sigma_n(\vartheta)}{Y_T}\right)^2 (\sigma_n(\vartheta) > 0), \tag{12}$$

$$f_{mc} = \left(\frac{\sigma_{nT}(\vartheta)}{S_T - \mu_T\sigma_n(\vartheta)}\right)^2 + \left(\frac{\sigma_{nL}(\vartheta)}{S_L - \mu_L\sigma_n(\vartheta)}\right)^2 (\sigma_n(\vartheta) < 0), \tag{13}$$

$$\mu_T = -\frac{1}{\tan(2\varphi_0)}, S_T = \frac{Y_c}{2\tan(\varphi_0)}, \mu_L = S_L\frac{\mu_T}{S_T}, \tag{14}$$

where S_T and S_L are the transverse shear failure strength and the longitudinal shear failure strength, respectively; μ_T and μ_L are the transverse and longitudinal friction coefficients, respectively; Y_T and Y_c are the transverse tensile strength and the compressive failure strength of composite laminates, respectively; and φ_0 is the through-thickness direction under transverse compressive load. For carbon-fiber-reinforced composites, the value of φ_0 determined by experiments is $\varphi_0 = 53^{\circ} \pm 2^{\circ}$ [47]. f_{mt} and f_{mc} are the matrix tensile damage index and the matrix compressive damage index, respectively; when $f_{mt} \geq 1$, matrix tensile failure occurs, and when $f_{mc} \geq 1$, matrix compressive failure occurs.

Following the prediction of a particular failure mode using the above criteria, an appropriate damage variable controlled the post-failure response. The damage variables used in this model included the fiber damage variable d_f and the matrix damage variable d_m. The fiber damage variable d_f includes the fiber tensile damage variable d_{ft} and the fiber compression damage variable d_{fc}. Similarly, the matrix damage variable d_m includes the matrix tensile damage variable d_{mt} and the matrix compression damage variable d_{mc}. The damage variable ranges from 0 to 1. When no damage occurs, the value is 0, and the value of 1 indicates the composite laminates are completely degraded [48]. The update method of damage variables adopts the form proposed by Donadon et al. [49]:

$$d^{t+\Delta t} = \max\left\{0, d^t, min\left\{1, 1 - \frac{\varepsilon^0}{\varepsilon}\left[1 + \kappa^2(2\kappa - 3)\right]\right\}\right\}, \tag{15}$$

$$\kappa = \frac{\varepsilon - \varepsilon^0}{\varepsilon^f - \varepsilon^0}, \tag{16}$$

where $d^{t+\Delta t}$ is the damage variable of the current incremental step, d^t is the damage variable of the previous incremental step, ε is the equivalent strain of the current incremental step, ε^0 is the equivalent strain at the beginning of damage, and ε^f is the equivalent strain at the final failure.

For fiber damage, the equivalent strain of the current incremental step is the strain component in the fiber direction. The calculations of the equivalent strain at the onset of damage and at the end of failure are as follows:

$$\varepsilon_{ft}^{0} = \frac{X_T}{E_1}, \tag{17}$$

$$\varepsilon_{fc}^{0} = \frac{X_C}{E_1}, \tag{18}$$

$$\varepsilon_{ft}^{f} = \frac{2G_{ft}}{X_T L}\left(f_{ft} \geq 1\right), \tag{19}$$

$$\varepsilon_{fc}^{f} = \frac{2G_{fc}}{X_C L}\left(f_{fc} \geq 1\right), \tag{20}$$

where G_{ft} and G_{fc} are the fracture energies of fiber tensile failure and fiber compressive failure, respectively, and L is the characteristic element length. These four equivalent strains are illustrated in Figure 6. Because there is no plastic deformation along the fiber direction, the parameters in Equations (17)–(20) do not vary with the progression of plastic deformation and damage. Under tensile load, stress softening is nonlinear, but it is linear under compression load. This is due to the fact that the equivalent strain at the onset of the damage is close to that at the end of the failure.

For the damage to the matrix, when either Equation (12) or Equation (13) is satisfied, the matrix is damaged and the equivalent stress and the equivalent strain are formulated as:

$$\sigma_m^0 = \sqrt{\sigma_n{}^2 + (\sigma_{nL})^2 + (\sigma_{nT})^2}\bigg|_{d_m^{t+\Delta t}=0}, \tag{21}$$

$$\varepsilon_m^0 = \sqrt{\varepsilon_n{}^2 + (\varepsilon_{nL})^2 + (\varepsilon_{nT})^2}\bigg|_{d_m^{t+\Delta t}=0}, \tag{22}$$

where, ε_n, ε_{nL}, and ε_{nT} are the normal traction and shear tractions on the fracture surface, respectively; $d_m^{t+\Delta t}$ is the matrix damage variable of the current incremental step; and x is the Macaulay bracket, defined as $\langle x \rangle = \max(0, x)$. After the damage, the equivalent strain of the matrix in the current incremental step becomes

$$\varepsilon_m = \sqrt{\varepsilon_n{}^2 + (\varepsilon_{nL})^2 + (\varepsilon_{nT})^2}\bigg|_{d_m^{t+\Delta t}>0} (f_{mt} \geq 1), \tag{23}$$

$$\varepsilon_m = \sqrt{(\varepsilon_{nL})^2 + (\varepsilon_{nT})^2}\bigg|_{d_m^{t+\Delta t}>0} (f_{mc} \geq 1), \tag{24}$$

For any mixed-modes failure scenario, the fracture energy for matrix failure is:

$$G_m = G_{Ic}\left\langle \frac{\sigma_n^d}{\sigma_m^0} \right\rangle^2 + G_{IIc}\left(\frac{\sigma_{nL}^d}{\sigma_m^0}\right)^2 + G_{IIc}\left(\frac{\sigma_{nT}^d}{\sigma_m^0}\right)^2, \tag{25}$$

where G_{Ic} and G_{IIc} are fracture energies for mode I and mode II fracture, respectively, and σ_n^d, σ_{nL}^d, and σ_{nT}^d are the tractions at the onset of matrix failure. The final failure strain of the matrix is defined as:

$$\varepsilon_m^f = \frac{2G_m}{\sigma_m^0 L}. \tag{26}$$

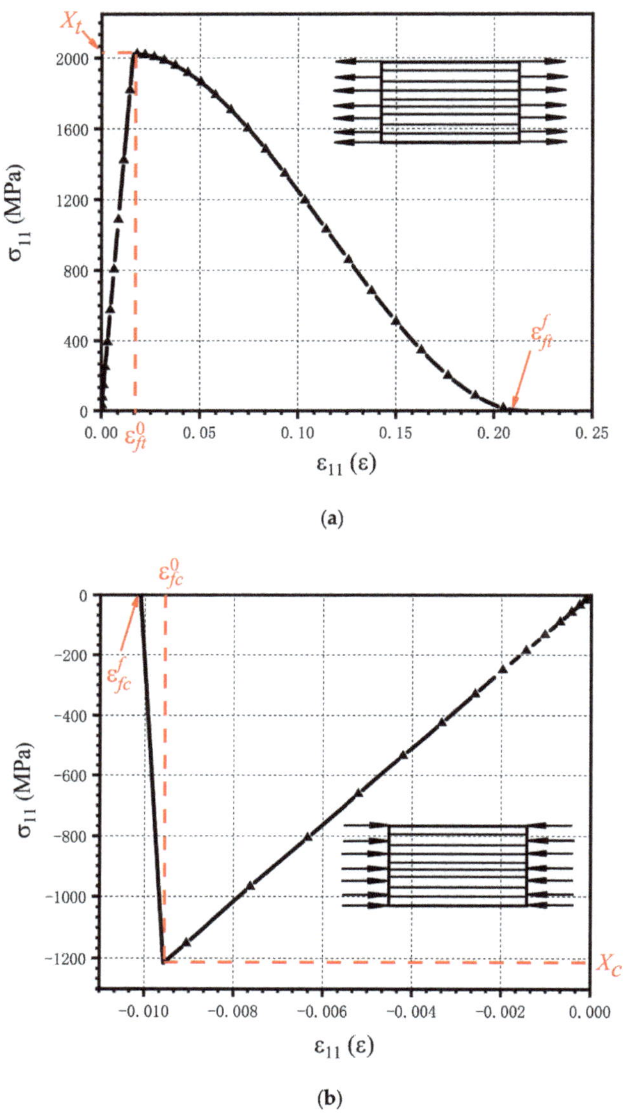

Figure 6. An illustration of the equivalent strain at the onset of damage and at the end of failure in the fiber direction: (**a**) the equivalent strain under tensile test and (**b**) the equivalent strain under compression test.

When matrix damage occurs in laminates, the stress tensor is rotated to the fracture plane by Equation (11) and the relevant tractions are softened according to Equation (27), yielding σ_n', σ_{nL}', and σ_{nT}'. To consider the effect of reverse load on crack closure, the normal traction is softened only when it is tensile.

$$\begin{cases} \sigma_n' = \sigma_n (1 - d_m)|_{\sigma_n > 0} \\ \sigma_{nL}' = \sigma_{nL}(1 - d_m) \\ \sigma_{nT}' = \sigma_{nT}(1 - d_m) \end{cases}, \tag{27}$$

After that, the tractions on the fracture surface are converted into stress in the global coordinate system through the inverse transformation of Equation (11), yielding $\sigma'_{ij}(i,j=1,2,3)$.

Matzenmiller et al. [50] assumed that fiber rupture damage, buckling, and compressive failure of fiber cause the damage evolution in the resin matrix. Based on this assumption, all stress components are softened:

$$\begin{bmatrix} \sigma''_{11} \\ \sigma''_{22} \\ \sigma''_{33} \\ \sigma''_{12} \\ \sigma''_{23} \\ \sigma''_{13} \end{bmatrix} = \left(1-d_f\right) \begin{bmatrix} \sigma'_{11} \\ \sigma'_{22} \\ \sigma'_{33} \\ \sigma'_{12} \\ \sigma'_{23} \\ \sigma'_{13} \end{bmatrix}, \quad (28)$$

where $\sigma''_{ij}(i,j=1,2,3)$ is the softened stress after fiber failure.

3.2. Calculation Procedure

According to the numerical algorithm of the 3D elastoplastic damage model, a user-defined material subroutine VUMAT based on ABAQUS was compiled. A calculation flow chart is shown in Figure 7. Ren et al. [51] proved that the forward Euler method and the backward Euler method can accurately solve the plastic strain. Considering that the algorithm used in this paper is an explicit dynamic algorithm and the stable time increment is 10^{-8} s, the forward Euler algorithm [52] was used to solve the plastic strain to minimize the computation time. The numerical simulation was carried out by step-by-step displacement loading. The calculation procedure in the n + 1 step was as follows:

(1) Initial conditions:

$$\Delta\varepsilon_{n+1}, \varepsilon_n, \varepsilon_n^p, \overline{\varepsilon_n^p},$$

where $\Delta\varepsilon_{n+1}$ is the strain tensor increment in step n + 1 and ε_n, ε_n^p, and $\overline{\varepsilon_n^p}$ are the total strain tensor, the plastic strain tensor, and the equivalent plastic strain, respectively, in step n.

(2) Assuming that the plastic strain tensor and the equivalent plastic strain of the current step are consistent with those of the previous step, the trial stress in the current step is calculated by:

$$\varepsilon_{n+1}^{p,trial} = \varepsilon_n^p, \overline{\varepsilon_{n+1}^{p,trial}} = \overline{\varepsilon_n^p}, \varepsilon_{n+1} = \varepsilon_n + \Delta\varepsilon_{n+1}, \sigma_{n+1}^{trial} = C : \left(\varepsilon_{n+1} - \varepsilon_{n+1}^{p,trial}\right),$$

where $\varepsilon_{n+1}^{p,trial}$ and $\overline{\varepsilon_{n+1}^{p,trial}}$ are the trial plastic strain tensor and the trial equivalent plastic strain, respectively, in step n + 1; ε_{n+1} is the strain tensor in step n + 1; and σ_{n+1}^{trial} is the trial stress tensor in step n + 1.

(3) Check the yield criterion:

$$F_{n+1}\left(\sigma_{n+1}^{trial}, \overline{\varepsilon_{n+1}^{p,trial}}\right) = F^p\left(\sigma_{n+1}^{trial}\right) - \kappa\left(\overline{\varepsilon_{n+1}^{p,trial}}\right).$$

If $F_{n+1}\left(\sigma_{n+1}^{trial}, \overline{\varepsilon_{n+1}^{p,trial}}\right) \leq 0$, then the plastic strain tensor ε_{n+1}^p, the equivalent strain $\overline{\varepsilon_{n+1}^p}$ and the stress tensor σ_{n+1} in step n + 1 are $\varepsilon_{n+1}^p = \varepsilon_n^p$, $\overline{\varepsilon_{n+1}^p} = \overline{\varepsilon_n^p}$, $\sigma_{n+1} = \sigma_{n+1}^{trial}$, respectively; otherwise, go to Equation (4) to calculate plastic strain increments.

(4) Calculate the plastic strain and the update stress:

(a) Calculate the scale factor increment:

$$\Delta\lambda_{n+1} = \frac{\frac{\partial F_{n+1}\left(\sigma_{n+1}^{trial}, \overline{\varepsilon_{n+1}^{p,trial}}\right)}{\partial \sigma_{n+1}^{trial}} : C : \Delta\varepsilon_{n+1}}{\frac{\partial F_{n+1}\left(\sigma_{n+1}^{trial}, \overline{\varepsilon_{n+1}^{p,trial}}\right)}{\partial \sigma_{n+1}^{trial}} : C : \frac{\partial F_{n+1}\left(\sigma_{n+1}^{trial}, \overline{\varepsilon_{n+1}^{p,trial}}\right)}{\partial \sigma_{n+1}^{trial}} - \frac{\partial F_{n+1}\left(\sigma_{n+1}^{trial}, \overline{\varepsilon_{n+1}^{p,trial}}\right)}{\partial \overline{\varepsilon_{n+1}^{p,trial}}}}.$$

(b) Update the plastic strain tensor and the equivalent plastic strain:

$$\varepsilon_{n+1}^{p} = \varepsilon_{n}^{p} + \Delta\lambda_{n+1} \frac{\partial F_{n+1}\left(\sigma_{n+1}^{trial}, \overline{\varepsilon_{n+1}^{p,trial}}\right)}{\partial \sigma_{n+1}^{trial}}, \overline{\varepsilon_{n+1}^{p}} = \overline{\varepsilon_{n}^{p}} + \Delta\lambda_{n+1}.$$

(c) Update the stress tensor:

$$\sigma_{n+1} = C : \left(\varepsilon_{n+1} - \varepsilon_{n+1}^{p}\right).$$

(5) Judge whether damage has been initiated and calculate the damage variable.

(a) Judge the positive and negative values of σ_{11}, and judge whether fiber damage has been initiated according to different fiber damage initiation criteria. If fiber damage occurs, the corresponding fiber damage variable d_f is calculated. Otherwise, $d_f = 0$.

(b) The stress is rotated to the fracture surface to determine whether the tractions on the fracture surface satisfy the damage initial criteria of the matrix. If matrix damage occurs, the corresponding matrix damage variable d_m is calculated. Otherwise, $d_m = 0$.

(6) Update the stress according to the calculated fiber damage variable and the matrix damage variable and skip to the next step.

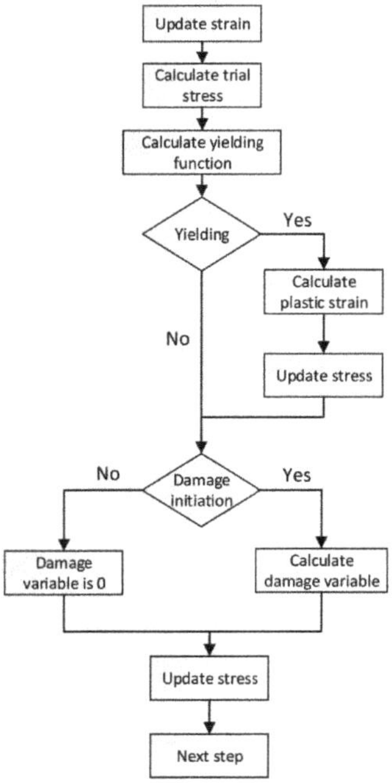

Figure 7. Calculation process of the 3D elastoplastic damage model.

3.3. Finite Element Model

The material properties, failure strength, and fracture energy of AS4/PEEK are shown in Tables 2 and 3 [45,51,53,54]. These material parameters were determined by a series of mechanical tests, including tensile tests, compression tests, off-axis tensile tests, and double cantilever beam tests. The elastic modulus and the Poisson ratio of the countersunk bolt were 110,000 MPa and 0.3, respectively, and the plastic deformation of the countersunk bolt was not considered in this paper.

Table 2. Material properties of AS4/PEEK.

E_1 (MPa)	E_2 (MPa)	E_3 (MPa)	G_{12} (MPa)	G_{13} (MPa)	G_{23} (MPa)
127,000	10,300	10,300	6000	6000	3450
v_{12}	v_{13}	v_{23}	a_{66}	β	n
0.32	0.32	0.49	1.50	292.67	0.135

Table 3. Strength and fracture energy of AS4/PEEK.

X_T (MPa)	X_C (MPa)	Y_T (MPa)	Y_c (MPa)	S_L (MPa)	G_{Ic} (kJ·m^{-2})
2023	1234	92.7	176	82.6	1.7
G_{IIc}(kJ·m^{-2})	G_{ft}(kJ·m^{-2})	G_{fc}(kJ·m^{-2})			
2.0	218	104			

The numerical simulation of tensile tests of AS4/PEEK laminates with holes was performed to investigate the effect of the mesh size on the 3D elastoplastic model. The test results data were obtained from Ref. [55], and there were three models in all. All models used linear, eight-node, three-dimensional, reduced-integration (C3D8R) element, and the total number of meshes were 2816, 9088, and 70,400, respectively. The model's dimension and boundary conditions were consistent with those of the test. As illustrated in Figure 8, the load–displacement curves acquired by the numerical simulation were compared to those obtained through testing. The stiffness and strength of the three models were close to the experimental values, suggesting that the 3D elastoplastic model is less dependent on mesh size. The above conclusions remained unchanged as the mesh was refined, which indicates that the influence of mesh size on calculation results can be ignored within a reasonable range of mesh size.

To reduce calculation time, the finite element model of the countersunk bolted joint did not consider the pads of the test specimens. We also ignored the effect of the screw thread. Thus, the countersunk bolt and nut were built as a whole model. The simplified finite element model is shown in Figure 9a, where the x–y–z coordinate system is the global coordinate system and the x-direction is parallel to the fiber direction. The analysis procedure was divided into two steps, where the first step implemented the tightening torque applied to the bolt. The tightening torques were 1.8 N·m, 3.6 N·m, 5.4 N·m, and 1.8 N·m, corresponding to the tightening torque of the test specimens. The bolt-tightening torque was simulated by applying an equivalent pressure load on both sides of the bolt. The normal pressure (p_w) acting on the bolt surface was by the equation $p_w = T/(\Lambda \times D \times A)$, where T is the applied bolt torque; Λ is the torque coefficient, which is commonly assumed to be 0.2 [56]; and D and A are the diameter and the area, respectively, of the upper and lower surfaces of the countersunk bolt. In the second step, the left end of the model was held fixed, while a 5 mm displacement with a smooth step amplitude curve along the x-displacements was applied to the right hand. The step time was by $t = 10/f^{1st}$, where f^{1st} is the fundamental frequency of the laminate [57]. The frequency analysis showed that the laminates had a fundamental frequency of 1116 Hz, and it was appropriate to set the time step to 0.01 s in both steps.

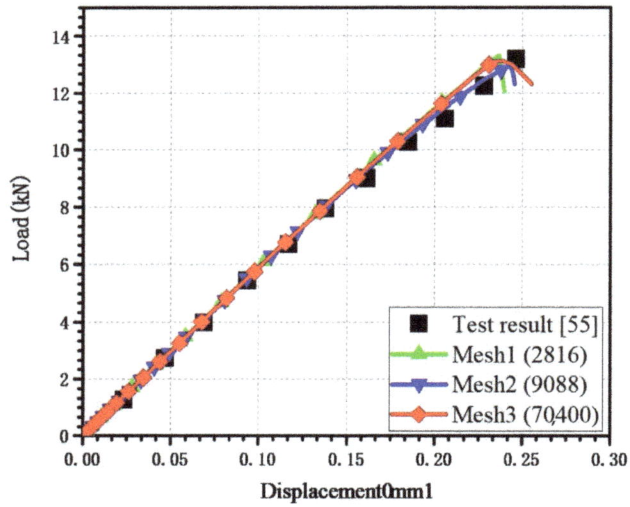

Figure 8. Comparison of load–displacement curves with different mesh densities.

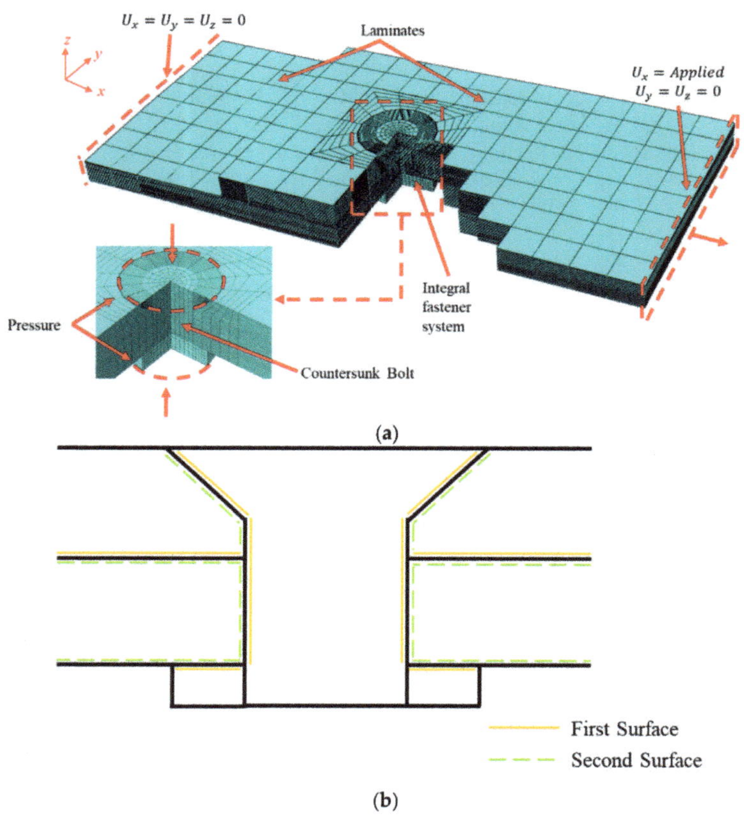

Figure 9. (**a**) Finite element model of single-lap thermoplastic composite joints bolted by countersunk with boundary conditions. (**b**) Surface definition for contact pairs around the countersunk bolt.

The contact properties included the normal behavior and the tangential behavior; the normal behavior uses hard contact, whereas the friction formulation in the tangential behavior uses the penalty formulation. Contact surfaces included multiple assembly-level interactions that used the friction coefficient of laminate–laminate interaction ($\mu = 0.7$) and bolt–laminate interaction ($\mu = 0.1$) [58]. The setting of the contact pair is shown in Figure 9b. The first surface and the second surface were the contact surfaces used in the general contact. The general contact algorithm of ABAQUS/Explicit uses a finite-sliding formulation and enforced-penalty-based, stick–slip interactions. The bolt-tightening torque was simulated by applying an equivalent pressure load on both sides of the bolt. In the model, the clearances between the laminate and the bolt were 0 μm, 60 μm, 120 μm, and 180 μm, which are 0%, 1%, 2%, and 3% of the bolt diameter, respectively, and 0 μm corresponds to the clearance of the test specimens. The above models with different clearances were named C0, C1, C2, and C3, respectively. The computation accuracy was set as "double precession" to reduce the accumulation of errors when running the simulation. The mesh type was C3D8R for the bolt and laminates, and each layer was separated into its elements in the through-thickness direction to get the stress distribution of and damage to each layer. At the same time, to improve the calculation accuracy of local stress, the mesh was refined at the hole edge. The mesh size of the bolt surface was consistent with that of the hole edge of the laminate in order to make the mesh of the bolt correspond to that of the hole edge of the laminate. The total mesh number was 23,340.

4. Numerical Simulation Results

4.1. Model Verification

A comparison of load–displacement curves of the single-lap countersunk bolted joints with three different stacking sequences obtained from simulation and tests is shown in Figure 10. It can be seen that the load–displacement curves obtained by numerical simulation were in good agreement with those from the tests, which indicates that the model is suitable for simulating mechanical behaviors of AS4/PEEK laminate single-lap countersunk bolted joints under tensile load.

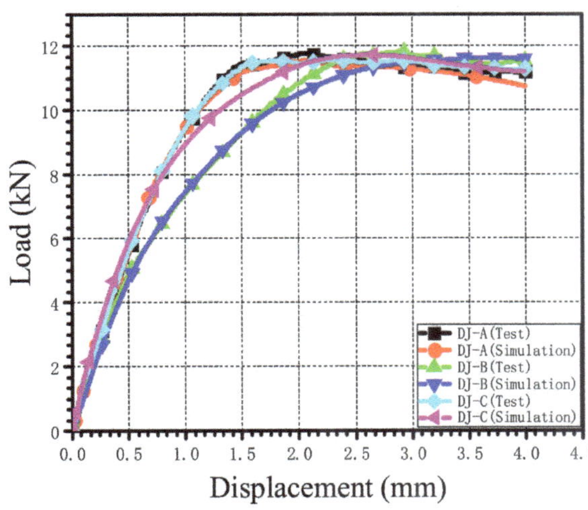

Figure 10. Comparison of load–displacement curves between numerical simulation and the test.

A comparison of the ultimate bearing strength and the 2% offset bearing strength obtained by numerical simulation and test is summarized in Table 4. The maximum error of the ultimate bearing strength and the 2% offset bearing strength was 5.23%, indicating that the model can accurately calculate the ultimate bearing strength and the 2% offset bearing strength of single-lap countersunk bolted joints of AS4/PEEK laminates under tensile load.

Taking the DJ-A model as an example, the displacements obtained from the numerical simulation and the specimen after the test are shown in Figure 11. The bending moment generated by the eccentric load path caused the laminates to bend, which was well simulated by the numerical simulation. As shown in the numerical simulation results, the highest displacement occurred near the left end of the lower laminate, which is consistent with the test results. The damage index and the damage variable are shown in Figures 12 and 13, respectively, where SDV29 is the matrix damage variable. The comparison in Figure 13 shows that the numerical simulation was accurate in simulating the damage to the countersunk bolted joint. The damage in the laminates was mainly matrix damage, and the matrix compressive damage mainly occurred in the upper part of the hole edge of the laminates, while the matrix tensile damage was distributed in the upper and lower parts of the hole edge. The fiber damage mainly occurred in the upper part of the hole edge of the laminates and was mainly fiber compressive damage.

Table 4. Comparison of the test and numerical simulation.

	Ultimate Bearing Strength (MPa)			2% Offset Bearing Strength (MPa)		
	Test	Simulation	Error *	Test	Simulation	Error *
DJ-A	560.50	578.94	−3.29%	542.20	570.58	−5.23%
DJ-B	562.36	546.83	2.76%	534.33	542.29	−1.49%
DJ-C	564.40	557.93	1.15%	561.40	557.67	0.66%

* Error = (Test − Simulation)/Test × 100%.

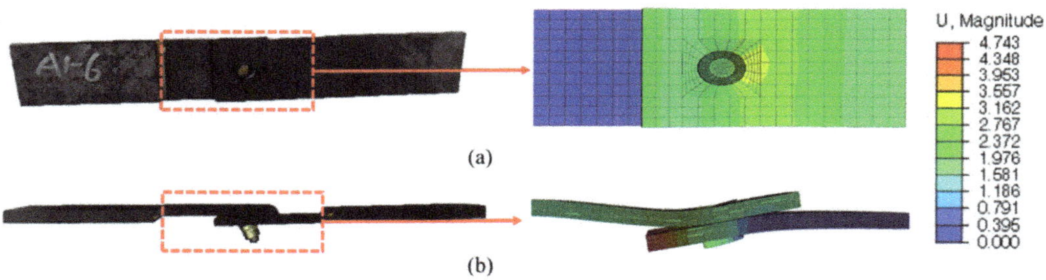

Figure 11. Comparison of displacement between the test specimen and numerical simulation: (**a**) nail-head side and (**b**) lateral side.

For ease of description, as illustrated in Figure 14, a polar coordinate system and the bolt–hole contact angle (θ) were defined to present the countersunk hole boundary stresses, which included the radial (r) stress, the tangential (t) stress, and the through-thickness (z) stress. The laminate plies were numbered 1–25, beginning at the top surface. The stresses were extracted from the integration points closest to the hole edge at an applied load level of 6 kN. The load level belonged to the elastoplastic region to ensure the hole was undamaged. The radial peak stress, the tangential peak stress, the through-thickness peak stress, and the ply number at different ply angles are summarized in Table 5, and the stress distribution of the corresponding layers is plotted in Figure 15. The peak values of radial stress and tangential stress mainly appeared in the 0° layer, while the peak values of radial stress and tangential stress in the 90° layer were the smallest. Therefore, we can conclude that the external load is mainly borne by the 0° layer, followed by the ±45° layer, and the load borne by the 90° layer is the smallest.

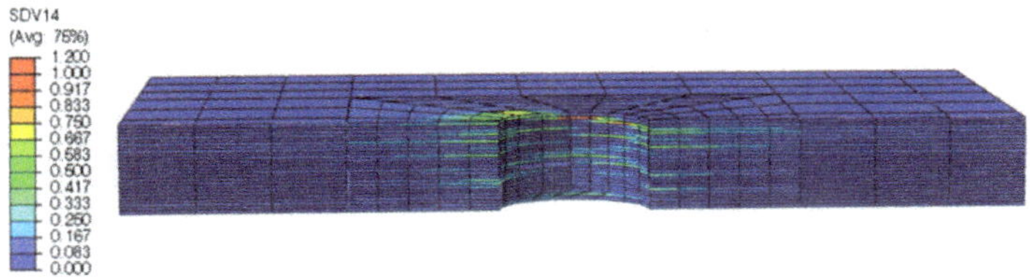

(**a**)

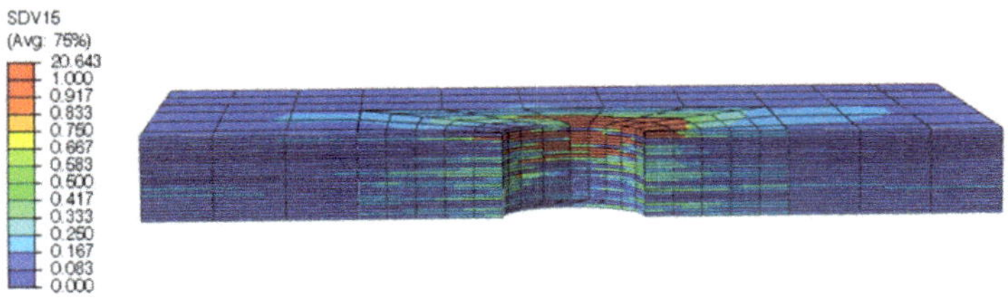

(**b**)

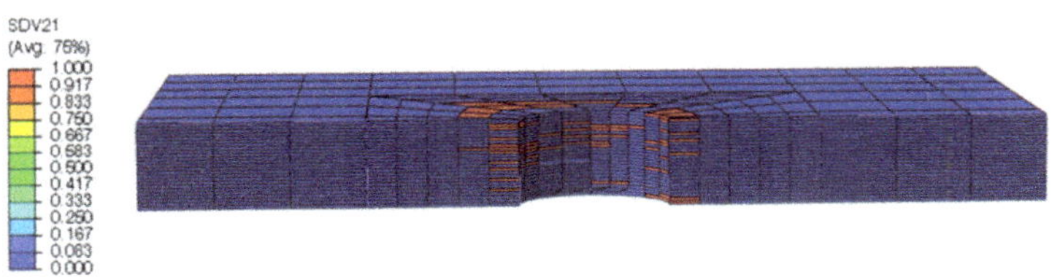

(**c**)

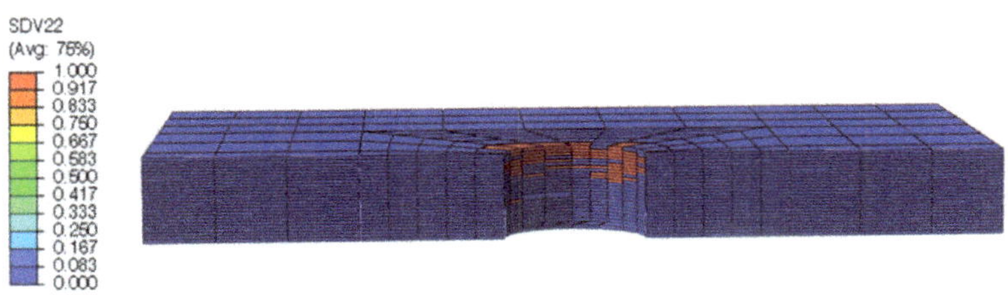

(**d**)

Figure 12. Damage index of the lower laminate around the hole. Damage occurs when the damage index is greater than or equal to 1. (**a**) Fiber tensile damage index, (**b**) fiber compressive damage index, (**c**) matrix tensile damage index, and (**d**) matrix compressive damage index.

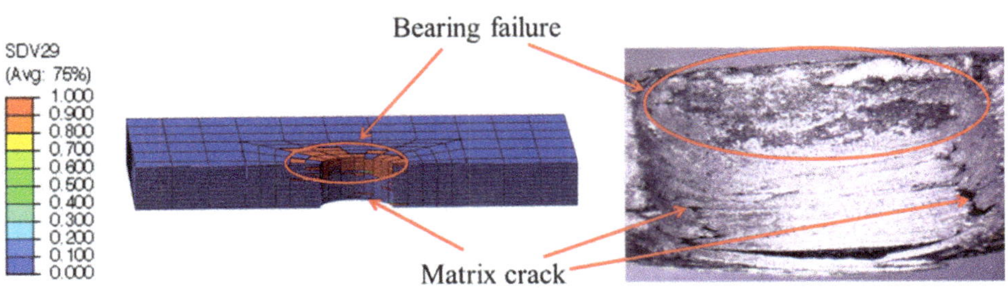

Figure 13. Comparison of damage modes between numerical simulation and the test.

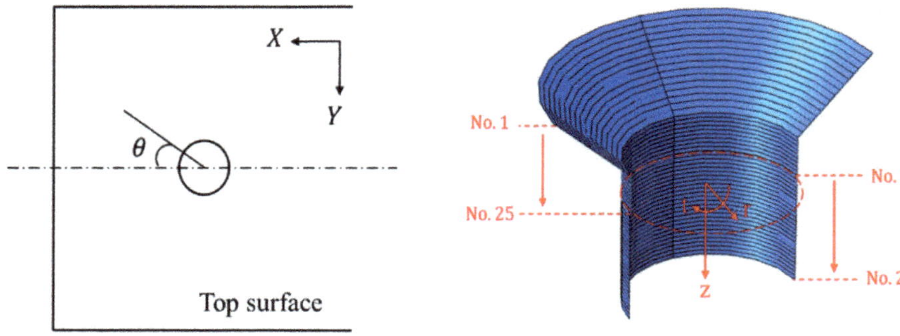

(a) (b)

Figure 14. Definition of a stress field around the bolt hole: (**a**) bolt–hole contact angle definition and (**b**) ply number definition.

Table 5. Peak stress of each ply angle and the number of each layer in the DJ-A model under 6 kN external load. The clearance is 0 μm, and the bolt-tightening torque is 1.8 N·m.

	Ply Angle	Radial Peak Stress (MPa)	Ply Number	Tangential Peak Stress (MPa)	Ply Number	Through-Thickness Peak Stress (MPa)	Ply Number
Upper laminate	−45°	−1253.22	No.21	1018.05	No.21	−104.62	No.17
	90°	−547.24	No.24	673.80	No.24	−67.76	No.2
	45°	−1370.35	No.23	1028.53	No.23	−93.54	No.3
	0°	−1367.04	No.22	1236.67	No.22	−123.04	No.16
Lower laminate	−45°	−1007.24	No.5	635.70	No.5	−52.66	No.25
	90°	−468.67	No.2	517.79	No.2	−47.93	No.13
	45°	−997.04	No.3	701.79	No.7	−52.80	No.11
	0°	−1053.13	No.4	835.62	No.3	−51.74	No.4

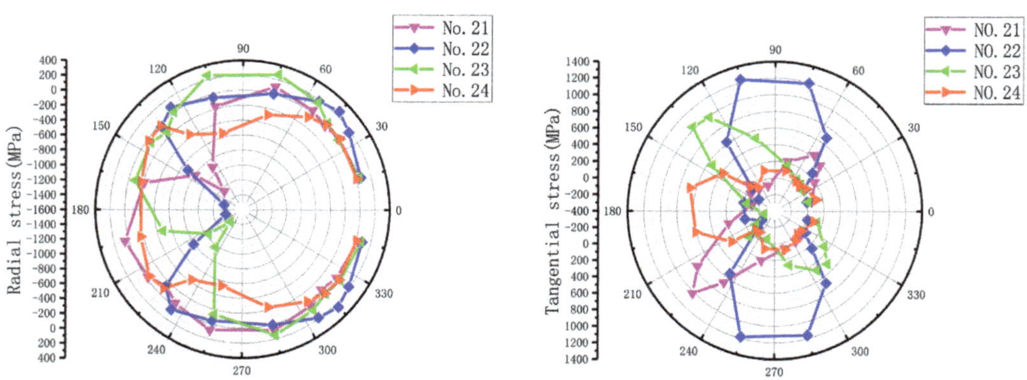

(a) (b)

Figure 15. *Cont.*

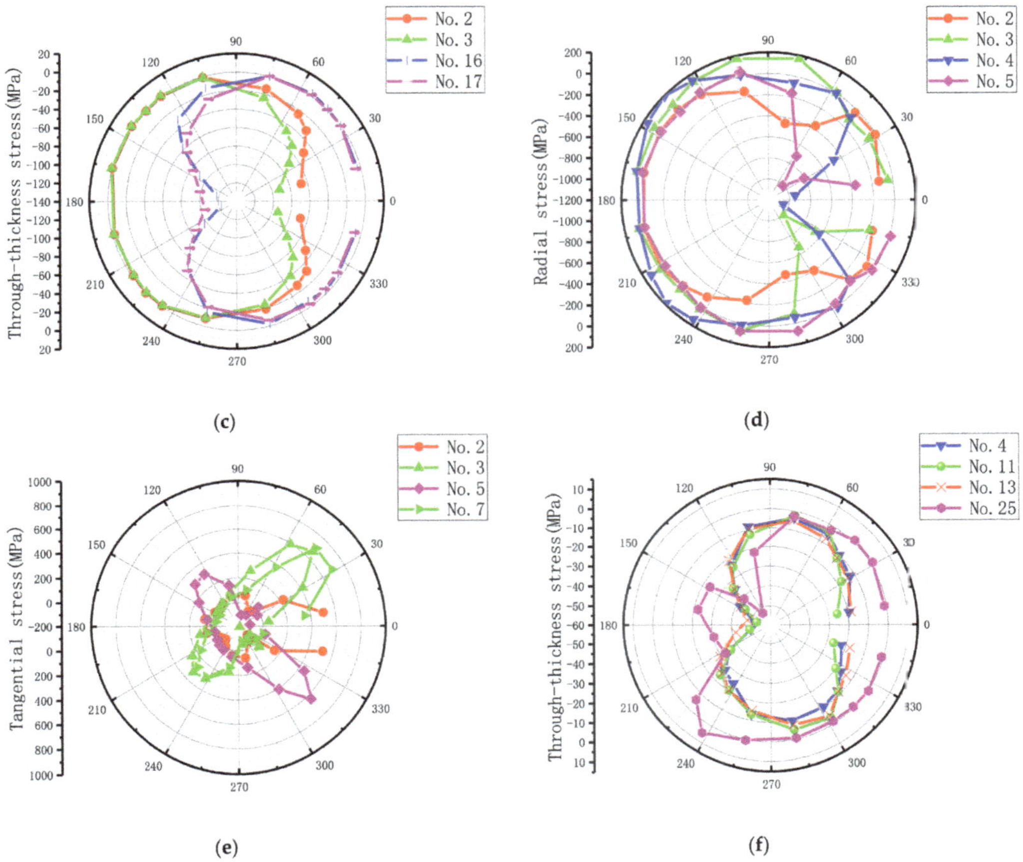

Figure 15. Stress distribution of the DJ-A model under 6 kN external load. (**a**) Radial stress in the upper plate, (**b**) tangential stress in upper laminates, (**c**) through-thickness stress in upper laminates, (**d**) radial stress in lower laminates, (**e**) tangential stress in lower laminates, and (**f**) through-thickness stress in lower laminates.

From the stress distribution in Figure 15, it can be clearly seen that the peak value of radial stress appeared in the position parallel to the fiber direction and was dominated by compressive stress. The radial stress distribution range around the hole circumference, which represents the bolt–hole contact arc, was approximately a semicircle. This was due to the absence of the bolt–hole clearance (0 μm) in this model. The peak tangential stress emerged at the location tangent to the fiber direction, which is due to the plies being the stiffest at these locations [13]. The peak tangential stress was predominantly tensile stress, but tangential compressive stresses existed at some sites owing to hole oval deformation, as found by Ref. [13]. The radial peak stress and the tangential peak stress were mainly distributed in the middle part of the countersunk bolted joint (i.e., the lower part of the upper plate and the upper part of the lower plate), which is in agreement with the results of Egan et al. [59] and Zhai et al. [60]. For the peak value of through-thickness stress, its value was only about 5–15% of the peak values of radial stress and tangential stress, and it was mainly compressive stress. This is because the through-thickness stress is mainly generated by the bolt-tightening torque.

4.2. Influence of Bolt-Tightening Torque

By changing the bolt-tightening torque applied on the bolt, the load–displacement curve obtained by a numerical simulation is shown in Figure 16. The ultimate bearing strength and the 2% offset bearing strength obtained by numerical simulations under different tightening torques are plotted in Figure 17. When the bolt-tightening torque increased from 1.8 N·m to 5.4 N·m, the ultimate bearing strength decreased by 6.34%, while the 2% offset bearing strength decreased by 8.01%.

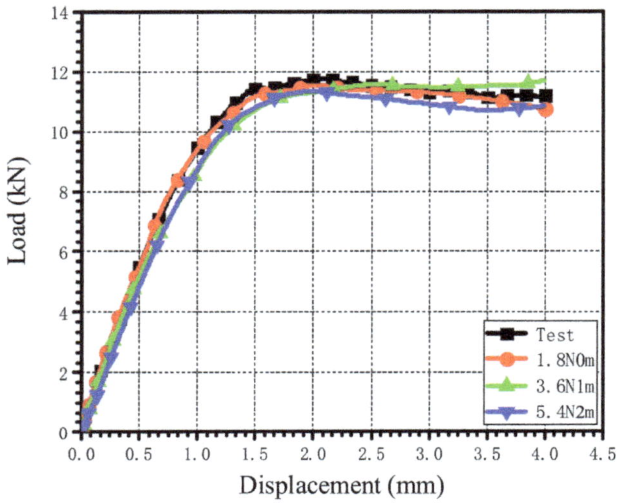

Figure 16. Load-displacement curves under different bolt-tightening torques when the bolt-hole clearance is 0 μm.

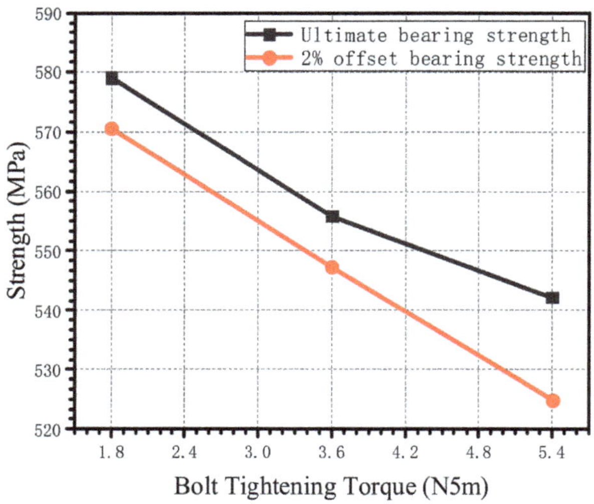

Figure 17. Ultimate bearing strength and 2% offset bearing strength under different bolt-tightening torques when the bolt–hole clearance is 0 μm.

The integral point stress on the laminate near the countersunk bolt was extracted when the bolt-tightening torque was 5.4 N·m and the external load was 6 kN. The peak stress at different ply angles and the layer number obtained by a numerical simulation are summarized in Table 6, and the stress distribution of the above layers is plotted in Figure 18. Compared to Figure 15, it can be clearly seen that increasing the external load did not change the stress distribution trend but changed the peak stress. The difference in peak stresses is shown in Table 7. It shows that when the bolt-tightening torque increased from 1.8 N·m to 5.4 N·m, the peak values of radial stress and tangential stress in the upper plate changed little, while in the lower plate, they changed greatly; the peak values decreased, and the maximum decline reached 18.92%. For the through-thickness peak stress, the upper laminate increased except for 0° ply, and the maximum increase reached 27.70%, while in the lower laminate, it decreased, and the maximum decrease reached 10.42%. Therefore, increasing the bolt-tightening torque within a certain range can effectively reduce the peak radial stress and the peak tangential stress. However, the increased through-thickness stress leads to laminate damage under low external load, resulting in a slight decrease in the ultimate bearing strength and the 2% offset bearing strength.

Table 6. Peak stress of each ply angle and ply number when the external load is 6 kN, the clearance remains 0 μm, and the bolt-tightening torque is 5.4 N·m.

	Ply Angle	Radial Peak Stress (MPa)	Ply Number	Tangential Peak Stress (Mpa)	Ply Number	Through-Thickness Peak Stress (Mpa)	Ply Number
Upper laminate	−45°	−1216.93	No.21	1013.53	No.21	−114.00	No.17
	90°	−532.98	No.24	700.77	No.24	−85.18	No.2
	45°	−1369.65	No.23	1039.45	No.23	−119.45	No.3
	0°	−1385.49	No.22	1266.79	No.22	−119.62	No.16
Lower laminate	−45°	−920.24	No.5	576.90	No.5	−52.22	No.25
	90°	−453.72	No.2	419.61	No.2	−45.34	No.13
	45°	−917.18	No.3	652.02	No.7	−47.30	No.7
	0°	−958.37	No.4	742.63	No.4	−49.03	No.4

Table 7. Comparison of the peak stress at a bolt-tightening torque of 5.4 N·m without clearance and at a bolt-tightening torque of 1.8 N·m with an external load of 6 kN.

	Ply Angle	Radial Peak Stress Difference	Tangential Peak Stress Difference	Through-Thickness Peak Stress Difference
Upper laminate	−45°	−2.90%	−0.44%	8.97%
	90°	−2.61%	4.00%	25.71%
	45°	−0.05%	1.06%	27.70%
	0°	1.35%	2.44%	−2.78%
Lower laminate	−45°	−8.64%	−9.25%	−0.84%
	90°	−3.19%	−18.92%	−5.40%
	45°	−8.01%	−7.09%	−10.42%
	0°	−9.00%	−11.13%	−5.24%

4.3. Influence of Clearance

The load–displacement curves under different clearances are shown in Figure 19a To study the influence of the clearance on the countersunk bolted joint when the displacement is small, the load–displacement curve when the displacement is less than 0.14 mm is plotted in Figure 19b. As shown in Figure 20, for the joint with clearances, it is obvious that there were three regions: the friction region, the transition region, and the load take-up region. The friction region was mainly produced by the friction force between the laminates; when the external load increased to a certain value, the upper and lower laminated plates began to slide rigidly, the slope was small, and the rigid displacement distance was about half of the clearance length. After the transition region, the laminates and the countersunk bolts were in complete contact, and the load began to increase gradually. The ultimate bearing strength and the 2% offset bearing strength corresponding to different clearances are plotted in Figure 21. When the clearance increased from 0 μm to 60 μm, the decrease in the ultimate bearing strength and the 2% offset bearing strength was the largest, at 11.47% and 10.78%, respectively. With the increase in the clearance, the ultimate bearing strength increased gradually and the 2% offset bearing strength increased first and then decreased.

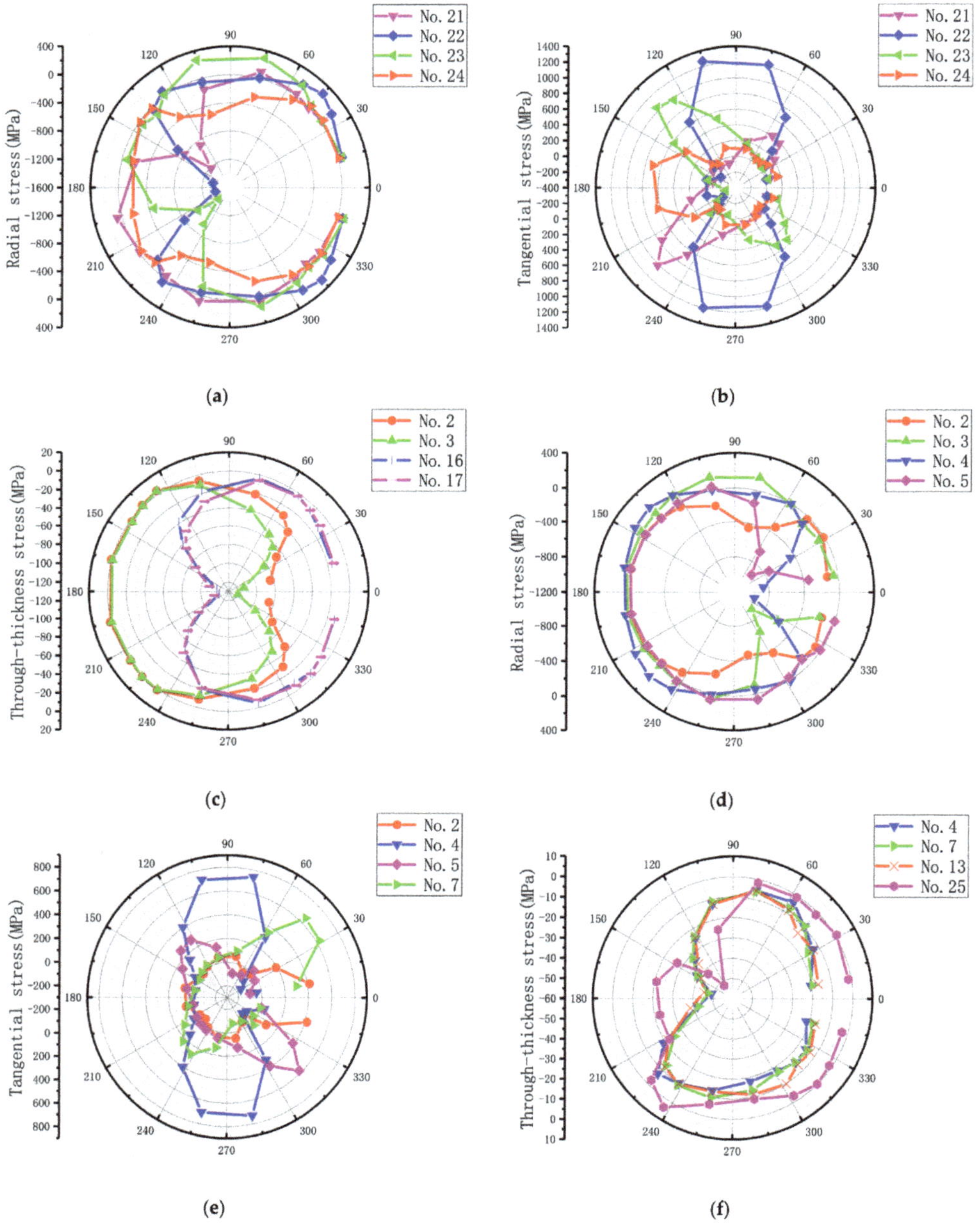

Figure 18. Stress distribution when the external load is 6 kN, the clearance remains 0 μm, and the bolt-tightening torque is 5.4 N·m. (**a**) Radial stress in the upper plate, (**b**) tangential stress in upper laminates, (**c**) through-thickness stress in upper laminates, (**d**) radial stress in lower laminates, (**e**) tangential stress in lower laminates, and (**f**) through-thickness stress in lower laminates.

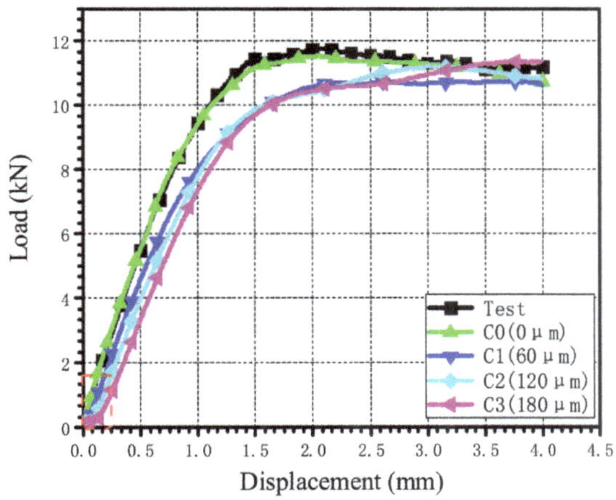

(**a**)

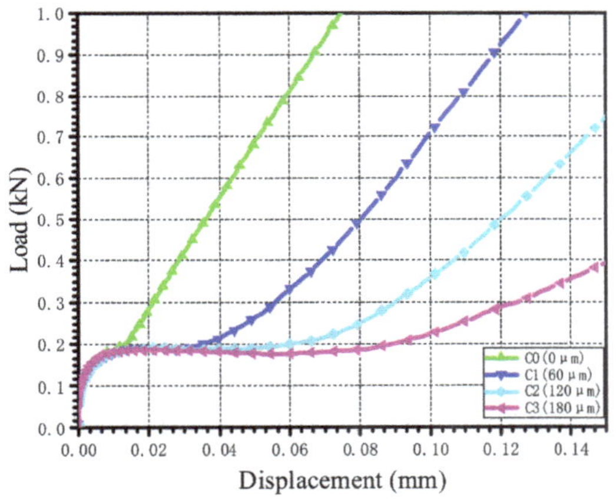

(**b**)

Figure 19. (**a**) Load–displacement curves under different clearances. (**b**) Local amplification of load–displacement curves under different clearances.

When the clearance was 180 μm and the external load was 6 kN, the integral point stresses on the laminate near the countersunk bolt were extracted. The peak stress of different ply angles and the corresponding ply number are summarized in Table 8, and the stress distribution of the above layer is drawn in Figure 22. In comparison to Figure 15, the peak stress positions and values changed; the radial stress distribution range in the counter-clockwise direction was approximately 105–255° in the upper laminate and 300–360° and 0–60° in the lower laminate; the contact area decreased by 16.7% in the upper laminate and 33.3% in the lower laminate. A decrease in the contact area led to an increase in the radial peak stress. This finding is consistent with that of Ref. [13]. The peak stress difference between the model and the model without clearance is summarized in Table 9. The peak value of radial stress increased except in the −45° layer of the upper layer and the 90° layer of the lower layer, and the maximum increase reached 90.94%. The tangential stress increased except in the −45° layer, and the maximum increase was 70.74%. For the peak stress in the through-thickness direction, except for the 45° layer of the upper laminate, the peak stress decreased and the maximum decrease was 57.33%. The decrease in the lower laminate was larger than that in the upper laminate. Therefore, the main influence of clearance on the single-lap countersunk bolted joint is to increase the peak values of radial stress and tangential stress, and it has a certain mitigation effect on the peak value of the through-thickness stress.

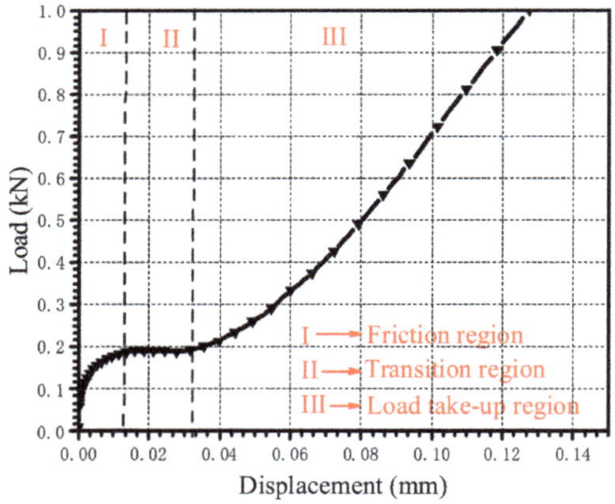

Figure 20. Three regions of the load–displacement curve.

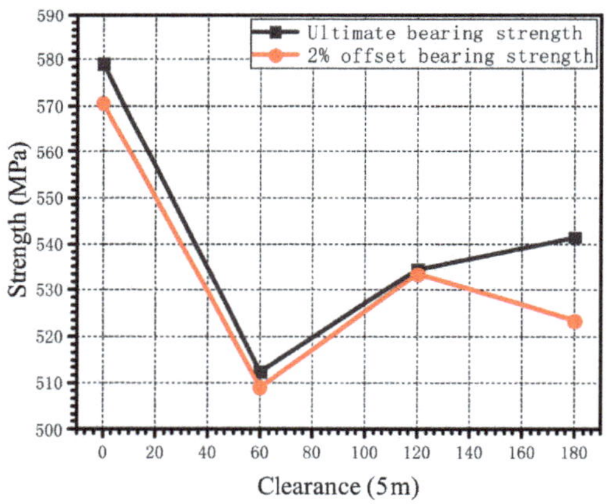

Figure 21. Ultimate bearing strength and 2% offset bearing strength under different clearances.

Table 8. Peak stress of each ply angle and ply number when the external load is 6 kN, the bolt-tightening torque remains 1.8 N·m, and the clearance is 180 μm.

	Ply Angle	Radial Peak Stress (MPa)	Ply Number	Tangential Peak Stress (MPa)	Ply Number	Through-Thickness Peak Stress (MPa)	Ply Number
Upper laminate	−45°	−1164.09	No.25	777.15	No.25	−86.44	No.17
	90°	−1044.92	No.24	913.51	No.24	−67.61	No.2
	45°	−2341.32	No.23	1756.09	No.23	−99.53	No.3
	0°	−2079.97	No.22	1377.03	No.22	−103.21	No.18
Lower laminate	−45°	−1057.99	No.1	613.82	No.5	−40.86	No.1
	90°	−311.20	No.2	537.24	No.2	−20.45	No.13
	45°	−1053.36	No.3	847.04	No.3	−29.16	No.7
	0°	−1449.25	No.4	1025.82	No.4	−26.81	No.10

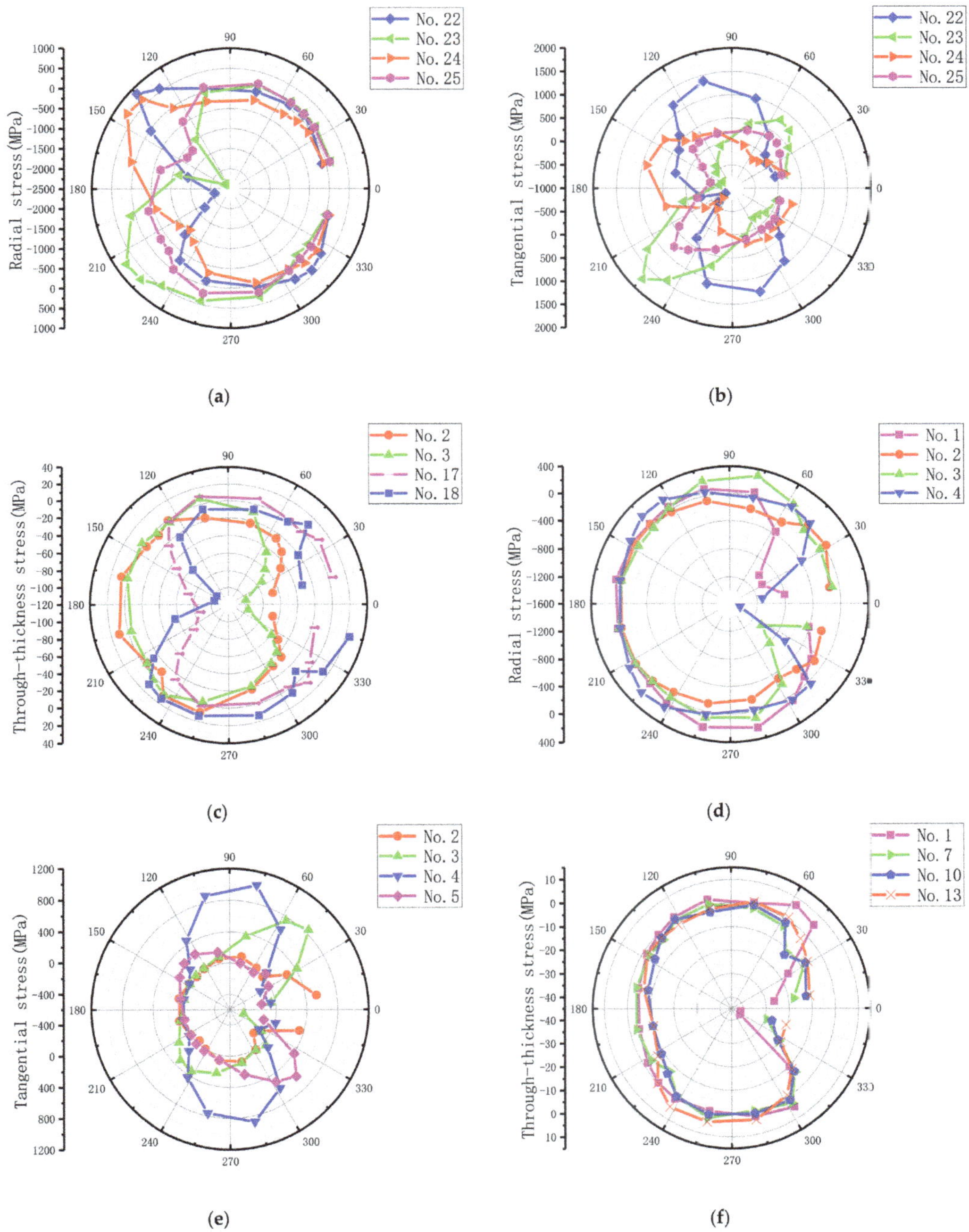

Figure 22. Stress distribution when the external load is 6 kN, the bolt-tightening torque remains 1.8 N·m, and the clearance is 180 µm. (**a**) Radial stress in the upper plate, (**b**) tangential stress in upper laminates, (**c**) through-thickness stress in upper laminates, (**d**) radial stress in lower laminates, (**e**) tangential stress in lower laminates, and (**f**) through-thickness stress in lower laminates.

Table 9. Comparison of peak stresses between 180 μm clearance and 0 μm clearance under 6 kN external load and constant bolt-tightening torque.

	Ply Angle	Radial Peak Stress Difference	Tangential Peak Stress Difference	Through-Thickness Peak Stress Difference
Upper laminate	−45°	−7.11%	−23.66%	−17.38%
	90°	90.94%	35.58%	−0.22%
	45°	70.86%	70.74%	6.40%
	0°	52.15%	11.35%	−16.12%
Lower laminate	−45°	5.04%	−3.44%	−22.41%
	90°	−33.60%	3.81%	−57.33%
	45°	5.65%	20.70%	−44.77%
	0°	37.61%	22.76%	−48.18%

5. Conclusions

In this paper, tensile tests were performed on single-lap countersunk bolt thermoplastic joints where AS4/PEEK laminates were stacked in three sequences. The ultimate failure of all test specimens occurred in the laminates around the countersunk bolt, and the effect of stacking sequences on both ultimate bearing strength and 2% offset bearing strength was limited. Furthermore, a 3D elastoplastic damage model was established and used in the numerical simulation of single-lap countersunk bolted joints. The numerical method of stiffness estimation was validated by comparing with the test stress–strain curves. The maximum error of predicted strength was only 5.23%, which proves that the present FEM can also accurately simulate the ultimate strength of AS4/PEEK laminate joints bolted by countersunk.

The proposed numerical model was used to analyze the effects of bolt-tightening torque and clearance on ultimate bearing strength and 2% offset bearing strength. When the bolt-tightening torque was increased from 1.8 N·m to 5.4 N·m, the ultimate bearing strength decreased by 6.34%, and the 2% offset bearing strength decreased by 8.01%. The analysis of the peak stress at different ply angles shows that increasing the bolt-tightening torque within a certain range can remarkably reduce the peak values of radial stress and tangential stress and the decrease in ultimate bearing strength and 2% offset bearing strength is mainly due to the increase in the through-thickness peak stress. When the clearance was increased from 0 μm to 60 μm with a constant bolt-tightening torque of 1.8 N·m, the ultimate bearing strength and the 2% offset bearing strength fell by 11.47% and 10.78%, to a minimum value, respectively. However, when the clearance was enlarged, the trend of ultimate bearing strength and 2% offset bearing strength was different: the ultimate bearing strength increased steadily, but the 2% offset bearing strength increased first and then decreased. In addition, the numerical results show that the size of the clearance will affect the value and position of peak radial and tangential stresses, which is mainly due to changes in the contact position and the contact area.

Author Contributions: Conceptualization, X.C., A.T. and Y.F.; methodology, X.C., Y.F. and J.Z.; software, X.C. and J.Z.; validation, X.C., A.T. and Y.F.; formal analysis, J.Z.; investigation, J.Z.; resources, X.C. and A.T.; data curation, J.Z.; writing—original draft preparation, J.Z.; writing—review and editing, X.C., Y.F. and A.T.; visualization, J.Z.; supervision, X.C., A.T. and Y.F.; project administration, X.C.; funding acquisition, X.C. and A.T. All authors have read and agreed to the published version of the manuscript.

Funding: This research received no external funding.

Institutional Review Board Statement: Not applicable.

Informed Consent Statement: Not applicable.

Data Availability Statement: The data presented in this study are available on request from the corresponding author.

Acknowledgments: The authors gratefully acknowledge Hui-Shen Shen in Shanghai Jiao Tong University for his guidance in the numerical simulation.

Conflicts of Interest: The authors declare no conflict of interest.

References

1. Hron, R.; Kadlec, M.; Růžek, R. Effect of the test procedure and thermoplastic composite resin type on the curved beam strength. *Materials* **2021**, *14*, 352. [CrossRef] [PubMed]
2. Lambiase, F.; Scipioni, S.I.; Lee, C.-J.; Ko, D.-C.; Liu, F. A state-of-the-art review on advanced joining processes for metal-composite and metal-polymer hybrid structures. *Materials* **2021**, *14*, 1890. [CrossRef] [PubMed]
3. Yao, S.-S.; Jin, F.-L.; Rhee, K.Y.; Hui, D.; Park, S.-J. Recent advances in carbon-fiber-reinforced thermoplastic composites: A review. *Compos. Part B Eng.* **2018**, *142*, 241–250. [CrossRef]
4. Yao, C.; Qi, Z.; Chen, W.; Zhang, C. Experimental study on CF/PEEK thermoplastic fastener: Effects of fastener matrix crystallinity and fibre content on the strength of single-lap joint. *Compos. Part B Eng.* **2021**, *213*, 108737. [CrossRef]
5. Zou, H.; Yin, W.; Cai, C.; Wang, B.; Liu, A.; Yang, Z.; Li, Y.; He, X. The out-of-plane compression behavior of cross-ply AS4/PEEK thermoplastic composite laminates at high strain rates. *Materials* **2018**, *11*, 2312. [CrossRef]
6. Wang, Y.; Li, Y.; Huan, D.; Li, L.; Liu, L.; Cheng, Y.; Liu, H. Experimental study and finite element analysis on the joining performance of GF/PP rivets. *J. Thermoplast. Compos. Mater.* **2020**, 0892705720939142. [CrossRef]
7. Yýlmaz, T.; Sýnmazçelik, T. Investigation of load bearing performances of pin connected carbon/polyphenylene sulphide composites under static loading conditions. *Mater. Des.* **2007**, *28*, 520–527. [CrossRef]
8. Thoppul, S.D.; Finegan, J.; Gibson, R.F. Mechanics of mechanically fastened joints in polymer–matrix composite structures—A review. *Compos. Sci. Technol.* **2009**, *69*, 301–329. [CrossRef]
9. Quinn, W.; Matthews, F. The effect of stacking sequence on the pin-bearing strength in glass fibre reinforced plastic. *J. Compos. Mater.* **1977**, *11*, 139–145. [CrossRef]
10. Wang, H.-S.; Hung, C.-L.; Chang, F.-K. Bearing failure of bolted composite joints. Part I: Experimental characterization. *J. Compos. Mater.* **1996**, *30*, 1284–1313. [CrossRef]
11. Park, H.-J. Effects of stacking sequence and clamping force on the bearing strengths of mechanically fastened joints in composite laminates. *Compos. Struct.* **2001**, *53*, 213–221. [CrossRef]
12. McCarthy, M.; Lawlor, V.; Stanley, W.; McCarthy, C. Bolt-hole clearance effects and strength criteria in single-bolt, single-lap, composite bolted joints. *Compos. Sci. Technol.* **2002**, *62*, 1415–1431. [CrossRef]
13. McCarthy, C.T.; McCarthy, M. Three-dimensional finite element analysis of single-bolt, single-lap composite bolted joints: Part II—Effects of bolt-hole clearance. *Compos. Struct.* **2005**, *71*, 159–175. [CrossRef]
14. McCarthy, C.; Gray, P. An analytical model for the prediction of load distribution in highly torqued multi-bolt composite joints. *Compos. Struct.* **2011**, *93*, 287–298. [CrossRef]
15. Khashaba, U.; Sallam, H.; Al-Shorbagy, A.; Seif, M. Effect of washer size and tightening torque on the performance of bolted joints in composite structures. *Compos. Struct.* **2006**, *73*, 310–317. [CrossRef]
16. Khashaba, U.; Sebaey, T.; Alnefaie, K. Failure and reliability analysis of pinned-joint composite laminates: Effects of pin–hole clearance. *J. Compos. Mater.* **2013**, *47*, 2287–2298. [CrossRef]
17. Khashaba, U.; Sebaey, T.; Alnefaie, K. Failure and reliability analysis of pinned-joints composite laminates: Effects of stacking sequences. *Compos. Part B Eng.* **2013**, *45*, 1694–1703. [CrossRef]
18. Hassan, J.; O'Higgins, R.M.; Feser, T.; Waimer, M.; McCarthy, C.T.; Toso, N.; McCarthy, M.A. Influence of layup, stacking sequence and loading rate on energy absorption of tension-absorber joints. *Compos. Struct.* **2021**, *261*, 113327. [CrossRef]
19. Giannopoulos, I.K.; Doroni-Dawes, D.; Kourousis, K.I.; Yasaee, M. Effects of bolt torque tightening on the strength and fatigue life of airframe FRP laminate bolted joints. *Compos. Part B Eng.* **2017**, *125*, 19–26. [CrossRef]
20. Mandal, B.; Chakrabarti, A. Numerical failure assessment of multi-bolt FRP composite joints with varying sizes and preloads of bolts. *Compos. Struct.* **2018**, *187*, 169–178. [CrossRef]
21. Kelly, G.; Hallström, S. Bearing strength of carbon fibre/epoxy laminates: Effects of bolt-hole clearance. *Compos. Part B Eng.* **2004**, *35*, 331–343. [CrossRef]
22. Liu, F.; Yao, W.; Zhao, L.; Wu, H.; Zhang, X.; Zhang, J. An improved 2D finite element model for bolt load distribution analysis of composite multi-bolt single-lap joints. *Compos. Struct.* **2020**, *253*, 112770. [CrossRef]
23. Belardi, V.G.; Fanelli, P.; Vivio, F. A novel composite bolted joint element: Application to a single-bolted joint. *Procedia Struct. Integr.* **2018**, *12*, 281–295. [CrossRef]
24. Gray, P.; McCarthy, C. A highly efficient user-defined finite element for load distribution analysis of large-scale bolted composite structures. *Compos. Sci. Technol.* **2011**, *71*, 1517–1527. [CrossRef]
25. Liu, F.; Zhang, J.; Zhao, L.; Xin, A.; Zhou, L. An analytical joint stiffness model for load transfer analysis in highly torqued multi-bolt composite joints with clearances. *Compos. Struct.* **2015**, *131*, 625–636. [CrossRef]
26. Gay, A.; Lefebvre, F.; Bergamo, S.; Valiorgue, F.; Chalandon, P.; Michel, P.; Bertrand, P. Fatigue performance of a self-piercing rivet joint between aluminum and glass fiber reinforced thermoplastic composite. *Int. J. Fatigue* **2016**, *83*, 127–134. [CrossRef]

27. Gay, A.; Roche, J.-M.; Lapeyronnie, P.; Valiorgue, F.; Bertrand, P. Non-destructive inspection of initial defects of PA6.6-GF50/aluminum self-piercing riveted joints and damage monitoring under mechanical static loading. *Int. J. Damage Mech.* **2017**, *26*, 1127–1146. [CrossRef]
28. Yao, C.; Qi, Z.; Chen, W.; Zhang, C. Effect of CF/PEEK plasticity behavior on the mechanical performance of interference-fit joint. *Polym. Compos.* **2021**, *42*, 2574–2588. [CrossRef]
29. Vieille, B.; Aucher, J.; Taleb, L. Influence of temperature on the behavior of carbon fiber fabrics reinforced PPS laminates. *Mater. Sci. Eng. A* **2009**, *517*, 51–60. [CrossRef]
30. Vieille, B.; Aucher, J.; Taleb, L. Woven ply thermoplastic laminates under severe conditions: Notched laminates and bolted joints. *Compos. Part B Eng.* **2011**, *42*, 341–349. [CrossRef]
31. Vieille, B.; Aucher, J.; Taleb, L. Comparative study on the behavior of woven-ply reinforced thermoplastic or thermosetting laminates under severe environmental conditions. *Mater. Des.* **2012**, *35*, 707–719. [CrossRef]
32. Vieille, B.; Albouy, W.; Taleb, L. Influence of stamping on the compressive behavior and the damage mechanisms of C/PEEK laminates bolted joints under severe conditions. *Compos. Part B Eng.* **2015**, *79*, 631–638. [CrossRef]
33. Sun, C.; Chen, J. A simple flow rule for characterizing nonlinear behavior of fiber composites. *J. Compos. Mater.* **1989**, *23*, 1009–1020. [CrossRef]
34. Pinho, S.; Darvizeh, R.; Robinson, P.; Schuecker, C.; Camanho, P. Material and structural response of polymer-matrix fibre-reinforced composites. *J. Compos. Mater.* **2012**, *46*, 2313–2341. [CrossRef]
35. Bažant, Z.P.; Oh, B.H. Crack band theory for fracture of concrete. *Mater. Struct.* **1983**, *16*, 155–177. [CrossRef]
36. ASTM International. *ASTM D5961-17 Standard Test Method for Bearing Response of Polymer Matrix Composite Laminates*; ASTM International: West Conshohocken, PA, USA, 2017.
37. Cao, Y.; Cao, Z.; Zhao, Y.; Zuo, D.; Tay, T. Damage progression and failure of single-lap thin-ply laminated composite bolted joints under quasi-static loading. *Int. J. Mech. Sci.* **2020**, *170*, 105360. [CrossRef]
38. Chen, J.; Morozov, E.; Shankar, K. A combined elastoplastic damage model for progressive failure analysis of composite materials and structures. *Compos. Struct.* **2012**, *94*, 3478–3489. [CrossRef]
39. Weeks, C.; Sun, C. Modeling non-linear rate-dependent behavior in fiber-reinforced composites. *Compos. Sci. Technol.* **1998**, *58*, 603–611. [CrossRef]
40. Wang, J.; Xiao, Y. Some improvements on Sun–Chen's one-parameter plasticity model for fibrous composites–part I: Constitutive modelling for tension–compression asymmetry response. *J. Compos. Mater.* **2017**, *51*, 405–418. [CrossRef]
41. Jang, J.; Jeon, S.-Y.; Choi, J.-H.; Shim, W.; Cho, J.-M.; Yoon, S.-J.; Choi, C.-H.; Yu, W.-R. Mechanical analysis of fiber-reinforced plastics using an elastoplastic-damage constitutive equation considering asymmetric material behavior. *Compos. Struct.* **2021**, *272*, 114268. [CrossRef]
42. Yoon, K.; Sun, C. Characterization of elastic-viscoplastic properties of an AS4/PEEK thermoplastic composite. *J. Compos. Mater.* **1991**, *25*, 1277–1296. [CrossRef]
43. Sun, C.; Yoon, K. Elastic-plastic analysis of AS4/PEEK composite laminate using a one-parameter plasticity model. *J. Compos. Mater.* **1992**, *26*, 293–308. [CrossRef]
44. Ding, S.; Tong, J.; Shen, M.; Huo, Y. Three-dimensional elastic-plastic analysis of the interlaminar stresses for the AS4/PEEK composite laminate with a circular hole. *Mech. Adv. Mater. Struct.* **2010**, *17*, 406–418. [CrossRef]
45. Liu, H.; Liu, J.; Ding, Y.; Hall, Z.E.; Kong, X.; Zhou, J.; Blackman, B.R.; Kinloch, A.J.; Dear, J.P. A three-dimensional elastic-plastic damage model for predicting the impact behaviour of fibre-reinforced polymer-matrix composites. *Compos. Part B Eng.* **2020**, *201*, 108389. [CrossRef]
46. Wang, X.; Guan, Z.; Du, S.; Han, G.; Li, Z. An accurate and easy to implement method for predicting matrix crack and plasticity of composites with an efficient search algorithm for LaRC05 criterion. *Compos. Part A Appl. Sci. Manuf.* **2020**, *131*, 105808. [CrossRef]
47. Pinho, S.; Iannucci, L.; Robinson, P. Physically-based failure models and criteria for laminated fibre-reinforced composites with emphasis on fibre kinking: Part I: Development. *Compos. Part A Appl. Sci. Manuf.* **2006**, *37*, 63–73. [CrossRef]
48. Egan, B. Finite element modelling and high speed testing of countersunk composite aircraft joints. Ph.D. Thesis, University of Limerick, Limerick, Ireland, 2014.
49. Donadon, M.V.; De Almeida, S.F.M.; Arbelo, M.A.; de Faria, A.R. A three-dimensional ply failure model for composite structures. *Int. J. Aerosp. Eng.* **2009**, *2009*, 486063. [CrossRef]
50. Matzenmiller, A.; Lubliner, J.; Taylor, R.L. A constitutive model for anisotropic damage in fiber-composites. *Mech. Mater.* **1995**, *20*, 125–152. [CrossRef]
51. Ren, R.; Le, G.; Zhong, J.; Ma, D.; He, Q. Numerical research on elasto-plastic behaviors of fiber-reinforced polymer based composite laminates. *Compos. Struct.* **2019**, *207*, 364–372. [CrossRef]
52. Simo, J.C.; Hughes, T.J. *Computational Inelasticity*; Springer Science & Business Media: New York, NY, USA, 2006; p. 33, ISBN 978-0-387-97520-7.
53. Liu, H.; Liu, J.; Kaboglu, C.; Chai, H.; Kong, X.; Blackman, B.R.; Kinloch, A.J.; Dear, J.P. Experimental and numerical studies on the behaviour of fibre-reinforced composites subjected to soft impact loading. *Procedia Struct. Integr.* **2019**, *17*, 992–1001. [CrossRef]
54. Turon, A.; Camanho, P.P.; Costa, J.; Dávila, C. A damage model for the simulation of delamination in advanced composites under variable-mode loading. *Mech. Mater.* **2006**, *38*, 1072–1089. [CrossRef]

55. Maa, R.-H.; Cheng, J.-H. A CDM-based failure model for predicting strength of notched composite laminates. *Compos. Part B Eng.* **2002**, *33*, 479–489. [CrossRef]
56. Yoon, D.; Kim, S.; Kim, J.; Doh, Y. Study on bearing strength and failure mode of a carbon-epoxy composite laminate for designing bolted joint structures. *Compos. Struct.* **2020**, *239*, 112023. [CrossRef]
57. Abaqus-Inc. Abaqus User Manual, Version 2019. Available online: https://help.3ds.com/2019/english/dssimulia_established/si.macaegsarefmap/simagsa-m-quasi-sb.htm?contextscope=all (accessed on 20 January 2022).
58. Egan, B.; McCarthy, M.; Frizzell, R.; Gray, P.; McCarthy, C. Modelling bearing failure in countersunk composite joints under quasi-static loading using 3D explicit finite element analysis. *Compos. Struct.* **2014**, *108*, 963–977. [CrossRef]
59. Egan, B.; McCarthy, C.T.; McCarthy, M.A.; Frizzell, R. Stress analysis of single-bolt, single-lap, countersunk composite joints with variable bolt-hole clearance. *Compos. Struct.* **2012**, *94*, 1038–1051. [CrossRef]
60. Zhai, Y.; Li, X.; Wang, L.; Li, D. Three-dimensional layer-by-layer stress analysis of single-lap, countersunk composite joints with varying joining interface conditions. *Compos. Struct.* **2018**, *202*, 1021–1031. [CrossRef]

MDPI AG
Grosspeteranlage 5
4052 Basel
Switzerland
Tel.: +41 61 683 77 34

Materials Editorial Office
E-mail: materials@mdpi.com
www.mdpi.com/journal/materials

www.ingramcontent.com/pod-product-compliance
Lightning Source LLC
Chambersburg PA
CBHW061825210926
51850CB00006B/811